$$\mathcal{R} = kT$$

Thermal
Physics

SECOND EDITION

Thermal Physics

Charles Kittel
Herbert Kroemer

University of California

W. H. Freeman and Company
New York

Sponsoring Editor: Peter Renz
Project Editor: Nancy Flight
Manuscript Editor: Ruth Veres
Designers: Gary A. Head and Sharon H. Smith
Production Coordinator: Frank Mitchell
Illustration Coordinator: Batyah Janowski
Artist: Felix Cooper
Compositor: Syntax International

Library of Congress Cataloging in Publication Data

Kittel, Charles.
  Thermal physics.

  Bibliography: p.
  Includes index.
  1. Statistical thermodynamics.   I. Kroemer,
Herbert, 1928–   joint author.   II. Title.
QC311.5.K52 1980   536′.7   79-16677
ISBN 0-7167-1088-9

Printed in the United State of America

Nineteenth printing, 1998

About the Authors

Charles Kittel has taught solid state physics at the University of California at Berkeley since 1951, having previously been at the Bell Laboratories. His undergraduate work in physics was done at M.I.T. and at the Cavendish Laboratory of Cambridge University. His Ph.D. research was in theoretical nuclear physics with Professor Gregory Breit at the University of Wisconsin. He has been awarded three Guggenheim fellowships, the Oliver Buckley Prize for Solid State Physics, and, for contributions to teaching, the Oersted Medal of the American Association of Physics Teachers. He is a member of the National Academy of Science and of the American Academy of Arts and Sciences. His research has been in magnetism, magnetic resonance, semiconductors, and the statistical mechanics of solids.

Herbert Kroemer is Professor of Electrical Engineering at the University of California at Santa Barbara. His background and training are in solid state physics. He received a Ph.D. in physics in 1952 from the University of Göttingen in Germany with a thesis on hot electron effects in the then new transistor. From 1952 through 1968 he worked in several semiconductor research laboratories in Germany and the United States. In 1968 he became Professor of Electrical Engineering at the University of Colorado; he came to UCSB in 1976. His research has been in the physics and technology of semiconductors and semiconductor devices, including high-frequency transistors, negative-mass effects in semiconductors, injection lasers, the Gunn effect, electron-hole drops, and semiconductor heterojunctions.

Preface

This book gives an elementary account of thermal physics. The subject is simple, the methods are powerful, and the results have broad applications. Probably no other physical theory is used more widely throughout science and engineering.

We have written for undergraduate students of physics and astronomy, and for electrical engineering students generally. These fields for our purposes have strong common bonds, most notably a concern with Fermi gases, whether in semiconductors, metals, stars, or nuclei. We develop methods (not original, but not easily accessible elsewhere) that are well suited to these fields. We wrote the book in the first place because we were delighted by the clarity of the "new" methods as compared to those we were taught when we were students ourselves.

The second edition is substantially rewritten and revised from the first edition, which, although warmly accepted, suffered from the concentration of abstract ideas at the beginning. In the new structure the free energy, the partition function, and the Planck distribution are developed before the chemical potential. Real problems can now be solved much earlier. We have added chapters on applications to semiconductors, binary mixtures, transport theory, cryogenics, and propagation. The treatment of heat and work is new and will be helpful to those concerned with energy conversion processes. Many more examples and problems are given, but we have not introduced problems where they do not contribute to the main line of advance. For this edition an instructor's guide is available from the publisher, upon request from the instructor.

This edition has been tested extensively over the past few years in classroom use. We have not emphasized several traditional topics, some because they are no longer useful and some because their reliance on classical statistical mechanics would make the course more difficult than we believe a first course should be. Also, we have avoided the use of combinatorial methods where they are unnecessary.

For a one quarter course for physics undergraduates, we suggest most of Chapters 1 through 10, plus 14. The Debye theory could be omitted from

Chapter 4 and the Boltzmann transport equation from Chapter 14. For a one quarter course for electrical engineers, we suggest Chapter 13 at any time after the discussion of the Fermi gas in Chapter 7. The material of Chapter 13 does not draw on Chapter 4. The scope of the book is ample for a one semester course, and here the pace can be relaxed.

Notation and units: We generally use the SI and CGS systems in parallel. We do not use the calorie. The kelvin temperature T is related to the fundamental temperature τ by $\tau = k_B T$, and the conventional entropy S is related to the fundamental entropy σ by $S = k_B \sigma$. The symbol log will denote natural logarithm throughout, simply because ln is less expressive when set in type. The notation (18) refers to Equation (18) of the current chapter, but (3.18) refers to Equation (18) of Chapter 3.

The book is the successor to course notes developed with the assistance of grants by the University of California. Edward M. Purcell contributed many ideas to the first edition. We benefited from review of the second edition by Seymour Geller, Paul L. Richards, and Nicholas Wheeler. Help was given by Ibrahim Adawi, Bernard Black, G. Domokos, Margaret Geller, Cameron Hayne, K. A. Jackson, S. Justi, Peter Kittel, Richard Kittler, Martin J. Klein, Ellen Leverenz, Bruce H. J. McKellar, F. E. O'Meara, Norman E. Phillips, B. Roswell Russell, T. M. Sanders, B. Stoeckly, John Verhoogen, John Wheatley, and Eyvind Wichmann. We thank Carol Tung for the typed manuscript and Sari Wilde for her help with the index.

An elementary treatment of the greenhouse effect in the Earth's atmosphere was added in 1994 on page 115, following an argument suggested by Professor Richard Muller.

Berkeley and Santa Barbara   *Charles Kittel*
                                               Herbert Kroemer

Note to the Student

For minimum coverage of the concepts presented in each chapter, the authors recommend the following exercises. Chapter 2: 1, 2, 3; Chapter 3: 1, 2, 3, 4, 8, 11; Chapter 4: 1, 2, 4, 5, 6, 8; Chapter 5: 1, 3, 4, 6, 8; Chapter 6: 1, 2, 3, 6, 12, 14, 15; Chapter 7: 2, 3, 5, 6, 7, 11; Chapter 8: 1, 2, 3, 5, 6, 7; Chapter 9: 1, 2, 3; Chapter 10: 1, 2, 3; Chapter 11: 1, 2, 3; Chapter 12: 3, 4, 5; Chapter 13: 1, 2, 3, 7, 8, 10; Chapter 14: 1, 3, 4, 5; Chapter 15: 2, 3, 4, 6.

Contents

Guide to Fundamental Definitions

General References

Thermodynamics

A. B. Pippard, *Elements of classical thermodynamics*, Cambridge University Press, 1966.

M. W. Zemansky and R. H. Dittman, *Heat and thermodynamics: an intermediate textbook*, 6th ed., McGraw-Hill, 1981.

Statistical Mechanics

B. K. Agarwal and M. Eisner, *Statistical mechanics*, Wiley, 1988.

T. L. Hill, *Statistical mechanics: principles and selected applications*, Dover Publications, 1987, c1956.

C. Kittel, *Elementary statistical physics*, Wiley, 1958. Parts 2 and 3 treat applications to noise and to elementary transport theory. Part 1 has been expanded into the present text.

R. Kubo, *Statistical mechanics*, North-Holland, 1990, c1965.

R. Kubo, M. Toda, N. Hashitsume, *Statistical physics II (Nonequilibrium)*, Springer, 1985.

L. D. Landau and E. M. Lifshitz, *Statistical physics*, 3rd ed. by E. M. Lifshitz and L. P. Pitaevskii, Pergamon, 1980, part 1.

Shang-Keng Ma, *Statistical mechanics*, World Scientific, 1985.

M. Toda, R. Kubo, N. Saito, *Statistical physics I (Equilibrium)*, Springer, 1983.

Mathematical tables

H. B. Dwight, *Tables of integrals and other mathematical data*, 4th ed., Macmillan, 1961. A widely useful small collection.

Applications

Astrophysics

R. J. Taylor, *The stars: their structure and evolution*, Springer, 1972.

S. Weinberg, *The first three minutes: a modern view of the origin of the universe*, new ed., Bantam Books, 1984.

Biophysics and macromolecules

T. L. Hill, *Cooperativity theory in biochemistry: steady state and equilibrium systems*, Springer, 1985.

Cryogenics and low temperature physics

G. K. White, *Experimental techniques in low-temperature physics*, 3rd ed., Oxford University Press, 1987, c1979 . . pa.

J. Wilks and D. S. Betts, *An introduction to liquid helium*, 2nd ed., Oxford Univesity Press, 1987.

Irreversible thermodynamics

J. A. McLennan, *Introduction to non-equilibrium statistical mechanics*, Prentice-Hall, 1989.

I. Prigogine and I. Stengers, *Order out of chaos: man's new dialog with nature*, Random House, 1984.

Kinetic theory and transport phenomena

S. G. Brush, *The kind of motion we call heat*, North-Holland, 1986, c1976.

H. Smith and H. H. Jensen, *Transport phenomena*, Oxford University Press, 1989.

Plasma physics

L. Spitzer, Jr., *Physical processes in the interstellar medium*, Wiley, 1978.

Phase transitions

P. Pfeuty and G. Toulouse, *Introduction to the renormalization group and to critical phenomena*, Wiley, 1977.

H. E. Stanley, *Introduction to phase transitions and critical phenomena*, Oxford University Press, 1987.

Metals and alloys

P. Haasen, *Physical metallurgy*, 2nd ed., Cambridge University Press, 1986. Superb.

Boundary value problems

H. S. Carslaw and J. C. Jaeger, *Conduction of heat in solids*, 2nd ed., Oxford University Press, 1986, c1959.

Semiconductor devices

R. Dalven, *Introduction to applied solid state physics*, Plenum, 1990.

K. Seeger, *Semiconductor physics: an introduction*, 5th ed., Springer, 1991.

S. M. Sze, *Physics of semiconductor devices*, 2nd ed., Wiley, 1981.

Solid state physics

C. Kittel, *Introduction to solid state physics*, 6th ed., Wiley, 1986. Referred to as *ISSP*.

Thermal
Physics

Introduction

Our approach to thermal physics differs from the tradition followed in beginning physics courses. Therefore we provide this introduction to set out what we are going to do in the chapters that follow. We show the main lines of the logical structure: in this subject all the physics comes from the logic. In order of their appearance, the leading characters in our story are the entropy, the temperature, the Boltzmann factor, the chemical potential, the Gibbs factor, and the distribution functions.

The entropy measures the number of quantum states accessible to a system. A closed system might be in any of these quantum states and (we assume) with equal probability. The fundamental statistical element, the fundamental logical assumption, is that quantum states are either accessible or inaccessible to the system, and the system is equally likely to be in any one accessible state as in any other accessible state. Given g accessible states, the entropy is defined as $\sigma = \log g$. The entropy thus defined will be a function of the energy U, the number of particles N, and the volume V of the system, because these parameters enter the determination of g; other parameters may enter as well. The use of the logarithm is a mathematical convenience: it is easier to write 10^{20} than $\exp(10^{20})$, and it is more natural for two systems to speak of $\sigma_1 + \sigma_2$ than of $g_1 g_2$.

When two systems, each of specified energy, are brought into thermal contact they may transfer energy; their total energy remains constant, but the constraints on their individual energies are lifted. A transfer of energy in one direction, or perhaps in the other, may increase the product $g_1 g_2$ that measures the number of accessible states of the combined systems. The fundamental assumption biases the outcome in favor of that allocation of the total energy that maximizes the number of accessible states: more is better, and more likely. This statement is the kernel of the law of increase of entropy, which is the general expression of the second law of thermodynamics.

We have brought two systems into thermal contact so that they may transfer energy. What is the most probable outcome of the encounter? One system will gain energy at the expense of the other, and meanwhile the total entropy of the two systems will increase. Eventually the entropy will reach a maximum for the given total energy. It is not difficult to show (Chapter 2) that the maximum is attained when the value of $(\partial\sigma/\partial U)_{N,V}$ for one system is equal to the value of the same quantity for the second system. This equality property for two systems in thermal contact is just the property we expect of the temperature. Accordingly, we define the fundamental temperature τ by the relation

$$\frac{1}{\tau} \equiv \left(\frac{\partial\sigma}{\partial U}\right)_{N,V}. \tag{1}$$

The use of $1/\tau$ assures that energy will flow from high τ to low τ; no more complicated relation is needed. It will follow that the Kelvin temperature T is directly proportional to τ, with $\tau = k_B T$, where k_B is the Boltzmann constant. The conventional entropy S is given by $S = k_B\sigma$.

Now consider a very simple example of the Boltzmann factor treated in Chapter 3. Let a small system with only two states, one at energy 0 and one at energy ε, be placed in thermal contact with a large system that we call the reservoir. The total energy of the combined systems is U_0; when the small system is in the state of energy 0, the reservoir has energy U_0 and will have $g(U_0)$ states accessible to it. When the small system is in the state of energy ε, the reservoir will have energy $U_0 - \varepsilon$ and will have $g(U_0 - \varepsilon)$ states accessible to it. By the fundamental assumption, the ratio of the probability of finding the small system with energy ε to the probability of finding it with energy 0 is

$$\frac{P(\varepsilon)}{P(0)} = \frac{g(U_0 - \varepsilon)}{g(U_0)} = \frac{\exp[\sigma(U_0 - \varepsilon)]}{\exp[\sigma(U_0)]}. \tag{2}$$

The reservoir entropy σ may be expanded in a Taylor series:

$$\sigma(U_0 - \varepsilon) \simeq \sigma(U_0) - \varepsilon(\partial\sigma/\partial U_0) = \sigma(U_0) - \varepsilon/\tau , \tag{3}$$

by the definition (1) of the temperature. Higher order terms in the expansion may be dropped. Cancellation of the term $\exp[\sigma(U_0)]$, which occurs in the numerator and denominator of (2) after the substitution of (3), leaves us with

$$P(\varepsilon)/P(0) = \exp(-\varepsilon/\tau). \tag{4}$$

This is Boltzmann's result. To show its use, we calculate the thermal average energy $\langle\varepsilon\rangle$ of the two state system in thermal contact with a reservoir at temperature τ:

$$\langle\varepsilon\rangle = \sum_i \varepsilon_i P(\varepsilon_i) = 0{\cdot}P(0) + \varepsilon P(\varepsilon) = \frac{\varepsilon\exp(-\varepsilon/\tau)}{1 + \exp(-\varepsilon/\tau)} , \tag{5}$$

where we have imposed the normalization condition on the sum of the probabilities:

$$P(0) + P(\varepsilon) = 1. \tag{6}$$

The argument can be generalized immediately to find the average energy of a harmonic oscillator at temperature τ, and we do this in Chapter 4 as the first step in the derivation of the Planck radiation law.

The most important extension of the theory is to systems that can transfer particles as well as energy with the reservoir. For two systems in diffusive and thermal contact, the entropy will be a maximum with respect to the transfer of particles as well as to the transfer of energy. Not only must $(\partial\sigma/\partial U)_{N,V}$ be equal for the two systems, but $(\partial\sigma/\partial N)_{U,V}$ must also be equal, where N refers to the number of particles of a given species. The new equality condition is the occasion for the introduction of a new quantity, the chemical potential μ:

$$-\frac{\mu}{\tau} = \left(\frac{\partial\sigma}{\partial N}\right)_{U,V}. \tag{7}$$

For two systems in thermal and diffusive contact, $\tau_1 = \tau_2$ and $\mu_1 = \mu_2$. The sign in (7) is chosen to ensure that the direction of particle flow as equilibrium is approached is from high chemical potential to low chemical potential.

The Gibbs factor of Chapter 5 is an extension of the Boltzmann factor and allows us to treat systems that can transfer particles. The simplest example is a system with two states, one with 0 particles and 0 energy, and one with 1 particle and energy ε. The system is in contact with a reservoir at temperature τ and chemical potential μ. We extend (3) for the reservoir entropy:

$$\sigma(U_0 - \varepsilon; N_0 - 1) = \sigma(U_0; N_0) - \varepsilon(\partial\sigma/\partial U_0) - 1 \cdot (\partial\sigma/\partial N_0)$$
$$= \sigma(U_0; N_0) - \varepsilon/\tau + \mu/\tau. \tag{8}$$

By analogy with (4), we have

$$P(1,\varepsilon)/P(0,0) = \exp[(\mu - \varepsilon)/\tau], \tag{9}$$

for the ratio of the probability the system is occupied by 1 particle at energy ε to the probability the system is unoccupied, with energy 0. The result (9) after normalization is readily expressed as

$$P(1,\varepsilon) = \frac{1}{\exp[(\varepsilon - \mu)/\tau] + 1}. \tag{10}$$

This particular result is known as the Fermi-Dirac distribution function and is used particularly in the theory of metals to describe the electron gas at low temperature and high concentration (Chapter 7).

The classical distribution function used in the derivation of the ideal gas law is just the limit of (10) when the occupancy $P(1,\varepsilon)$ is much less than 1:

$$P(1,\varepsilon) \simeq \exp[(\mu - \varepsilon)/\tau]. \qquad (11)$$

The properties of the ideal gas are developed from this result in Chapter 6.

The Helmholtz free energy $F \equiv U - \tau\sigma$ appears as an important computational function, because the relation $(\partial F/\partial \tau)_{N,V} = -\sigma$ offers the easiest method for finding the entropy, once we have found out how to calculate F from the energy eigenvalues (Chapter 3). Other powerful tools for the calculation of thermodynamic functions are developed in the text. Most of the remainder of the text concerns applications that are useful in their own right and that illuminate the meaning and utility of the principal thermodynamic functions.

Thermal physics connects the world of everyday objects, of astronomical objects, and of chemical and biological processes with the world of molecular, atomic, and electronic systems. It unites the two parts of our world, the microscopic and the macroscopic.

Chapter 1

States of a Model System

But although, as a matter of history, statistical mechanics owes its origin to investigations in thermodynamics, it seems eminently worthy of an independent development, both on account of the elegance and simplicity of its principles, and because it yields new results and places old truths in a new light in departments quite outside of thermodynamics.

J. W. Gibbs

A theory is the more impressive the greater the simplicity of its premises, the more different kinds of things it relates, and the more extended its area of applicability. Therefore the deep impression that classical thermodynamics made upon me. It is the only physical theory of universal content which I am convinced will never be overthrown, within the framework of applicability of its basic concepts.

A. Einstein

Thermal physics is the fruit of the union of statistical and mechanical principles. Mechanics tells us the meaning of work; thermal physics tells us the meaning of heat. There are three new quantities in thermal physics that do not appear in ordinary mechanics: entropy, temperature, and free energy. We shall motivate their definitions in the first three chapters and deduce their consequences thereafter.

Our point of departure for the development of thermal physics is the concept of the stationary quantum states of a system of particles. When we can count the quantum states accessible to a system, we know the entropy of the system, for the entropy is defined as the logarithm of the number of states (Chapter 2). The dependence of the entropy on the energy of the system defines the temperature. From the entropy, the temperature, and the free energy we find the pressure, the chemical potential, and all other thermodynamic properties of the system.

For a system in a stationary quantum state, all observable physical properties such as the energy and the number of particles are independent of the time. For brevity we usually omit the word stationary; the quantum states that we treat are stationary except when we discuss transport processes in Chapters 14–15. The systems we discuss may be composed of a single particle or, more often, of many particles. The theory is developed to handle general systems of interacting particles, but powerful simplifications can be made in special problems for which the interactions may be neglected.

Each quantum state has a definite energy. States with identical energies are said to belong to the same energy level. The **multiplicity** or degeneracy of an energy level is the number of quantum states with very nearly the same energy. $\longrightarrow E \to E + dE$ It is the number of quantum states that is important in thermal physics, not the number of energy levels. We shall frequently deal with sums over all quantum states. Two states at the same energy must always be counted as two states, not as one level.

Let us look at the quantum states and energy levels of several atomic systems. The simplest is hydrogen, with one electron and one proton. The low-lying energy levels of hydrogen are shown in Figure 1.1. The zero of energy in the figure is taken at the state of lowest energy. The number of quantum states belonging to the same energy level is in parentheses. In the figure we overlook that the proton has a spin of $\frac{1}{2}\hbar$ and has two independent orientations, parallel

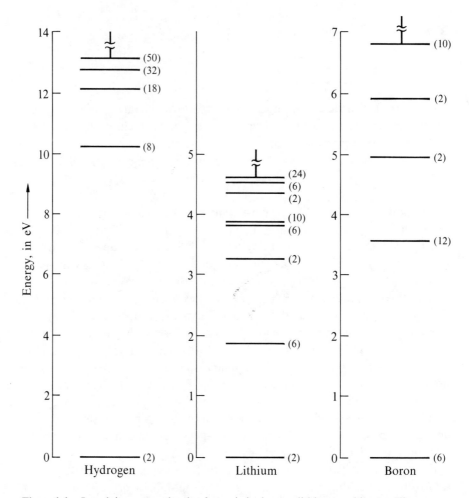

Figure 1.1   Low-lying energy levels of atomic hydrogen, lithium, and boron. The energies are given in electron volts, with 1 eV = 1.602×10^{-12} erg. The numbers in parentheses give the number of quantum states having the same energy, with no account taken of the spin of the nucleus. The zero of energy in the figure is taken for convenience at the lowest energy state of each atom.

or antiparallel to the direction of an arbitrary external axis, such as the direction of a magnetic field. To take account of the two orientations we should double the values of the multiplicities shown for atomic hydrogen.

An atom of lithium has three electrons which move about the nucleus. Each electron interacts with the nucleus, and each electron also interacts with all the

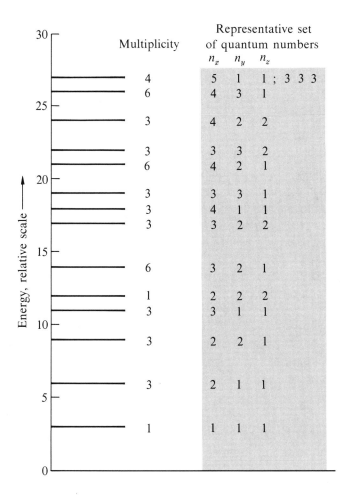

Energy	Multiplicity	Representative set of quantum numbers n_x n_y n_z		
	4	5	1	1 ; 3 3 3
	6	4	3	1
	3	4	2	2
	3	3	3	2
	6	4	2	1
	3	3	3	1
	3	4	1	1
	3	3	2	2
	6	3	2	1
	1	2	2	2
	3	3	1	1
	3	2	2	1
	3	2	1	1
	1	1	1	1

Figure 1.2   Energy levels, multiplicities, and quantum numbers n_x, n_y, n_z of a particle confined to a cube.

other electrons. The energies of the levels of lithium shown in the figure are the collective energies of the entire system. The energy levels shown for boron, which has five electrons, are also the energies of the entire system.

The energy of a system is the total energy of all particles, kinetic plus potential, with account taken of interactions between particles. A quantum state of the system is a state of all particles. Quantum states of a one-particle system are called **orbitals**. The low-lying energy levels of a single particle of mass M confined to a cube of side L are shown in Figure 1.2. We shall find in Chapter 3 that an orbital of a free particle can be characterized by three positive integral quantum numbers n_x, n_y, n_z. The energy is

$$\varepsilon = \frac{\hbar^2}{2M}\left(\frac{\pi}{L}\right)^2 (n_x^{\,2} + n_y^{\,2} + n_z^{\,2}).\qquad (1)$$

The multiplicities of the levels are indicated in the figure. The three orbitals with (n_x, n_y, n_z) equal to (4,1,1), (1,4,1), and (1,1,4) all have $n_x^{\,2} + n_y^{\,2} + n_z^{\,2} = 18$; the corresponding energy level has the multiplicity 3.

To describe the statistical properties of a system of N particles, it is essential to know the set of values of the energy $\varepsilon_s(N)$, where ε is the energy of the quantum state s of the N particle system. Indices such as s may be assigned to the quantum states in any convenient arbitrary way, but two different states should not be assigned the same index.

It is a good idea to start our program by studying the properties of simple model systems for which the energies $\varepsilon_s(N)$ can be calculated exactly. We choose as a model a simple binary system because the general statistical properties found for the model system are believed to apply equally well to any realistic physical system. This assumption leads to predictions that always agree with experiment. What general statistical properties are of concern will become clear as we go along.

BINARY MODEL SYSTEMS

The binary model system is illustrated in Figure 1.3. We assume there are N separate and distinct sites fixed in space, shown for convenience on a line. Attached to each site is an elementary magnet that can point only up or down, corresponding to magnetic moments $\pm m$. To understand the system means to count the states. This requires no knowledge of magnetism: an element of the system can be any site capable of two states, labeled as yes or no, red or blue, occupied or unoccupied, zero or one, plus one or minus one. The sites are numbered, and sites with different numbers are supposed not to overlap in physical space. You might even think of the sites as numbered parking spaces in a car parking lot, as in Figure 1.4. Each parking space has two states, vacant or occupied by one car.

Whatever the nature of our objects, we may designate the two states by arrows that can only point straight up or straight down. If the magnet points up, we say that the magnetic moment is $+m$. If the magnet points down, the magnetic moment is $-m$.

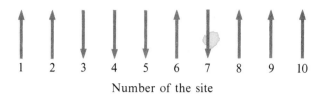

Number of the site

Figure 1.3 Model system composed of 10 elementary magnets at fixed sites on a line, each having magnetic moment $\pm m$. The numbers shown are attached to the sites; each site has its own magnet. We assume there are no interactions among the magnets and there is no external magnetic field. Each magnetic moment may be oriented in two ways, up or down, so that there are 2^{10} distinct arrangements of the 10 magnetic moments shown in the figure. If the arrangements are selected in a random process, the probability of finding the particular arrangement shown is $1/2^{10}$.

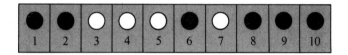

Figure 1.4 State of a parking lot with 10 numbered parking spaces. The ●'s denote spaces occupied by a car; the ○'s denote vacant spaces. This particular state is equivalent to that shown in Figure 1.3.

Now consider N different sites, each of which bears a moment that may assume the values $\pm m$. Each moment may be oriented in two ways with a probability independent of the orientation of all other moments. The total number of arrangements of the N moments is $2 \times 2 \times 2 \times \cdots 2 = 2^N$. A state of the system is specified by giving the orientation of the moment on each site; there are 2^N states. We may use the following simple notation for a single state of the system of N sites:

$$\uparrow\uparrow\downarrow\downarrow\uparrow\downarrow\uparrow\uparrow\uparrow \cdots . \tag{2}$$

Figure 1.5 The four different states of a system of two elements numbered 1 and 2, where each element can have two conditions. The element is a magnet which can be in condition ↑ or condition ↓.

The sites themselves are assumed to be arranged in a definite order. We may number them in sequence from left to right, as we did in Figure 1.3. According to this convention the state (2) also can be written as

$$\uparrow_1\uparrow_2\downarrow_3\downarrow_4\downarrow_5\uparrow_6\downarrow_7\uparrow_8\uparrow_9\uparrow_{10}\cdots. \tag{3}$$

Both sets of symbols (2) and (3) denote the same state of the system, the state in which the magnetic moment on site 1 is $+m$; on site 2, the moment is $+m$; on site 3, the moment is $-m$; and so forth.

It is not hard to convince yourself that every distinct state of the system is contained in a symbolic product of N factors:

$$(\uparrow_1 + \downarrow_1)(\uparrow_2 + \downarrow_2)(\uparrow_3 + \downarrow_3)\cdots(\uparrow_N + \downarrow_N). \tag{4}$$

The multiplication rule is defined by

$$(\uparrow_1 + \downarrow_1)(\uparrow_2 + \downarrow_2) = \uparrow_1\uparrow_2 + \uparrow_1\downarrow_2 + \downarrow_1\uparrow_2 + \downarrow_1\downarrow_2. \tag{5}$$

The function (4) on multiplication generates a sum of 2^N terms, one for each of the 2^N possible states. Each term is a product of N individual magnetic moment symbols, with one symbol for each elementary magnet on the line. Each term denotes an independent state of the system and is a simple product of the form $\uparrow_1\uparrow_2\downarrow_3\cdots\uparrow_N$, for example.

For a system of two elementary magnets, we multiply $(\uparrow_1 + \downarrow_1)$ by $(\uparrow_2 + \downarrow_2)$ to obtain the four possible states of Figure 1.5:

$$(\uparrow_1 + \downarrow_1)(\uparrow_2 + \downarrow_2) = \uparrow_1\uparrow_2 + \uparrow_1\downarrow_2 + \downarrow_1\uparrow_2 + \downarrow_1\downarrow_2. \tag{6}$$

The sum is not a state but is a way of listing the four possible states of the system. The product on the left-hand side of the equation.is called a generating function: it generates the states of the system.

The generating function for the states of a system of three magnets is

$$(\uparrow_1 + \downarrow_1)(\uparrow_2 + \downarrow_2)(\uparrow_3 + \downarrow_3).$$

This expression on multiplication generates $2^3 = 8$ different states:

Three magnets up:		$\uparrow_1\uparrow_2\uparrow_3$	
Two magnets up:	$\uparrow_1\uparrow_2\downarrow_3$	$\uparrow_1\downarrow_2\uparrow_3$	$\downarrow_1\uparrow_2\uparrow_3$
One magnet up:	$\uparrow_1\downarrow_2\downarrow_3$	$\downarrow_1\uparrow_2\downarrow_3$	$\downarrow_1\downarrow_2\uparrow_3$
None up:		$\downarrow_1\downarrow_2\downarrow_3.$	

$$m_i = \mu s_i$$

$$M = \sum_{i=1}^{N} m_i$$

The total magnetic moment of our model system of N magnets each of magnetic moment m will be denoted by M, which we will relate to the energy in a magnetic field. The value of M varies from Nm to $-Nm$. The set of possible values is given by

$$M = Nm, \quad (N-2)m, \quad (N-4)m, \quad (N-6)m, \quad \cdots, \quad -Nm. \tag{7}$$

The set of possible values of M is obtained if we start with the state for which all magnets are up $(M = Nm)$ and reverse one at a time. We may reverse N magnets to obtain the ultimate state for which all magnets are down $(M = -Nm)$.

There are $N + 1$ possible values of the total moment, whereas there are 2^N states. When $N \gg 1$, we have $2^N \gg N + 1$. There are many more states than values of the total moment. If $N = 10$, there are $2^{10} = 1024$ states distributed among 11 different values of the total magnetic moment. For large N many different states of the system may have the same value of the total moment M. We will calculate in the next section how many states have a given value of M.

Only one state of a system has the moment $M = Nm$; that state is

$$\uparrow\uparrow\uparrow\uparrow \cdots \uparrow\uparrow\uparrow\uparrow. \tag{8}$$

There are N ways to form a state with one magnet down:

$$\downarrow\uparrow\uparrow\uparrow \cdots \uparrow\uparrow\uparrow\uparrow \tag{9}$$

is one such state; another is

$$\uparrow\downarrow\uparrow\uparrow \cdots \uparrow\uparrow\uparrow\uparrow, \tag{10}$$

and the other states with one magnet down are formed from (8) by reversing any single magnet. The states (9) and (10) have total moment $M = Nm - 2m$.

Enumeration of States and the Multiplicity Function

We use the word spin as a shorthand for elementary magnet. It is convenient to assume that N is an even number. We need a mathematical expression for the number of states with $N_\uparrow = \frac{1}{2}N + s$ magnets up and $N_\downarrow = \frac{1}{2}N - s$ magnets down, where s is an integer. When we turn one magnet from the up to the down orientation, $\frac{1}{2}N + s$ goes to $\frac{1}{2}N + s - 1$ and $\frac{1}{2}N - s$ goes to $\frac{1}{2}N - s + 1$. The difference (number up − number down) changes from $2s$ to $2s - 2$. The difference

$$N_\uparrow - N_\downarrow = 2s \tag{11}$$

is called the **spin excess**. The spin excess of the 4 states in Figure 1.5 is $2, 0, 0, -2$, from left to right. The factor of 2 in (11) appears to be a nuisance at this stage, but it will prove to be convenient.

The product in (4) may be written symbolically as

$$(\uparrow + \downarrow)^N.$$

We may drop the site labels (the subscripts) from (4) when we are interested only in how many of the magnets in a state are up or down, and not in which particular sites have magnets up or down. It we drop the labels and neglect the order in which the arrows appear in a given product, then (5) becomes

$$(\uparrow + \downarrow)^2 = \uparrow\uparrow + 2\uparrow\downarrow + \downarrow\downarrow;$$

further,

$$(\uparrow + \downarrow)^3 = \uparrow\uparrow\uparrow + 3\uparrow\uparrow\downarrow + 3\uparrow\downarrow\downarrow + \downarrow\downarrow\downarrow.$$

We find $(\uparrow + \downarrow)^N$ for arbitrary N by the binomial expansion

$$\begin{aligned}
(x + y)^N &= x^N + Nx^{N-1}y + \tfrac{1}{2}N(N-1)x^{N-2}y^2 + \cdots + y^N \\
&= \sum_{t=0}^{N} \frac{N!}{(N-t)!\,t!} x^{N-t}y^t.
\end{aligned} \tag{12}$$

We may write the exponents of x and y in a slightly different, but equivalent, form by replacing t with $\frac{1}{2}N - s$:

$$n = t = \frac{1}{2}N - s$$

$$(x + y)^N = \sum_{s=-\frac{1}{2}N}^{\frac{1}{2}N} \frac{N!}{(\frac{1}{2}N + s)! \, (\frac{1}{2}N - s)!} \, x^{\frac{1}{2}N+s} y^{\frac{1}{2}N-s}. \tag{13}$$

With this result the symbolic expression $(\uparrow + \downarrow)^N$ becomes

$$(\uparrow + \downarrow)^N \equiv \sum_s \frac{N!}{(\frac{1}{2}N + s)! \, (\frac{1}{2}N - s)!} \, \uparrow^{\frac{1}{2}N+s} \downarrow^{\frac{1}{2}N-s}. \tag{14}$$

The coefficient of the term in $\uparrow^{\frac{1}{2}N+s} \downarrow^{\frac{1}{2}N-s}$ is the number of states having $N_\uparrow = \frac{1}{2}N + s$ magnets up and $N_\downarrow = \frac{1}{2}N - s$ magnets down. This class of states has spin excess $N_\uparrow - N_\downarrow = 2s$ and net magnetic moment $2sm$. Let us denote the number of states in this class by $g(N,s)$, for a system of N magnets:

$$g(N,s) = \frac{N!}{(\frac{1}{2}N + s)! \, (\frac{1}{2}N - s)!} = \frac{N!}{N_\uparrow! \, N_\downarrow!}. \tag{15}$$

Thus (14) is written as

$$(\uparrow + \downarrow)^N = \sum_{s=-\frac{1}{2}N}^{\frac{1}{2}N} g(N,s) \, \uparrow^{\frac{1}{2}N+s} \downarrow^{\frac{1}{2}N-s}. \tag{16}$$

We shall call $g(N,s)$ the **multiplicity function**; it is the number of states having the same value of s. The reason for our definition emerges when a magnetic field is applied to the spin system: in a magnetic field, states of different values of s have different values of the energy, so that our g is equal to the multiplicity of an energy level in a magnetic field. Until we introduce a magnetic field, all states of the model system have the same energy, which may be taken as zero. Note from (16) that the total number of states is given by

$$\sum_{s=-\frac{1}{2}N}^{s=\frac{1}{2}N} g(N,s) = (1 + 1)^N = 2^N. \tag{17}$$

Examples related to $g(N,s)$ for $N = 10$ are given in Figures 1.6 and 1.7. For a coin, "heads" could stand for "magnet up" and "tails" could stand for "magnet down."

Figure 1.6   Number of distinct arrangements of $5 + s$ spins up and $5 - s$ spins down. Values of $g(N,s)$ are for $N = 10$, where $2s$ is the spin excess $N\uparrow - N\downarrow$. The total number of states is

$$2^{10} = \sum_{s=-5}^{5} g(10,s).$$

The values of the g's are taken from a table of the binomial coefficients.

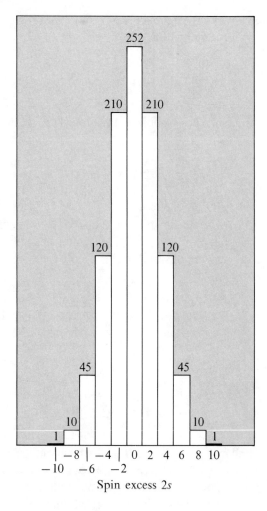

Spin excess $2s$

Binary Alloy System

To illustrate that the exact nature of the two states on each site is irrelevant to the result, we consider an alternate system—an alloy crystal with N distinct sites, numbered from 1 through 12 in Figure 1.8. Each site is occupied by either an atom of chemical species A or an atom of chemical species B, with no provision for vacant sites. In brass, A could be copper and B zinc. In analogy to (3), a single state of the alloy system can be written as

$$A_1B_2B_3A_4B_5A_6B_7B_8B_9A_{10}A_{11}A_{12} \cdots . \tag{18}$$

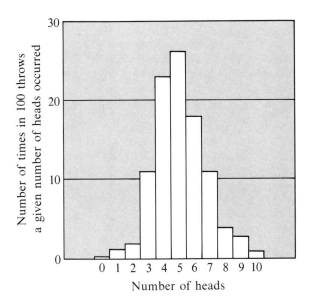

Figure 1.7   An experiment was done in which 10 pennies were thrown 100 times. The number of heads in each throw was recorded.

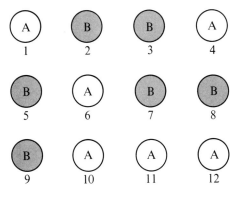

Figure 1.8   A binary alloy system of two chemical components A and B, whose atoms occupy distinct numbered sites.

Every distinct state of a binary alloy system on N sites is contained in the symbolic product of N factors:

$$(A_1 + B_1)(A_2 + B_2)(A_3 + B_3) \cdots (A_N + B_N) , \tag{19}$$

in analogy to (4). The average composition of a binary alloy is specified conventionally by the chemical formula $A_{1-x}B_x$, which means that out of a total of N atoms, the number of A atoms is $N_A = (1 - x)N$ and the number of B atoms is $N_B = xN$. Here x lies between 0 and 1.

The symbolic expression

$$(A + B)^N = \sum_{t=0}^{N} \frac{N!}{(N - t)! \, t!} A^{N-t} B^t \tag{20}$$

is analogous to the result (12). The coefficient of the term in $A^{N-t} B^t$ gives the number $g(N,t)$ of possible arrangements or states of $N - t$ atoms A and t atoms B on N sites:

$$g(N,t) = \frac{N!}{(N - t)! \, t!} = \frac{N!}{N_A! \, N_B!} , \tag{21}$$

which is identical to the result (15) for the spin model system, except for notation.

Sharpness of the Multiplicity Function

We know from common experience that systems held at constant temperature usually have well-defined properties; this stability of physical properties is a major prediction of thermal physics. The stability follows as a consequence of the exceedingly sharp peak in the multiplicity function and of the steep variation of that function away from the peak. We can show explicitly that for a very large system, the function $g(N,s)$ defined by (15) is peaked very sharply about the value $s = 0$. We look for an approximation that allows us to examine the form of $g(N,s)$ versus s when $N \gg 1$ and $|s| \ll N$. We cannot look up these values in tables: common tables of factorials do not go above $N = 100$, and we may be interested in $N \approx 10^{20}$, of the order of the number of atoms in a solid specimen big enough to be seen and felt. An approximation is clearly needed, and a good one is available.

It is convenient to work with $\log g$. Except where otherwise specified, all logarithms are understood to be log base e, written here as log. The international standard usage is ln for log base e, but it is clearer to write log when there is no ambiguity whatever. When you confront a very, very large number such as

2^N, where $N = 10^{20}$, it is a simplification to look at the logarithm of the number. We take the logarithm of both sides of (15) to obtain

$$\log g(N,s) = \log N! - \log(\tfrac{1}{2}N + s)! - \log(\tfrac{1}{2}N - s)! , \qquad (22)$$

by virtue of the characteristic property of the logarithm of a product:

$$\log xy = \log x + \log y; \qquad \log(x/y) = \log x - \log y. \qquad (23)$$

With the notation

$$N_\uparrow = \tfrac{1}{2}N + s; \qquad N_\downarrow = \tfrac{1}{2}N - s \qquad (24)$$

for the number of magnets up and down, (22) appears as

$$\log g(N,s) = \log N! - \log N_\uparrow! - \log N_\downarrow!. \qquad (25)$$

We evaluate the logarithm of $N!$ in (25) by use of the **Stirling approximation**, according to which

$$N! \simeq (2\pi N)^{1/2} N^N \exp[-N + 1/(12N) + \cdots] , \qquad (26)$$

for $N \gg 1$. This result is derived in Appendix A. For sufficiently large N, the terms $1/(12N) + \cdots$ in the argument may be neglected in comparison with N. We take the logarithm of both sides of (26) to obtain

$$\log N! \cong \tfrac{1}{2}\log 2\pi + (N + \tfrac{1}{2})\log N - N. \qquad (27)$$

Similarly

$$\log N_\uparrow! \cong \tfrac{1}{2}\log 2\pi + (N_\uparrow + \tfrac{1}{2})\log N_\uparrow - N_\uparrow; \qquad (28)$$

$$\log N_\downarrow! \cong \tfrac{1}{2}\log 2\pi + (N_\downarrow + \tfrac{1}{2})\log N_\downarrow - N_\downarrow. \qquad (29)$$

After rearrangement of (27),

$$\log N! \cong \tfrac{1}{2}\log(2\pi/N) + (N_\uparrow + \tfrac{1}{2} + N_\downarrow + \tfrac{1}{2})\log N - (N_\uparrow + N_\downarrow) , \qquad (30)$$

where we have used $N = N_\uparrow + N_\downarrow$. We subtract (28) and (29) from (30) to obtain for (25):

$$\log g \cong \tfrac{1}{2}\log(1/2\pi N) - (N_\uparrow + \tfrac{1}{2})\log(N_\uparrow/N) - (N_\downarrow + \tfrac{1}{2})\log(N_\downarrow/N). \qquad (31)$$

This may be simplified because

$$\log(N_\uparrow/N) = \log\tfrac{1}{2}(1 + 2s/N) = -\log 2 + \log(1 + 2s/N)$$
$$\cong -\log 2 + (2s/N) - (2s^2/N^2) \tag{32}$$

by virtue of the expansion $\log(1 + x) = x - \tfrac{1}{2}x^2 + \cdots$, valid for $x \ll 1$. Similarly,

$$\log(N_\downarrow/N) = \log\tfrac{1}{2}(1 - 2s/N) \simeq -\log 2 - (2s/N) - (2s^2/N^2). \tag{33}$$

On substitution in (31) we obtain

$$\log g \cong \tfrac{1}{2}\log(2/\pi N) + N \log 2 - 2s^2/N. \tag{34}$$

We write this result as

$$\boxed{g(N,s) \cong g(N,0)\exp(-2s^2/N) ,} \tag{35}$$

where

$$g(N,0) \simeq (2/\pi N)^{1/2}2^N. \tag{36}$$

Such a distribution of values of s is called a **Gaussian distribution**. The integral* of (35) over the range $-\infty$ to $+\infty$ for s gives the correct value 2^N for the total number of states. Several useful integrals are treated in Appendix A.

The exact value of $g(N,0)$ is given by (15) with $s = 0$:

$$g(N,0) = \frac{N!}{(\tfrac{1}{2}N)!\,(\tfrac{1}{2}N)!}. \tag{37}$$

---

* The replacement of a sum by an integral, such as $\sum_s (\ldots)$ by $\int(\ldots)ds$, usually does not introduce significant errors. For example, the ratio of

$$\sum_{s=0}^{N} s = \tfrac{1}{2}(N^2 + N) \qquad \text{to} \qquad \int_0^N s\,ds = \tfrac{1}{2}N^2$$

is equal to $1 + (1/N)$, which approaches 1 as N approaches ∞.

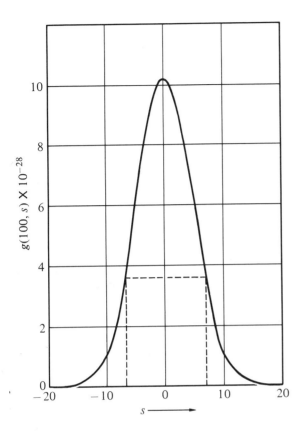

Figure 1.9　The Gaussian approximation to the binomial coefficients $g(100,s)$ plotted on a linear scale. On this scale it is not possible to distinguish on the drawing the approximation from the exact values over the range of s plotted. The entire range of s is from -50 to $+50$. The dashed lines are drawn from the points at $1/e$ of the maximum value of g.

For $N = 50$, the value of $g(50,0)$ is 1.264×10^{14}, from (37). The approximate value from (36) is 1.270×10^{14}. The distribution plotted in Figure 1.9 is centered in a maximum at $s = 0$. When $s^2 = \frac{1}{2}N$, the value of g is reduced to e^{-1} of the maximum value. That is, when

$$s/N = (1/2N)^{1/2} \; , \tag{38}$$

the value of g is e^{-1} of $g(N,0)$. The quantity $(1/2N)^{1/2}$ is thus a reasonable measure of the fractional width of the distribution. For $N \approx 10^{22}$, the fractional width is of the order of 10^{-11}. When N is very large, the distribution is exceedingly sharply defined, in a relative sense. It is this sharp peak and the continued sharp variation of the multiplicity function far from the peak that will lead to a prediction that the physical properties of systems in thermal equilibrium are well defined. We now consider one such property, the mean value of s^2.

AVERAGE VALUES

The average value, or mean value, of a function $f(s)$ taken over a probability distribution function $P(s)$ is defined as

$$\langle f \rangle = \sum_s f(s)\, P(s) \,, \tag{39}$$

provided that the distribution function is normalized to unity:

$$\sum_s P(s) = 1. \tag{40}$$

The binomial distribution (15) has the property (17) that

$$\sum_s g(N,s) = 2^N \,, \tag{41}$$

and is not normalized to unity. If all states are equally probable, then $P(s) = g(N,s)/2^N$, and we have $\sum P(s) = 1$. The average of $f(s)$ over this distribution will be

$$\langle f \rangle = \sum_s f(s)\, P(N,s). \tag{42}$$

Consider the function $f(s) = s^2$. In the approximation that led to (35) and (36), we replace in (42) the sum $\sum$ over s by an integral $\int \cdots ds$ between $-\infty$ and $+\infty$. Then

$$\langle s^2 \rangle = \frac{(2/\pi N)^{1/2}\, 2^N \int ds\, s^2 \exp(-2s^2/N)}{2^N} \,,$$

$$= (2/\pi N)^{1/2}\, (N/2)^{3/2} \int_{-\infty}^{\infty} dx\, x^2 e^{-x^2}$$

$$= (2/\pi N)^{1/2}\, (N/2)^{3/2}\, (\pi/4)^{1/2} \,,$$

whence

$$\langle s^2 \rangle = \tfrac{1}{4}N; \qquad \langle (2s)^2 \rangle = N. \tag{43}$$

The quantity $\langle (2s)^2 \rangle$ is the mean square spin excess. The root mean square spin excess is

$$\langle (2s)^2 \rangle^{1/2} = \sqrt{N} \,, \tag{44}$$

and the fractional fluctuation in $2s$ is defined as

$$\mathscr{F} \equiv \frac{\langle (2s)^2 \rangle^{1/2}}{N} = \frac{1}{\sqrt{N}}. \tag{45}$$

The larger N is, the smaller is the fractional fluctuation. This means that the central peak of the distribution function becomes relatively more sharply defined as the size of the system increases, the size being measured by the number of sites N. For 10^{20} particles, $\mathscr{F} = 10^{-10}$, which is very small.

Energy of the Binary Magnetic System

The thermal properties of the model system become physically relevant when the elementary magnets are placed in a magnetic field, for then the energies of the different states are no longer all equal. If the energy of the system is specified, then only the states having this energy may occur. The energy of interaction of a single magnetic moment **m** with a fixed external magnetic field **B** is

$$U = -\mathbf{m} \cdot \mathbf{B}. \tag{46}$$

This is the potential energy of the magnet **m** in the field **B**.

For the model system of N elementary magnets, each with two allowed orientations in a uniform magnetic field **B**, the total potential energy U is

$$U = \sum_{i=1}^{N} U_i = -\mathbf{B} \cdot \sum_{i=1}^{N} \mathbf{m}_i = -2smB = -MB , \tag{47}$$

using the expression M for the total magnetic moment $2sm$. In this example the spectrum of values of the energy U is discrete. We shall see later that a continuous or quasi-continuous spectrum will create no difficulty. Furthermore, the spacing between adjacent energy levels of this model is constant, as in Figure 1.10. Constant spacing is a special feature of the particular model, but this feature will not restrict the generality of the argument that is developed in the following sections.

The value of the energy for moments that interact only with the external magnetic field is completely determined by the value of s. This functional dependence is indicated by writing $U(s)$. Reversing a single moment lowers $2s$ by -2, lowers the total magnetic moment by $-2m$, and raises the energy by $2mB$. The energy difference between adjacent levels is denoted by $\Delta\varepsilon$, where

$$\Delta\varepsilon = U(s) - U(s + 1) = 2mB. \tag{48}$$

s		$U(s)/mB$	$g(s)$	$\log g(s)$
-5	——————	$+10$	1	0
-4	——————	$+8$	10	2.30
-3	——————	$+6$	45	3.81
-2	——————	$+4$	120	4.79
-1	——————	$+2$	210	5.35
0	——————	0	252	5.53
$+1$	——————	-2	210	5.35
$+2$	——————	-4	120	4.79
$+3$	——————	-6	45	3.81
$+4$	——————	-8	10	2.30
$+5$	——————	-10	1	0

Figure 1.10   Energy levels of the model system of 10 magnetic moments m in a magnetic field B. The levels are labeled by their s values, where $2s$ is the spin excess and $\frac{1}{2}N + s = 5 + s$ is the number of up spins. The energies $U(s)$ and multiplicities $g(s)$ are shown. For this problem the energy levels are spaced equally, with separation $\Delta\varepsilon = 2mB$ between adjacent levels.

Example: Multiplicity function for harmonic oscillators.   The problem of the binary model system is the simplest problem for which an exact solution for the multiplicity function is known. Another exactly solvable problem is the harmonic oscillator, for which the solution was originally given by Max Planck. The original derivation is often felt to be not entirely simple. The beginning student need not worry about this derivation. The modern way to do the problem is given in Chapter 4 and is simple.

The quantum states of a harmonic oscillator have the energy eigenvalues

$$\varepsilon_s = s\hbar\omega \ , \tag{49}$$

where the quantum number s is a positive integer or zero, and ω is the angular frequency of the oscillator. The number of states is infinite, and the multiplicity of each is one. Now consider a system of N such oscillators, all of the same frequency. We want to find the number of ways in which a given total excitation energy

$$\varepsilon = \sum_{i=1}^{N} s_i\hbar\omega = n\hbar\omega \tag{50}$$

can be distributed among the oscillators. That is, we want the multiplicity function $g(N,n)$ for the N oscillators. The oscillator multiplicity function is not the same as the spin multiplicity function found earlier.

We begin the analysis by going back to the multiplicity function for a single oscillator, for which $g(1,n) = 1$ for all values of the quantum number s, here identical to n. To solve the problem of (53) below, we need a function to represent or generate the series

$$\sum_{n=0}^{\infty} g(1,n)t^n = \sum_{n=0}^{\infty} t^n. \tag{51}$$

All $\sum$ run from 0 to ∞. Here t is just a temporary tool that will help us find the result (53), but t does not appear in the final result. The answer is

$$\frac{1}{1-t} = \sum_{n=0}^{\infty} t^n , \tag{52}$$

provided we assume $|t| < 1$. For the problem of N oscillators, the generating function is

$$\left(\frac{1}{1-t}\right)^N = \left(\sum_{s=0}^{\infty} t^s\right)^N = \sum_{n=0}^{\infty} g(N,n)t^n , \tag{53}$$

because the number of ways a term t^n can appear in the N-fold product is precisely the number of ordered ways in which the integer n can be formed as the sum of N non-negative integers.

We observe that

$$g(N,n) = \lim_{t \to 0} \frac{1}{n!}\left(\frac{d}{dt}\right)^n \sum_{s=0}^{\infty} g(N,s)t^s$$

$$= \lim_{t \to 0} \frac{1}{n!}\left(\frac{d}{dt}\right)^n (1-t)^{-N}$$

$$= \frac{1}{n!} N(N+1)(N+2)\cdots(N+n-1). \tag{54}$$

Thus for the system of oscillators,

$$g(N,n) = \frac{(N+n-1)!}{n!\,(N-1)!}. \tag{55}$$

This result will be needed in solving a problem in the next chapter.

SUMMARY

1. The multiplicity function for a system of N magnets with spin excess $2s = N_\uparrow - N_\downarrow$ is

$$g(N,s) = \frac{N!}{(\frac{1}{2}N + s)!\,(\frac{1}{2}N - s)!} = \frac{N!}{N_\uparrow!\,N_\downarrow!}.$$

In the limit $s/N \ll 1$, with $N \gg 1$, we have the Gaussian approximation

$$g(N,s) \simeq (2/\pi N)^{1/2} 2^N \exp(-2s^2/N).$$

2. If all states of the model spin system are equally likely, the average value of s^2 is

$$\langle s^2 \rangle = \int_{-\infty}^{\infty} ds\, s^2 g(N,s) \Big/ \int_{-\infty}^{\infty} ds\, g(N,s) = \tfrac{1}{4}N \ ,$$

in the Gaussian approximation.

3. The fractional fluctuation of s^2 is defined as $\langle s^2 \rangle^{1/2}/N$ and is equal to $1/2N^{1/2}$.

4. The energy of the model spin system in a state of spin excess $2s$ is

$$U(s) = -2smB \ ,$$

where m is the magnetic moment of one spin and B is the magnetic field.

Chapter 2

Entropy and Temperature

Note on problems: The method of this chapter can be used to solve some problems, as illustrated by Problems 1, 2, and 3. Because much simpler methods are developed in Chapter 3 and later, we do not emphasize problem solving at this stage.

One should not imagine that two gases in a 0.1 liter container, initially unmixed, will mix, then again after a few days separate, then mix again, and so forth. On the contrary, one finds . . . that not until a time enormously long compared to $10^{10^{10}}$ years will there by any noticeable unmixing of the gases. One may recognize that this is practically equivalent to never. . . .

L. Boltzmann

If we wish to find in rational mechanics an a priori foundation for the principles of thermodynamics, we must seek mechanical definitions of temperature and entropy.

J. W. Gibbs

The general connection between energy and temperature may only be established by probability considerations. [Two systems] are in statistical equilibrium when a transfer of energy does not increase the probability.

M. Planck

We start this chapter with a definition of probability that enables us to define the average value of a physical property of a system. We then consider systems in thermal equilibrium, the definition of entropy, and the definition of temperature. The second law of thermodynamics will appear as the law of increase of entropy. This chapter is perhaps the most abstract in the book. The chapters that follow will apply the concepts to physical problems.

FUNDAMENTAL ASSUMPTION

The **fundamental assumption** of thermal physics is that a closed system is equally likely to be in any of the quantum states accessible to it. All accessible quantum states are assumed to be equally probable—there is no reason to prefer some accessible states over other accessible states.

A **closed system** will have constant energy, a constant number of particles, constant volume, and constant values of all external parameters that may influence the system, including gravitational, electric, and magnetic fields.

A quantum state is **accessible** if its properties are compatible with the physical specification of the system: the energy of the state must be in the range within which the energy of the system is specified, and the number of particles must be in the range within which the number of particles is specified. With large systems we can never know either of these exactly, but it will suffice to have $\delta U/U \ll 1$ and $\delta N/N \ll 1$.

Unusual properties of a system may sometimes make it impossible for certain states to be accessible during the time the system is under observation. For example, the states of the crystalline form of SiO_2 are inaccessible at low temperatures in any observation that starts with the glassy or amorphous form: fused silica will not convert to quartz in our lifetime in a low-temperature experiment. You will recognize many exclusions of this type by common sense. We treat all quantum states as accessible unless they are excluded by the specification of the system (Figure 2.1) and the time scale of the measurement process. States that are not accessible are said to have zero probability.

Of course, it is possible to specify the configuration of a closed system to a point that its statistical properties as such are of no interest. If we specify that the

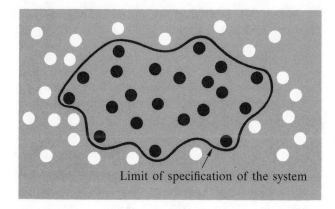

Limit of specification of the system

Figure 2.1   A purely symbolic diagram: each solid spot represents an accessible quantum state of a closed system. The fundamental assumption of statistical physics is that a closed system is equally likely to be in any of the quantum states accessible to it. The empty circles represent some of the states that are not accessible because their properties do not satisfy the specification of the system.

system is exactly in a stationary quantum state s, no statistical aspect is left in the problem.

PROBABILITY

Suppose we have a closed system that we know is equally likely to be in any of the g accessible quantum states. Let s be a general state label (and not one-half the spin excess). The probability $P(s)$ of finding the system in this state is

$$P(s) = 1/g \tag{1}$$

if the state s is accessible and $P(s) = 0$ otherwise, consistent with the fundamental assumption. We shall be concerned later with systems that are not closed, for which the energy U and particle number N may vary. For these systems $P(s)$ will not be a constant as in (1), but will have a functional dependence on U and on N.

The sum $\sum P(s)$ of the probability over all states is always equal to unity, because the total probability that the system is in some state is unity:

$$\sum_s P(s) = 1. \tag{2}$$

The probabilities defined by (1) lead to the definition of the average value of any physical property. Suppose that the physical property X has the value $X(s)$ when the system is in the state s. Here X might denote magnetic moment, energy, square of the energy, charge density near a point $\mathbf{r}$, or any property that can be observed when the system is in a quantum state. Then the average of the observations of the quantity X taken over a system described by the probabilities $P(s)$ is

$$\boxed{\langle X \rangle = \sum_s X(s)P(s).} \tag{3}$$

This equation defines the average value of X. Here $P(s)$ is the probability that the system is in the state s. The angular brackets $\langle \cdots \rangle$ are used to denote average value.

For a closed system, the average value of X is

$$\langle X \rangle = \sum_s X(s)(1/g) , \tag{4}$$

because now all g accessible states are equally likely, with $P(s) = 1/g$. The average in (4) is an elementary example of what may be called an **ensemble average**: we imagine g similar systems, one in each accessible quantum state. Such a group of systems constructed alike is called an ensemble of systems. The average of any property over the group is called the ensemble average of that property.

An ensemble of systems is composed of many systems, all constructed alike. Each system in the ensemble is a replica of the actual system in one of the quantum states accessible to the system. If there are g accessible states, then there will be g systems in the ensemble, one system for each state. Each system in the ensemble is equivalent for all practical purposes to the actual system. Each system satisfies all external requirements placed on the original system and in this sense is "just as good" as the actual system. Every quantum state

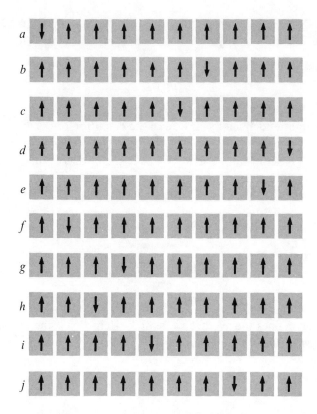

Figure 2.2   This ensemble a through j represents a system of 10 spins with energy $-8mB$ and spin excess $2s = 8$. The multiplicity $g(N,s)$ is $g(10,4) = 10$, so that the representative ensemble must contain 10 systems. The order in which the various systems in the ensemble are listed has no significance.

accessible to the actual system is represented in the ensemble by one system in a stationary quantum state, as in Figure 2.2. We assume that the ensemble represents the real system—this is implied in the fundamental assumption.

Example: Construction of an ensemble.   We construct in Figure 2.3 an ensemble to represent a closed system of five spins, each system with spin excess $2s = 1$. The energy of each in a magnetic field is $-mB$. (Do not confuse the use of s in spin excess with our frequent use of s as a state index or label.) Each system represents one of the multiples of

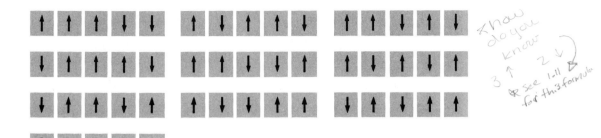

Figure 2.3   The ensemble represents a system with $N = 5$ spins and spin excess $2s = 1$.

Figure 2.4   With $N = 5$ and $2s = 5$, a single system may represent the ensemble. This is not a typical situation.

quantum states at this energy. The number of such states is given by the multiplicity function (1.15):

$$g(5,\tfrac{1}{2}) = \frac{5!}{3!\,2!} = 10.$$

The 10 systems shown in Figure 2.3 make up the ensemble.

If the energy in the magnetic field were such that $2s = 5$, then a single system comprises the ensemble, as in Figure 2.4. In zero magnetic field, all energies of all $2^N = 2^5 = 32$ states are equal, and the new ensemble must represent 32 systems, of which 1 system has $2s = 5$; 5 systems have $2s = 3$; 10 systems have $2s = 1$; 10 systems have $2s = -1$; 5 systems have $2s = -3$; and 1 system has $2s = -5$.

Most Probable Configuration

Let two systems $\mathcal{S}_1$ and $\mathcal{S}_2$ be brought into contact so that energy can be transferred freely from one to the other. This is called **thermal contact** (Figure 2.5). The two systems in contact form a larger closed system $\mathcal{S} = \mathcal{S}_1 + \mathcal{S}_2$ with constant energy $U = U_1 + U_2$. What determines whether there will be a net flow of energy from one system to another? The answer leads to the concept of temperature. The direction of energy flow is not simply a matter of whether the energy of one system is greater than the energy of the other, because the

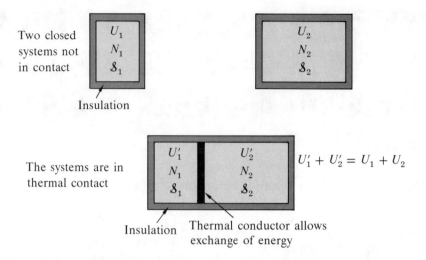

Figure 2.5 Establishment of thermal contact between two systems $\mathcal{S}_1$ and $\mathcal{S}_2$.

systems can be different in size and constitution. A constant total energy can be shared in many ways between two systems.

The most probable division of the total energy is that for which the combined system has the maximum number of accessible states. We shall enumerate the accessible states of two model systems and then study what characterizes the systems when in thermal contact. We first solve in detail the problem of thermal contact between two spin systems, 1 and 2, in a magnetic field which is introduced in order to define the energy. The numbers of spins N_1, N_2 may be different, and the values of the spin excess $2s_1$, $2s_2$ may be different for the two systems. All spins have magnetic moment m. The actual exchange of energy might take place via some weak (residual) coupling between the spins near the interface between the two systems. We assume that the quantum states of the total system $\mathcal{S}$ can be represented accurately by a combination of any state of $\mathcal{S}_1$ with any state of $\mathcal{S}_2$. We keep N_1, N_2 constant, but the values of the spin excess are allowed to change. The spin excess of a state of the combined system will be denoted by $2s$, where $s = s_1 + s_2$. The energy of the combined system is directly proportional to the total spin excess:

$$U(s) = U_1(s_1) + U_2(s_2) = -2mB(s_1 + s_2) = -2mBs. \qquad (5)$$

The total number of particles is $N = N_1 + N_2$.

We assume that the energy splittings between adjacent energy levels are equal to $2mB$ in both systems, so that the magnetic energy given up by system 1 when one spin is reversed can be taken up by the reversal of one spin of system 2 in the opposite sense. Any large physical system will have enough diverse modes of energy storage so that energy exchange with another system is always possible. The value of $s = s_1 + s_2$ is constant because the total energy is constant, but when the two systems are brought into thermal contact a redistribution is permitted in the separate values of s_1, s_2 and thus in the energies U_1, U_2.

The multiplicity function $g(N,s)$ of the combined system $\mathcal{S}$ is related to the product of the multiplicity functions of the individual systems $\mathcal{S}_1$ and $\mathcal{S}_2$ by the relation:

$$g(N,s) = \sum_{s_1} g_1(N_1,s_1)g_2(N_2,s - s_1) \;, \tag{6}$$

where the multiplicity functions g_1, g_2 are given by expressions of the form of (1.15). The range of s_1 in the summation is from $-\tfrac{1}{2}N_1$ to $\tfrac{1}{2}N_1$, if $N_1 < N_2$. To see how (6) comes about, consider first that configuration of the combined system for which the first system has spin excess $2s_1$ and the second system has spin excess $2s_2$. A **configuration** is defined as the set of all states with specified values of s_1 and s_2. The first system has $g_1(N_1,s_1)$ accessible states, each of which may occur together with any of the $g_2(N_2,s_2)$ accessible states of the second system. The total number of states in one configuration of the combined system is given by the product $g_1(N_1,s_1)g_2(N_2,s_2)$ of the multiplicity functions of $\mathcal{S}_1$ and $\mathcal{S}_2$. Because $s_2 = s - s_1$, the product of the g's may be written as

$$g_1(N_1,s_1)g_2(N_2,s - s_1). \tag{7}$$

This product forms one term of the sum (6).

Different configurations of the combined system are characterized by different values of s_1. We sum over all possible values of s_1 to obtain the total number of states of all the configurations with fixed s or fixed energy. We thus obtain (6), where $g(N,s)$ is the number of accessible states of the combined system. In the sum we hold s, N_1, and N_2 constant, as part of the specification of thermal contact.

The result (6) is a sum of products of the form (7). Such a product will be a maximum for some value of s_1, say $\hat{s}_1$, to be read as "s_1 hat" or "s_1 caret". The configuration for which g_1g_2 is a maximum is called the **most probable configuration**; the number of states in it is

$$g_1(N_1,\hat{s}_1)g_2(N_2,s - \hat{s}_1). \tag{8}$$

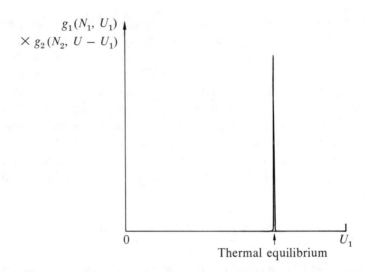

Figure 2.6 Schematic representation of the dependence of the configuration multiplicity on the division of the total energy between two systems, $\mathcal{S}_1$ and $\mathcal{S}_2$.

If the systems are large, the maximum with respect to changes in s_1 will be extremely sharp, as in Figure 2.6. A relatively small number of configurations will dominate the statistical properties of the combined system. The most probable configuration alone will describe many of these properties.

Such a sharp maximum is a property of every realistic type of large system for which exact solutions are available; we postulate that it is a general property of all large systems. From the sharpness property it follows that fluctuations about the most probable configuration are small, in a sense that we will define.

The important result follows that the values of the average physical properties of a large system in thermal contact with another large system are accurately described by the properties of the most probable configuration, the configuration for which the number of accessible states is a maximum. Such average values (used in either of these two senses) are called **thermal equilibrium values**.

Because of the sharp maximum, we may replace the average of a physical quantity over all accessible configurations (6) by an average over only the most probable configuration (8). In the example below we estimate the error involved in such a replacement and find the error to be negligible.

Example: Two spin systems in thermal contact. We investigate for the model spin system the sharpness of the product (7) near the maximum (8) as follows. We form the product of the multiplicity functions for $g_1(N_1,s_1)$ and $g_2(N_2,s_2)$, both of the form of (1.35):

$$g_1(N_1,s_1)g_2(N_2,s_2) = g_1(0)g_2(0)\exp\left(-\frac{2s_1^2}{N_1} - \frac{2s_2^2}{N_2}\right),\qquad (9)$$

where $g_1(0)$ denotes $g_1(N_1,0)$ and $g_2(0)$ denotes $g_2(N_2,0)$. We replace s_2 by $s - s_1$:

$$g_1(N_1,s_1)g_2(N_2,s - s_1) = g_1(0)g_2(0)\exp\left(-\frac{2s_1^2}{N_1} - \frac{2(s - s_1)^2}{N_2}\right).\qquad (10)$$

This product* gives the number of states accessible to the combined system when the spin excess of the combined system is $2s$, and the spin excess of the first system is $2s_1$.

We find the maximum value of (10) as a function of s_1 when the total spin excess $2s$ is held constant; that is, when the energy of the combined systems is constant. It is convenient to use the property that the maximum of $\log y(x)$ occurs at the same value of x as the maximum of $y(x)$. The calculation can be done either way. From (10),

$$\log g_1(N_1,s_1)g_2(N_2,s - s_1) = \log g_1(0)g_2(0) - \frac{2s_1^2}{N_1} - \frac{2(s - s_1)^2}{N_2}.\qquad (11)$$

This quantity is an extremum when the first derivative with respect to s_1 is zero. An extremum may be a maximum, a minimum, or a point of inflection. The extremum is a maximum if the second derivative of the function is negative, so that the curve bends downward.

At the extremum the first derivative is

$$\frac{\partial}{\partial s_1}\{\log g_1(N_1,s_1)g_2(N_2,s - s_1)\} = -\frac{4s_1}{N_1} + \frac{4(s - s_1)}{N_2} = 0,\qquad (12)$$

where N_1, N_2, and s are held constant as s_1 is varied. The second derivative $\partial^2/\partial s_1^2$ of Equation (11) is

$$-4\left(\frac{1}{N_1} + \frac{1}{N_2}\right)$$

---

* The product function of two Gaussian functions is always a Gaussian.

and is negative, so that the extremum is a maximum. Thus the most probable configuration of the combined system is that for which (12) is satisfied:

$$\frac{s_1}{N_1} = \frac{s - s_1}{N_2} = \frac{s_2}{N_2}. \tag{13}$$

The two systems are in equilibrium with respect to interchange of energy when the fractional spin excess of system 1 is equal to the fractional spin excess of system 2.

We prove that nearly all the accessible states of the combined systems satisfy or very nearly satisfy (13). If $\hat{s}_1$ and $\hat{s}_2$ denote the values of s_1 and s_2 at the maximum, then (13) is written as

$$\frac{\hat{s}_1}{N_1} = \frac{\hat{s}_2}{N_2} = \frac{s}{N}. \tag{14}$$

To find the number of states in the most probable configuration, we insert (14) in (9) to obtain

$$(g_1 g_2)_{\text{max}} \equiv g_1(\hat{s}_1) g_2(s - \hat{s}_1) = g_1(0) g_2(0) \exp(-2s^2/N). \tag{15}$$

To investigate the sharpness of the maximum of $g_1 g_2$ at a given value of s, introduce δ such that

$$s_1 = \hat{s}_1 + \delta; \qquad s_2 = \hat{s}_2 - \delta. \tag{16}$$

Here δ measures the deviation of s_1, s_2 from their values $\hat{s}_1, \hat{s}_2$ at the maximum of $g_1 g_2$. Square s_1, s_2 to form

$$s_1{}^2 = \hat{s}_1{}^2 + 2\hat{s}_1\delta + \delta^2; \qquad s_2{}^2 = \hat{s}_2{}^2 - 2\hat{s}_2\delta + \delta^2 ,$$

which we substitute in (9) and (15) to obtain the number of states

$$g_1(N_1,s_1) g_2(N_2,s_2) = (g_1 g_2)_{\text{max}} \exp\left(-\frac{4\hat{s}_1\delta}{N_1} - \frac{2\delta^2}{N_1} + \frac{4\hat{s}_2\delta}{N_2} - \frac{2\delta^2}{N_2}\right).$$

We know from (14) that $\hat{s}_1/N_1 = \hat{s}_2/N_2$, so that the number of states in a configuration of deviation δ from equilibrium is

$$g_1(N_1,\hat{s}_1 + \delta) g_2(N_2,\hat{s}_2 - \delta) = (g_1 g_2)_{\text{max}} \exp\left(-\frac{2\delta^2}{N_1} - \frac{2\delta^2}{N_2}\right). \tag{17}$$

As a numerical example in which the fractional deviation from equilibrium is very small, let $N_1 = N_2 = 10^{22}$ and $\delta = 10^{12}$; that is, $\delta/N_1 = 10^{-10}$. Then $2\delta^2/N_1 = 200$, and the product g_1g_2 is reduced to $e^{-400} \approx 10^{-174}$ of its maximum value. This is an extremely large reduction, so that g_1g_2 is truly a very sharply peaked function of s_1. The probability that the fractional deviation will be 10^{-10} *or larger* is found by integrating (17) from $\delta = 10^{12}$ out to a value of the order of s or of N, thereby including the area under the wings of the probability distribution. This is the subject of Problem 6. An upper limit to the integrated probability is given by $N \times 10^{-174} = 10^{-152}$, still very small. When two systems are in thermal contact, the values of s_1, s_2 that occur most often will be very close to the values of $\hat{s}_1, \hat{s}_2$ for which the product g_1g_2 is a maximum. It is extremely rare to find systems with values of s_1, s_2 perceptibly different from $\hat{s}_1, \hat{s}_2$.

What does it mean to say that the probability of finding the system with a fractional deviation larger than $\delta/N_1 = 10^{-10}$ is only 10^{-152} of the probability of finding the system in equilibrium? We mean that the system will never be found with a deviation as much as 1 part in 10^{10}, small as this deviation seems. We would have to sample 10^{152} similar systems to have a reasonable chance of success in such an experiment. If we sample one system every 10^{-12} s, which is pretty fast work, we would have to sample for 10^{140} s. The age of the universe is only 10^{18} s. Therefore we say with great surety that the deviation described will never be observed. The estimate is rough, but the message is correct. The quotation from Boltzmann given at the beginning of this chapter is relevant here.

We may expect to observe substantial fractional deviations only in the properties of a *small* system in thermal contact with a large system or reservoir. The energy of a small system, say a system of 10 spins, in thermal contact with a large reservoir may undergo fluctuations that are large in a fractional sense, as have been observed in experiments on the Brownian motion of small particles in suspension in liquids. The average energy of a small system in contact with a large system can always be determined accurately by observations at one time on a large number of identical small systems or by observations on one small system over a long period of time.

THERMAL EQUILIBRIUM

The result for the number of accessible states of two model spin systems in thermal contact may be generalized to any two systems in thermal contact, with constant total energy $U = U_1 + U_2$. By direct extension of the earlier argument, the multiplicity $g(N,U)$ of the combined system is:

$$g(N,U) = \sum_{U_1} g_1(N_1,U_1)g_2(N_2,U - U_1) , \qquad (18)$$

summed over all values of $U_1 \leq U$. Here $g_1(N_1,U_1)$ is the number of accessible states of system 1 at energy U_1. A configuration of the combined system is specified by the value of U_1, together with the constants U, N_1, N_2. The number of accessible states in a configuration is the product $g_1(N_1,U_1)g_2(N_2,U - U_1)$. The sum over all configurations gives $g(N,U)$.

The largest term in the sum in (18) governs the properties of the total system in thermal equilibrium. For an extremum it is necessary that the differential* of $g(N,U)$ be zero for an infinitesimal exchange of energy:

$$dg = \left(\frac{\partial g_1}{\partial U_1}\right)_{N_1} g_2 \, dU_1 + g_1 \left(\frac{\partial g_2}{\partial U_2}\right)_{N_2} dU_2 = 0; \qquad dU_1 + dU_2 = 0. \quad (19)$$

We divide by $g_1 g_2$ and use the result $dU_2 = -dU_1$ to obtain the thermal equilibrium condition:

$$\frac{1}{g_1}\left(\frac{\partial g_1}{\partial U_1}\right)_{N_1} = \frac{1}{g_2}\left(\frac{\partial g_2}{\partial U_2}\right)_{N_2}, \qquad\qquad (20a)$$

which we may write as

$$\left(\frac{\partial \log g_1}{\partial U_1}\right)_{N_1} = \left(\frac{\partial \log g_2}{\partial U_2}\right)_{N_2}. \qquad\qquad (20b)$$

We define the quantity σ, called the **entropy**, by

$$\boxed{\sigma(N,U) \equiv \log g(N,U) \;,} \qquad\qquad (21)$$

where σ is the Greek letter sigma. We now write (20) in the final form

$$\boxed{\left(\frac{\partial \sigma_1}{\partial U_1}\right)_{N_1} = \left(\frac{\partial \sigma_2}{\partial U_2}\right)_{N_2}.} \qquad\qquad (22)$$

---

* The notation

$$\left(\frac{\partial g_1}{\partial U_1}\right)_{N_1}$$

means that N_1 is held constant in the differentiation of $g_1(N_1, U_1)$ with respect to U_1. That is, the partial derivative with respect to U_1 is defined as

$$\left(\frac{\partial g_1}{\partial U_1}\right)_{N_1} = \lim_{\Delta U_1 \to 0} \frac{g_1(N_1, U_1 + \Delta U_1) - g_1(N_1, U_1)}{\Delta U_1}.$$

For example, if $g(x,y) = 3x^4 y$, then $(\partial g/\partial x)_y = 12x^3 y$ and $(\partial g/\partial y)_x = 3x^4$.

This is the condition for thermal equilibrium for two systems in thermal contact. Here N_1 and N_2 may symbolize not only the numbers of particles, but all constraints on the systems.

TEMPERATURE

The last equality (22) leads us immediately to the concept of temperature. We know the everyday rule: in thermal equilibrium the temperatures of the two systems are equal:

$$T_1 = T_2. \tag{23}$$

This rule must be equivalent to (22), so that T must be a function of $(\partial\sigma/\partial U)_N$. If T denotes the absolute temperature in kelvin, this function is simply the inverse relationship

$$\frac{1}{T} = k_B \left(\frac{\partial\sigma}{\partial U} \right)_N. \tag{24}$$

The proportionality constant k_B is a universal constant called the **Boltzmann constant**. As determined experimentally,

$$k_B = 1.381 \times 10^{-23} \text{ joules/kelvin}$$
$$= 1.381 \times 10^{-16} \text{ ergs/kelvin}. \tag{25}$$

We defer the discussion to Appendix B because we prefer to use a more natural temperature scale: we define the **fundamental temperature** τ by

$$\boxed{\frac{1}{\tau} = \left(\frac{\partial\sigma}{\partial U} \right)_N.} \tag{26}$$

This temperature differs from the Kelvin temperature by the scale factor, k_B:

$$\boxed{\tau = k_B T.} \tag{27}$$

Because σ is a pure number, the fundamental temperature τ has the dimensions of energy. We can use as a temperature scale the energy scale, in whatever unit may be employed for the latter—joule or erg. This procedure is much simpler than the introduction of the Kelvin scale in which the unit of temperature is arbitrarily selected so that the triple point of water is exactly 273.16 K. The triple point of water is the unique temperature at which water, ice, and water vapor coexist.

Historically, the conventional scale dates from an age in which it was possible to build accurate thermometers even though the relation of temperature to quantum states was as yet not understood. Even at present, it is still possible to measure temperatures with thermometers calibrated in kelvin to a higher precision than the accuracy with which the conversion factor k_B itself is known— about 32 parts per million. Questions of practical thermometry are discussed in Appendix B.

Comment.   In (26) we defined the reciprocal of τ as the partial derivative $(\partial\sigma/\partial U)_N$. It is permissible to take the reciprocal of both sides to write

$$\tau = (\partial U/\partial\sigma)_N. \tag{28}$$

The two expressions (26) and (28) have a slightly different meaning. In (26), the entropy σ was given as a function of the independent variables U and N as $\sigma = \sigma(U,N)$. Hence τ determined from (26) has the same independent variables, $\tau = \tau(U,N)$. In (28), however, differentiation of U with respect to σ with N constant implies $U = U(\sigma,N)$, so that $\tau = \tau(\sigma,N)$. The definition of temperature is the same in both cases, but it is expressed as a function of different independent variables. The question "What are the independent variables?" arises frequently in thermal physics because in some experiments we control some variables, and in other experiments we control other variables.

ENTROPY

The quantity $\sigma \equiv \log g$ was introduced in (21) as the entropy of the system: **the entropy is defined as the logarithm of the number of states accessible to the system.** As defined, the entropy is a pure number. In classical thermodynamics the entropy S is defined by

$$\frac{1}{T} = \left(\frac{\partial S}{\partial U}\right)_N. \tag{29}$$

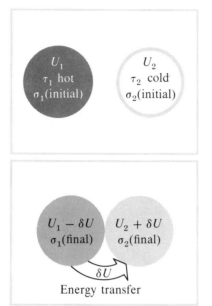

Figure 2.7   If the temperature τ_1 is higher than τ_2, the transfer of a positive amount of energy δU from system 1 to system 2 will increase the total entropy $\sigma_1 + \sigma_2$ of the combined systems over the initial value σ_1(initial) $+ \sigma_2$(initial). In other words, the final system will be in a more probable condition if energy flows from the warmer body to the cooler body when thermal contact is established. This is an example of the law of increasing entropy.

σ_1(final) $+ \sigma_2$(final) $> \sigma_1$(initial) $+ \sigma_2$(initial)

As a consequence of (24), we see that S and σ are connected by a scale factor:

$$S = k_B \sigma. \tag{30}$$

We will call S the conventional entropy.

The more states that are accessible, the greater the entropy. In the definition of $\sigma(N,U)$ we have indicated a functional dependence of the entropy on the number of particles in the system and on the energy of the system. The entropy may depend on additional independent variables: the entropy of a gas (Chapter 3) depends on the volume.

In the early history of thermal physics the physical significance of the entropy was not known. Thus the author of the article on thermodynamics in the *Encyclopaedia Britannica*, 11th ed. (1905), wrote: "The utility of the conception of entropy . . . is limited by the fact that it does not correspond directly to any directly measurable physical property, but is merely a mathematical function of the definition of absolute temperature." We now know what absolute physical property the entropy measures. An example of the comparison of the experimental determination and theoretical calculation of the entropy is discussed in Chapter 6.

Consider the total entropy change $\Delta\sigma$ when we remove a positive amount of energy ΔU from 1 and add the same amount of energy to 2, as in Figure 2.7.

The total entropy change is

$$\Delta\sigma = \left(\frac{\partial\sigma_1}{\partial U_1}\right)_{N_1}(-\Delta U) + \left(\frac{\partial\sigma_2}{\partial U_2}\right)_{N_2}(\Delta U) = \left(-\frac{1}{\tau_1} + \frac{1}{\tau_2}\right)\Delta U. \quad (31)$$

When $\tau_1 > \tau_2$ the quantity in parentheses on the right-hand side is positive, so that the total change of entropy is positive when the direction of energy flow is from the system with the higher temperature to the system with the lower temperature.

Example: Entropy increase on heat flow.   This example makes use of the reader's previous familiarity with heat and specific heat.

(a) Let a 10-g specimen of copper at a temperature of 350 K be placed in thermal contact with an identical specimen at a temperature of 290 K. Let us find the quantity of energy transferred when the two specimens are placed in contact and come to equilibrium at the final temperature T_f. The specific heat of metallic copper over the temperature range 15°C to 100°C is approximately $0.389\,\mathrm{J\,g^{-1}\,K^{-1}}$, according to a standard handbook.

The energy increase of the second specimen is equal to the energy loss of the first; thus the energy increase of the second specimen is, in joules,

$$\Delta U = (3.89\,\mathrm{J\,K^{-1}})(T_f - 290\,\mathrm{K}) = (3.89\,\mathrm{J\,K^{-1}})(350\,\mathrm{K} - T_f) ,$$

where the temperatures are in kelvin. The final temperature after contact is

$$T_f = \tfrac{1}{2}(350 + 290)\mathrm{K} = 320\,\mathrm{K}.$$

Thus

$$\Delta U_1 = (3.89\,\mathrm{J\,K^{-1}})(-30\,\mathrm{K}) = -11.7\,\mathrm{J} ,$$

and

$$\Delta U_2 = -\Delta U_1 = 11.7\,\mathrm{J}.$$

(b) What is the change of entropy of the two specimens when a transfer of 0.1 J has taken place, almost immediately after initial contact? Notice that this transfer is a small fraction of the final energy transfer as calculated above. Because the energy transfer considered is small, we may suppose the specimens are approximately at their initial temperatures of 350 and 290 K. The entropy of the first body is decreased by

$$\Delta S_1 = \frac{-0.1\,\mathrm{J}}{350\,\mathrm{K}} = -2.86 \times 10^{-4}\,\mathrm{J\,K^{-1}}.$$

The entropy of the second body is increased by

$$\Delta S_2 = \frac{0.1\,\mathrm{J}}{290\,\mathrm{K}} = 3.45 \times 10^{-4}\,\mathrm{J\,K^{-1}}.$$

The total entropy increases by

$$\Delta S_1 + \Delta S_2 = (-2.86 + 3.45) \times 10^{-4}\,\mathrm{J\,K^{-1}} = 0.59 \times 10^{-4}\,\mathrm{J\,K^{-1}}.$$

In fundamental units the increase of entropy is

$$\Delta \sigma = \frac{0.59 \times 10^{-4}}{k_B} = \frac{0.59 \times 10^{-4}\,\mathrm{J\,K^{-1}}}{1.38 \times 10^{-23}\,\mathrm{J\,K^{-1}}} = 0.43 \times 10^{19}\,, \tag{32}$$

where k_B is the Boltzmann constant. This result means that the number of accessible states of the two systems increases by the factor $\exp(\Delta\sigma) = \exp(0.43 \times 10^{19})$.

Law of Increase of Entropy

We can show that the total entropy always increases when two systems are brought into thermal contact. We have just demonstrated this in a special case. If the total energy $U = U_1 + U_2$ is constant, the total multiplicity after the systems are in thermal contact is

$$g(U) = \sum_{U_1} g_1(U_1)g_2(U - U_1)\,, \tag{33}$$

by (18). This expression contains the term $g_1(U_{10})g_2(U - U_{10})$ for the initial multiplicity before contact and many other terms besides. Here U_{10} is the initial energy of system 1 and $U - U_{10}$ is the initial energy of system 2. Because all terms in (33) are positive numbers, the multiplicity is always increased by establishment of thermal contact between two systems. This is a proof of the law of increase of entropy for a well-defined operation.

The significant effect of contact, the effect that stands out even after taking the logarithm of the multiplicity, is not just that the number of terms in the summation is large, but that the largest single term in the summation may be very, very much larger than the initial multiplicity. That is,

$$(g_1 g_2)_{\mathrm{max}} \equiv g_1(\hat{U}_1)g_2(U - \hat{U}_1) \tag{34}$$

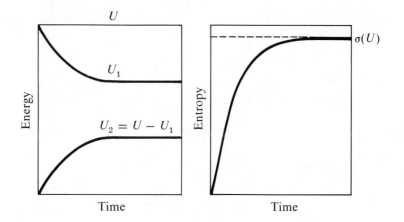

Figure 2.8   A system with two parts, 1 and 2, is prepared at zero time with $U_2 = 0$ and $U_1 = U$. Exchange of energy takes place between two parts and presently the system will be found in or close to the most probable configuration. The entropy increases as the system attains configurations of increasing multiplicity or probability. The entropy eventually reaches the entropy $\sigma(U)$ of the most probable configuration.

may be very, very much larger than the initial term

$$g_1(U_{10})g_2(U - U_{10}). \tag{35}$$

Here $\hat{U}_1$ denotes the value of U_1 for which the product g_1g_2 is a maximum. The essential effect is that the systems after contact evolve from their initial configurations to their final configurations. The fundamental assumption implies that evolution in this operation will always take place, with all accessible final states equally probable.

The statement

$$\sigma_{\text{final}} \simeq \log(g_1g_2)_{\text{max}} \geq \sigma_{\text{initial}} = \log(g_1g_2)_0 \tag{36}$$

is a statement of the **law of increase of entropy**: the entropy of a closed system tends to remain constant or to increase when a constraint internal to the system is removed. The operation of establishing thermal contact is equivalent to the removal of the constraint that U_1, U_2 each be constant; after contact only $U_1 + U_2$ need be constant.

The evolution of the combined system towards the final thermal equilibrium configuration takes a certain time. If we separate the two systems before they

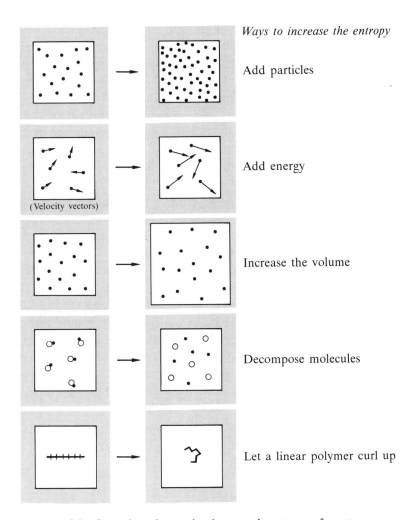

Ways to increase the entropy

Add particles

Add energy

(Velocity vectors)

Increase the volume

Decompose molecules

Let a linear polymer curl up

Figure 2.9   Operations that tend to increase the entropy of a system.

reach this configuration, we will obtain an intermediate configuration with intermediate energies and an intermediate entropy. It is therefore meaningful to view the entropy as a function of the time that has elapsed since removal of the constraint, called the time of evolution in Figure 2.8.

Processes that tend to increase the entropy of a system are shown in Figure 2.9; the arguments in support of each process will be developed in the chapters that follow.

For a large system* (in thermal contact with another large system) there will never occur spontaneously significant differences between the actual value of the entropy and the value of the entropy of the most probable configuration of the system. We showed this for the model spin system in the argument following (17); we used "never" in the sense of not once in the entire age of the universe, 10^{18} s. We can only find a significant difference between the actual entropy and the entropy of the most probable configuration of the macroscopic system very shortly after we have changed the nature of the contact between two systems, which implies that we had prepared the system initially in some special way. Special preparation could consist of lining up all the spins in one system parallel to one another or collecting all the molecules in the air of the room into the system formed by a small volume in one corner of the room. Such extreme situations never arise naturally in systems left undisturbed, but arise from artificial operations performed on the system.

Consider the gas in a room: the gas in one half of the room might be prepared initially with a low value of the average energy per molecule, while the gas in the other half of the room might be prepared initially with a higher value of the average energy per molecule. If the gas in the two halves is now allowed to interact by removal of a partition, the gas molecules will come very quickly[†] to a most probable configuration in which the molecules in both halves of the room have the same average energy. Nothing else will ever be observed to happen. We will never observe the system to leave the most probable configuration and reappear later in the initial specially prepared configuration. This is true even though the equations of motion of physics are reversible in time and do not distinguish past and future.

LAWS OF THERMODYNAMICS

When thermodynamics is studied as a nonstatistical subject, four postulates are introduced. These postulates are called the laws of thermodynamics. In essence, these laws are contained within our statistical formulation of thermal physics, but it is useful to exhibit them as separate statements.

Zeroth law. If two systems are in thermal equilibrium with a third system, they must be in thermal equilibrium with each other. This law is a consequence

---

* A large or macroscopic system may be taken to be one with more than 10^{10} or 10^{15} atoms.
† The calculation of the time required for the process is largely a problem in hydrodynamics.

of the condition (20b) for equilibrium in thermal contact:

$$\left(\frac{\partial \log g_1}{\partial U_1}\right)_{N_1} = \left(\frac{\partial \log g_3}{\partial U_3}\right)_{N_3}; \qquad \left(\frac{\partial \log g_2}{\partial U_2}\right)_{N_2} = \left(\frac{\partial \log g_3}{\partial U_3}\right)_{N_3}.$$

In other words, $\tau_1 = \tau_3$ and $\tau_2 = \tau_3$ imply $\tau_1 = \tau_2$.

First law. Heat is a form of energy. This law is no more than a statement of the principle of conservation of energy. Chapter 8 discusses what form of energy heat is.

Second law. There are many equivalent statements of the second law. We shall use the statistical statement, which we have called the law of increase of entropy, applicable when a constraint internal to a closed system is removed. The commonly used statement of the law of increase of entropy is: "If a closed system is in a configuration that is not the equilibrium configuration, the most probable consequence will be that the entropy of the system will increase monotonically in successive instants of time." This is a looser statement than the one we gave with Eq. (36) above.

The traditional thermodynamic statement is the Kelvin-Planck formulation of second law of thermodynamics: "It is impossible for any cyclic process to occur whose sole effect is the extraction of heat from a reservoir and the performance of an equivalent amount of work." An engine that violates the second law by extracting the energy of one heat reservoir is said to be performing perpetual motion of the second kind. We will see in Chapter 8 that the Kelvin-Planck formulation is a consequence of the statistical statement.

Third law. The entropy of a system approaches a constant value as the temperature approaches zero. The earliest statement of this law, due to Nernst, is that at the absolute zero the entropy difference disappears between all those configurations of a system which are in internal thermal equilibrium. The third law follows from the statistical definition of the entropy, provided that the ground state of the system has a well-defined multiplicity. If the ground state multiplicity is $g(0)$, the corresponding entropy is $\sigma(0) = \log g(0)$ as $\tau \to 0$. From a quantum point of view, the law does not appear to say much that is not implicit in the definition of entropy, provided, however, that the system is in its lowest set of quantum states at absolute zero. Except for glasses, there would not be any objection to affirming that $g(0)$ is a small number and $\sigma(0)$ is essentially zero. Glasses have a frozen-in disorder, and for them $\sigma(0)$ can be substantial, of the order of the number of atoms N. What the third law tells us in real life is that curves of many reasonable physical quantities plotted against τ must come in flat as τ approaches 0.

Entropy as a Logarithm

Several useful properties follow from the definition of the entropy as the logarithm of the number of accessible states, instead of as the number of accessible states itself. First, the entropy of two independent systems is the sum of the separate entropies.

Second, the entropy is entirely insensitive—for all practical purposes—to the precision δU with which the energy of a closed system is defined. We have never meant to imply that the system energy is known exactly, a circumstance that for a discrete spectrum of energy eigenvalues would make the number of accessible states depend erratically on the energy. We have simply not paid much attention to the precision, whether it be determined by the uncertainty principle $\delta U \, \delta(\text{time}) \sim \hbar$, or determined otherwise. Define $\mathfrak{D}(U)$ as the number of accessible states per unit energy range; $\mathfrak{D}(U)$ can be a suitable smoothed average centered at U. Then $g(U) = \mathfrak{D}(U)\delta U$ is the number of accessible states in the range δU at U. The entropy is

$$\sigma(U) = \log \mathfrak{D}(U)\delta U = \log \mathfrak{D}(U) + \log \delta U. \qquad (37)$$

Typically, as for the system of N spins, the total number of states will be of the order of 2^N. If the total energy is of the order of N times some average one-particle energy Δ, then $\mathfrak{D}(U) \sim 2^N/N\Delta$. Thus

$$\sigma(U) = N \log 2 - \log N\Delta + \log \delta U. \qquad (38)$$

Let $N = 10^{20}$; $\Delta = 10^{-14}$ erg; and $\delta U = 10^{-1}$ erg.

$$\sigma(U) = 0.69 \times 10^{20} - 13.82 - 2.3. \qquad (39)$$

We see from this example that the value of the entropy is dominated overwhelmingly by the value of N; the precision δU is without perceptible effect on the result. In the problem of N free particles in a box, the number of states is proportional to something like $U^N \delta U$, whence $\sigma \sim N \log U + \log \delta U$. Again the term in N is dominant, a conclusion independent of even the system of units used for the energy.

Example: Perpetual motion of the second kind.   Early in our study of physics we came to understand the impossibility of a perpetual motion machine, a machine that will give forth more energy than it absorbs.

Equally impossible is a perpetual motion machine of the second kind, as it is called, in which heat is extracted from part of the environment and delivered to another part of the environment, the difference in temperature thus established being used to power a heat engine that delivers mechanical work available for any purpose at no cost to us. In brief, we cannot propel a ship by cooling the surrounding ocean to extract the energy necessary to propel the ship. The spontaneous transfer of energy from the low temperature ocean to a higher temperature boiler on the ship would decrease the total entropy of the combined systems and would thus be in violation of the law of increase of entropy.

SUMMARY

1. The fundamental assumption is that a closed system is equally likely to be in any of the quantum states accessible to it.

2. If $P(s)$ is the probability that a system is in the state s, the average value of a quantity X is

$$\langle X \rangle = \sum_s X(s)P(s).$$

3. An ensemble of systems is composed of very many systems, all constructed alike.

4. The number of accessible states of the combined systems 1 and 2 is

$$g(s) = \sum_s g_1(s_1)g_2(s - s_1) \, ,$$

where $s_1 + s_2 = s$.

5. The entropy $\sigma(N,U) \equiv \log g(N,U)$. The relation $S = k_B \sigma$ connects the conventional entropy S with the fundamental entropy σ.

6. The fundamental temperature τ is defined by

$$1/\tau \equiv (\partial \sigma / \partial U)_{N,V}.$$

The relation $\tau = k_B T$ connects the fundamental temperature and the conventional temperature.

7. The law of increase of entropy states that the entropy of a closed system tends to remain constant or to increase when a constraint internal to the system is removed.

8. The thermal equilibrium values of the physical properties of a system are defined as averages over all states accessible when the system is in contact with a large system or reservoir. If the first system also is large, the thermal equilibrium properties are given accurately by consideration of the states in the most probable configuration alone.

PROBLEMS

1. Entropy and temperature.   Suppose $g(U) = CU^{3N/2}$, where C is a constant and N is the number of particles. (a) Show that $U = \frac{3}{2}N\tau$. (b) Show that $(\partial^2\sigma/\partial U^2)_N$ is negative. This form of $g(U)$ actually applies to an ideal gas.

2. Paramagnetism.   Find the equilibrium value at temperature τ of the fractional magnetization

$$M/Nm = 2\langle s \rangle/N$$

of the system of N spins each of magnetic moment m in a magnetic field B. The spin excess is $2s$. Take the entropy as the logarthithm of the multiplicity $g(N,s)$ as given in (1.35):

$$\sigma(s) \simeq \log g(N,0) - 2s^2/N \ , \tag{40}$$

for $|s| \ll N$. *Hint*: Show that in this approximation

$$\sigma(U) = \sigma_0 - U^2/2m^2B^2N \ , \tag{41}$$

with $\sigma_0 = \log g(N,0)$. Further, show that $1/\tau = -U/m^2B^2N$, where U denotes $\langle U \rangle$, the thermal average energy.

3. Quantum harmonic oscillator.   (a) Find the entropy of a set of N oscillators of frequency ω as a function of the total quantum number n. Use the multiplicity function (1.55) and make the Stirling approximation $\log N! \simeq N \log N - N$. Replace $N - 1$ by N. (b) Let U denote the total energy $n\hbar\omega$ of the oscillators. Express the entropy as $\sigma(U,N)$. Show that the total energy at temperature τ is

$$U = \frac{N\hbar\omega}{\exp(\hbar\omega/\tau) - 1}. \tag{42}$$

This is the Planck result; it is derived again in Chapter 4 by a powerful method that does not require us to find the multiplicity function.

4. The meaning of "never." It has been said* that "six monkeys, set to strum unintelligently on typewriters for millions of years, would be bound in time to write all the books in the British Museum." This statement is nonsense, for it gives a misleading conclusion about very, very large numbers. Could all the monkeys in the world have typed out a single specified book in the age of the universe?[†]

Suppose that 10^{10} monkeys have been seated at typewriters throughout the age of the universe, 10^{18} s. This number of monkeys is about three times greater than the present human population[‡] of the earth. We suppose that a monkey can hit 10 typewriter keys per second. A typewriter may have 44 keys; we accept lowercase letters in place of capital letters. Assuming that Shakespeare's *Hamlet* has 10^5 characters, will the monkeys hit upon *Hamlet*?

(a) Show that the probability that any given sequence of 10^5 characters typed at random will come out in the correct sequence (the sequence of *Hamlet*) is of the order of

$$(\tfrac{1}{44})^{100\ 000} = 10^{-164\ 345}\ ,$$

where we have used $\log_{10} 44 = 1.64345$.

(b) Show that the probability that a *monkey-Hamlet* will be typed in the age of the universe is approximately $10^{-164\ 316}$. The probability of *Hamlet* is therefore zero in any operational sense of an event, so that the original statement at the beginning of this problem is nonsense: one book, much less a library, will never occur in the total literary production of the monkeys.

5. Additivity of entropy for two spin systems. Given two systems of $N_1 \simeq N_2 = 10^{22}$ spins with multiplicity functions $g_1(N_1,s_1)$ and $g_2(N_2, s - s_1)$, the product $g_1 g_2$ as a function of s_1 is relatively sharply peaked at $s_1 = \hat{s}_1$. For $s_1 = \hat{s}_1 + 10^{12}$, the product $g_1 g_2$ is reduced by 10^{-174} from its peak value. Use the Gaussian approximation to the multiplicity function; the form (17) may be useful.

(a) Compute $g_1 g_2/(g_1 g_2)_{max}$ for $s_1 = \hat{s}_1 + 10^{11}$ and $s = 0$.

(b) For $s = 10^{20}$, by what factor must you multiply $(g_1 g_2)_{max}$ to make it equal to $\sum_{s_1} g_1(N_1,s_1) g_2(N_2,s - s_1)$; give the factor to the nearest order of magnitude.

---

* J. Jeans, *Mysterious universe*, Cambridge University Press, 1930, p. 4. The statement is attributed to Huxley.
[†] For a related mathematico-literary study, see "The Library of Babel," by the fascinating Argentine writer Jorge Luis Borges, in *Ficciones*, Grove Press, Evergreen paperback, 1962, pp. 79–88.
[‡] For every person now alive, some thirty persons have once lived. This figure is quoted by A. C. Clarke in *2001*. We are grateful to the Population Reference Bureau and to Dr. Roger Revelle for explanations of the evidence. The cumulative number of man-seconds is 2×10^{20}, if we take the average lifetime as 2×10^9 s and the number of lives as 1×10^{11}. The cumulative number of man-seconds is much less than the number of monkey-seconds (10^{28}) taken in the problem.

(c) How large is the fractional error in the entropy when you ignore this factor?

6. Integrated deviation.   For the example that gave the result (17), calculate approximately the probability that the fractional deviation from equilibrium δ/N_1 is 10^{-10} or larger. Take $N_1 = N_2 = 10^{22}$. You will find it convenient to use an asymptotic expansion for the complementary error function. When $x \gg 1$,

$$2x\exp(x^2)\int_x^\infty \exp(-t^2)\,dt \approx 1 + \text{small terms.}$$

Chapter 3

Boltzmann Distribution and Helmholtz Free Energy

Units: Thermodynamic results can easily be translated from fundamental units to conventional units. The only quantity that will cause difficulty is the heat capacity, defined below in (17a) as $C(\text{fund.}) \equiv \tau(\partial\sigma/\partial\tau)$ in fundamental units and as $C(\text{conv.}) = T(\partial S/\partial T)$ in conventional units. These two quantities are not equal, for $C(\text{conv.}) = k_B C(\text{fund.})$.

The laws of thermodynamics may easily be obtained from the principles of statistical mechanics, of which they are the incomplete expression.

Gibbs

We are able to distinguish in mechanical terms the thermal action of one system on another from that which we call mechanical in the narrower sense . . . so as to specify cases of thermal action and cases of mechanical action.

Gibbs

In this chapter we develop the principles that permit us to calculate the values of the physical properties of a system as a function of the temperature. We assume that the system $\mathcal{S}$ of interest to us is in thermal equilibrium with a very large system $\mathcal{R}$, called the **reservoir**. The system and the reservoir will have a common temperature τ because they are in thermal contact.

The total system $\mathcal{R} + \mathcal{S}$ is a closed system, insulated from all external influences, as in Figure 3.1. The total energy $U_0 = U_\mathcal{R} + U_\mathcal{S}$ is constant. In particular, if the system is in a state of energy ε_s, then $U_0 - \varepsilon_s$ is the energy of the reservoir.

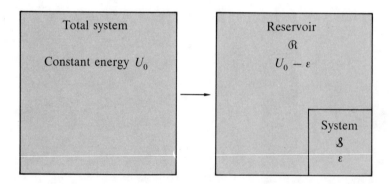

Figure 3.1 Representation of a closed total system decomposed into a reservoir $\mathcal{R}$ in thermal contact with a system $\mathcal{S}$.

BOLTZMANN FACTOR

A central problem of thermal physics is to find the probability that the system $\mathcal{S}$ will be in a specific quantum state s of energy ε_s. This probability is proportional to the Boltzmann factor.

When we specify that $\mathcal{S}$ should be in the state s, the number of accessible states of the total system is reduced to the number of accessible states of the reservoir $\mathcal{R}$, at the appropriate energy. That is, the number $g_{\mathcal{R}+\mathcal{S}}$ of states

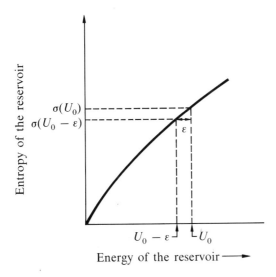

Figure 3.2 The change of entropy when the reservoir transfers energy ε to the system. The fractional effect of the transfer on the reservoir is small when the reservoir is large, because a large reservoir will have a high entropy.

accessible to $\mathcal{R} + \mathcal{S}$ is

$$g_{\mathcal{R}} \times 1 = g_{\mathcal{R}} , \qquad (1)$$

because for our present purposes we have specified the state of $\mathcal{S}$.

If the system energy is ε_s, the reservoir energy is $U_0 - \varepsilon_s$. The number of states accessible to the reservoir in this condition is $g_{\mathcal{R}}(U_0 - \varepsilon_s)$, as in Figure 3.2. The ratio of the probability that the system is in quantum state 1 at energy ε_1 to the probability that the system is in quantum state 2 at energy ε_2 is the ratio of the two multiplicities:

$$\frac{P(\varepsilon_1)}{P(\varepsilon_2)} = \frac{\text{Multiplicity of } \mathcal{R} \text{ at energy } U_0 - \varepsilon_1}{\text{Multiplicity of } \mathcal{R} \text{ at energy } U_0 - \varepsilon_2} = \frac{g_{\mathcal{R}}(U_0 - \varepsilon_1)}{g_{\mathcal{R}}(U_0 - \varepsilon_2)}. \qquad (2)$$

This result is a direct consequence of what we have called the fundamental assumption. The two situations are shown in Figure 3.3. Although questions about the system depend on the constitution of the reservoir, we shall see that the dependence is only on the temperature of the reservoir.

If the reservoirs are very large, the multiplicities are very, very large numbers. We write (2) in terms of the entropy of the reservoir:

$$\frac{P(\varepsilon_1)}{P(\varepsilon_2)} = \frac{\exp[\sigma_{\mathcal{R}}(U_0 - \varepsilon_1)]}{\exp[\sigma_{\mathcal{R}}(U_0 - \varepsilon_2)]} = \exp[\sigma_{\mathcal{R}}(U_0 - \varepsilon_1) - \sigma_{\mathcal{R}}(U_0 - \varepsilon_2)]. \qquad (3)$$

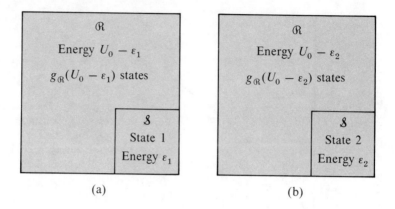

Figure 3.3   The system in (a), (b) is in quantum state 1, 2. The reservoir has $g_{\mathcal{R}}(U_0 - \varepsilon_1)$, $g_{\mathcal{R}}(U_0 - \varepsilon_2)$ accessible quantum states, in (a) and (b) respectively.

With

$$\Delta\sigma_{\mathcal{R}} \equiv \sigma_{\mathcal{R}}(U_0 - \varepsilon_1) - \sigma_{\mathcal{R}}(U_0 - \varepsilon_2) \; , \tag{4}$$

the probability ratio for the two states 1, 2 of the system is simply

$$\frac{P(\varepsilon_1)}{P(\varepsilon_2)} = \exp(\Delta\sigma_{\mathcal{R}}). \tag{5}$$

Let us expand the entropies in (4) in a Taylor series expansion about $\sigma_{\mathcal{R}}(U_0)$. The Taylor series expansion of $f(x)$ about $f(x_0)$ is

$$f(x_0 + a) = f(x_0) + a\left(\frac{df}{dx}\right)_{x=x_0} + \frac{1}{2!}a^2\left(\frac{d^2f}{dx^2}\right)_{x=x_0} + \cdots. \tag{6}$$

Thus

$$\sigma\,(U_0 - \varepsilon) = \sigma_{\mathcal{R}}(U_0) - \varepsilon(\partial\sigma_{\mathcal{R}}/\partial U)_{V,N} + \cdots$$
$$= \sigma_{\mathcal{R}}(U_0) - \varepsilon/\tau + \cdots \; , \tag{7}$$

where $1/\tau \equiv (\partial\sigma_{\mathcal{R}}/\partial U)_{V,N}$ gives the temperature. The partial derivative is taken at energy U_0. The higher order terms in the expansion vanish in the limit of an infinitely large reservoir.*

Therefore $\Delta\sigma_{\mathfrak{R}}$ defined by (4) becomes

$$\Delta\sigma_{\mathfrak{R}} = -(\varepsilon_1 - \varepsilon_2)/\tau. \tag{8}$$

The final result of (5) and (8) is

$$\frac{P(\varepsilon_1)}{P(\varepsilon_2)} = \frac{\exp(-\varepsilon_1/\tau)}{\exp(-\varepsilon_2/\tau)}. \tag{9}$$

A term of the form $\exp(-\varepsilon/\tau)$ is known as a **Boltzmann factor**. This result is of vast utility. It gives the ratio of the probability of finding the system in a single quantum state 1 to the probability of finding the system in a single quantum state 2.

Partition Function

It is helpful to consider the function

$$Z(\tau) = \sum_s \exp(-\varepsilon_s/\tau) , \tag{10}$$

called the **partition function**. The summation is over the Boltzmann factor $\exp(-\varepsilon_s/\tau)$ for all states s of the system. The partition function is the proportionality factor between the probability $P(\varepsilon_s)$ and the Boltzmann factor $\exp(-\varepsilon_s/\tau)$:

$$P(\varepsilon_s) = \frac{\exp(-\varepsilon_s/\tau)}{Z}. \tag{11}$$

We see that $\sum P(\varepsilon_s) = Z/Z = 1$: the sum of all probabilities is unity.

The result (11) is one of the most useful results of statistical physics. The average energy of the system is $U = \langle\varepsilon\rangle = \sum \varepsilon_s P(\varepsilon_s)$, or

$$U = \frac{\sum \varepsilon_s \exp(-\varepsilon_s/\tau)}{Z} = \tau^2(\partial \log Z/\partial\tau). \tag{12}$$

---

* We expand $\sigma(U_0 - \varepsilon)$ and not $g(U_0 - \varepsilon)$ because the expansion of the latter quantity immediately gives convergence difficulties.

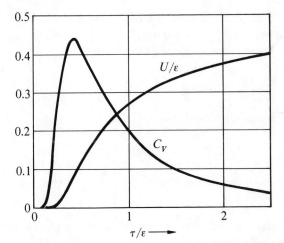

Figure 3.4   Energy and heat capacity of a two state system as functions of the temperature τ. The energy is plotted in units of ε.

The average energy refers to those states of a system that can exchange energy with a reservoir. The notation $\langle \cdots \rangle$ denotes such an average value and is called the **thermal average** or **ensemble average**. In (12) the symbol U is used for $\langle \varepsilon \rangle$ in conformity with common practice; U will now refer to the system and not, as earlier, to the system + reservoir.

Example: Energy and heat capacity of a two state system.   We treat a system of one particle with two states, one of energy 0 and one of energy ε. The particle is in thermal contact with a reservoir at temperature τ. We want to find the energy and the heat capacity of the system as a function of the temperature τ. The partition function for the two states of the particle is

$$Z = \exp(-0/\tau) + \exp(-\varepsilon/\tau) = 1 + \exp(-\varepsilon/\tau). \tag{13}$$

The average energy is

$$U \equiv \langle \varepsilon \rangle = \frac{\varepsilon \exp(-\varepsilon/\tau)}{Z} = \varepsilon \, \frac{\exp(-\varepsilon/\tau)}{1 + \exp(-\varepsilon/\tau)}. \tag{14}$$

This function is plotted in Figure 3.4.

If we shift the zero of energy and take the energies of the two states as $-\tfrac{1}{2}\varepsilon$ and $+\tfrac{1}{2}\varepsilon$, instead of as 0 and ε, the results appear differently. We have

$$Z = \exp(\varepsilon/2\tau) + \exp(-\varepsilon/2\tau) = 2\cosh(\varepsilon/2\tau) \text{ ,} \tag{15}$$

and

$$\langle \varepsilon \rangle = \frac{(-\tfrac{1}{2}\varepsilon)\exp(\varepsilon/2\tau) + (\tfrac{1}{2}\varepsilon)\exp(-\varepsilon/2\tau)}{Z} = -\varepsilon\frac{\sinh(\varepsilon/2\tau)}{2\cosh(\varepsilon/2\tau)}$$

$$= -\tfrac{1}{2}\varepsilon\tanh(\varepsilon/2\tau). \tag{16}$$

The **heat capacity** C_V of a system at constant volume is defined as

$$C_V \equiv \tau(\partial\sigma/\partial\tau)_V , \tag{17a}$$

which by the thermodynamic identity (34a) derived below is equivalent to the alternate definition

$$C_V \equiv (\partial U/\partial\tau)_V. \tag{17b}$$

We hold V constant because the values of the energy are calculated for a system at a specified volume. From (14) and (17b),

$$C_V = \varepsilon\frac{\partial}{\partial\tau}\frac{1}{\exp(\varepsilon/\tau) + 1} = \left(\frac{\varepsilon}{\tau}\right)^2\frac{\exp(\varepsilon/\tau)}{[\exp(\varepsilon/\tau) + 1]^2}. \tag{18a}$$

The same result follows from (16).

In conventional units C_V is defined as $T(\partial S/\partial T)_V$ or $(\partial U/\partial T)_V$, whence

$$c_V = C_V/m$$

(conventional) $$C_V = k_B\left(\frac{\varepsilon}{k_B T}\right)^2\frac{\exp(\varepsilon/k_B T)}{[\exp(\varepsilon/k_B T) + 1]^2}. \tag{18b}$$

In fundamental units the heat capacity is dimensionless; in conventional units it has the dimensions of energy per kelvin. The **specific heat** is defined as the heat capacity per unit mass.

The hump in the plot of heat capacity versus temperature in Figure 3.4 is called a Schottky anomaly. For $\tau \gg \varepsilon$ the heat capacity (18a) becomes

$$C_V \simeq (\varepsilon/2\tau)^2. \tag{19}$$

Notice that $C_V \propto \tau^{-2}$ in this high temperature limit. In the low temperature limit the temperature is small in comparison with the energy level spacing ε. For $\tau \ll \varepsilon$ we have

$$C_V \simeq (\varepsilon/\tau)^2\exp(-\varepsilon/\tau). \tag{20}$$

The exponential factor $\exp(-\varepsilon/\tau)$ reduces C_V rapidly as τ decreases, because $\exp(-1/x) \to 0$ as $x \to 0$.

Definition: Reversible process.   A process is reversible if carried out in such a way that the system is always infinitesimally close to the equilibrium condition. For example, if the entropy is a function of the volume, any change of volume must be carried out so slowly that the entropy at any volume V is closely equal to the equilibrium entropy $\sigma(V)$. Thus, the entropy is well defined at every stage of a reversible process, and by reversing the direction of the change the system will be returned to its initial condition. In reversible processes, the condition of the system is well defined at all times, in contrast to irreversible processes, where usually we will not know what is going on during the process. We cannot apply the mathematical methods of thermal physics to systems whose condition is undefined.

A volume change that leaves the system in the same quantum state is an example of an isentropic reversible process. If the system always remains in the same state the entropy change will be zero between any two stages of the process, because the number of states in an ensemble (p. 31) of similar systems does not change. Any process in which the entropy change vanishes is an isentropic reversible process. But reversible processes are not limited to isentropic processes, and we shall have a special interest also in isothermal reversible processes.

PRESSURE

Consider a system in the quantum state s of energy ε_s. We assume ε_s to be a function of the volume of the system. The volume is decreased slowly from V to $V - \Delta V$ by application of an external force. Let the volume change take place sufficiently slowly that the system remains in the same quantum state s throughout the compression. The "same" state may be characterized by its quantum numbers (Figure 3.5) or by the number of zeros in the wavefunction.

The energy of the state s after the reversible volume change is

$$\varepsilon_s(V - \Delta V) = \varepsilon_s(V) - (d\varepsilon_s/dV)\Delta V + \cdots. \tag{21}$$

Consider a pressure p_s applied normal to all faces of a cube. The mechanical work done on the system by the pressure in a contraction (Figure 3.6) of the cube volume from V to $V - \Delta V$ appears as the change of energy of the system:

$$U(V - \Delta V) - U(V) = \Delta U = -(d\varepsilon_s/dV)\Delta V. \tag{22}$$

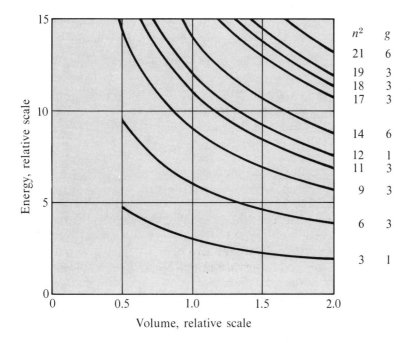

n^2	g
21	6
19	3
18	3
17	3
14	6
12	1
11	3
9	3
6	3
3	1

Figure 3.5 Dependence of energy on volume, for the energy levels of a free particle confined to a cube. The curves are labeled by $n^2 = n_x^2 + n_y^2 + n_z^2$, as in Figure 1.2. The multiplicities g are also given. The volume change here is isotropic: a cube remains a cube. The energy range $\delta\varepsilon$ of the states represented in an ensemble of systems will increase in a reversible compression, but we know from the discussion in Chapter 2 that the width of the energy range itself is of no practical importance. It is the change in the average energy that is important.

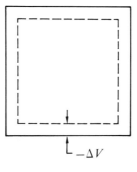

Figure 3.6 Volume change $-\Delta V$ in uniform compression of a cube.

Here U denotes the energy of the system. Let A be the area of one face of the cube; then

$$A(\Delta x + \Delta y + \Delta z) = \Delta V ,\tag{23}$$

if all increments ΔV and $\Delta x = \Delta y = \Delta z$ are taken as positive in the compression. The work done in the compression is

$$\Delta U = p_s A(\Delta x + \Delta y + \Delta z) = p_s \Delta V ,\tag{24}$$

so that, on comparison with (22),

$$p_s = -d\varepsilon_s/dV\tag{25}$$

is the pressure on a system in the state s.

We average (25) over all states of the ensemble to obtain the average pressure $\langle p \rangle$, usually written as p:

$$\boxed{p = -\left(\frac{\partial U}{\partial V}\right)_\sigma ,}\tag{26}$$

where $U \equiv \langle \varepsilon \rangle$. The entropy σ is held constant in the derivative because the number of states in the ensemble is unchanged in the reversible compression we have described. We have a collection of systems, each in some state, and each remains in this state in the compression.

The result (26) corresponds to our mechanical picture of the pressure on a system that is maintained in some specific state. Appendix D discusses the result more deeply. For applications we shall need also the later result (50) for the pressure on a system maintained at constant temperature.

We look for other expressions for the pressure. The number of states and thus the entropy depend only on U and on V, for a fixed number of particles, so that only the two variables U and V describe the system. The differential of the entropy is

$$d\sigma(U,V) = \left(\frac{\partial \sigma}{\partial U}\right)_V dU + \left(\frac{\partial \sigma}{\partial V}\right)_U dV.\tag{27}$$

This gives the differential change of the entropy for arbitrary independent differential changes dU and dV. Assume now that we select dU and dV interdependently, in such a way that the two terms on the right-hand side of (27) cancel. The overall entropy change $d\sigma$ will be zero. If we denote these inter-dependent values of dU and dV by $(\delta U)_\sigma$ and $(\delta V)_\sigma$, the entropy change will be zero:

$$0 = \left(\frac{\partial \sigma}{\partial U}\right)_V (\delta U)_\sigma + \left(\frac{\partial \sigma}{\partial V}\right)_U (\delta V)_\sigma. \tag{28}$$

After division by $(\delta V)_\sigma$,

$$0 = \left(\frac{\partial \sigma}{\partial U}\right)_V \frac{(\delta U)_\sigma}{(\delta V)_\sigma} + \left(\frac{\partial \sigma}{\partial V}\right)_U. \tag{29}$$

But the ratio $(\delta U)_\sigma/(\delta V)_\sigma$ is the partial derivative of U with respect to V at constant σ:

$$(\delta U)_\sigma/(\delta V)_\sigma \equiv (\partial U/\partial V)_\sigma. \tag{30}$$

With this and the definition $1/\tau \equiv (\partial \sigma/\partial U)_V$, Eq. (29) becomes

$$\left(\frac{\partial U}{\partial V}\right)_\sigma = -\tau \left(\frac{\partial \sigma}{\partial V}\right)_U. \tag{31}$$

By (26) the left-hand side of (31) is equal to $-p$, whence

$$p = \tau \left(\frac{\partial \sigma}{\partial V}\right)_U. \tag{32}$$

Thermodynamic Identity

Consider again the differential (27) of the entropy; substitute the new result for the pressure and the definition of τ to obtain

$$d\sigma = \frac{1}{\tau} dU + \frac{p}{\tau} dV , \tag{33}$$

or

$$\tau d\sigma = dU + pdV. \tag{34a}$$

This useful relation will be called the **thermodynamic identity.** The form with N variable will appear in (5.38). A simple transposition gives

$$dU = \tau d\sigma - pdV, \quad \text{or} \quad dU = TdS - pdV. \tag{34b}$$

If the actual process of change of state of the system is reversible, we can identify $\tau d\sigma$ as the heat added to the system and $-pdV$ as the work done on the system. The increase of energy is caused in part by mechanical work and in part by the transfer of heat. **Heat** is defined as the transfer of energy between two systems brought into thermal contact (Chapter 8).

HELMHOLTZ FREE ENERGY

The function

$$\boxed{F \equiv U - \tau\sigma} \tag{35}$$

is called the Helmholtz free energy. This function plays the part in thermal physics at constant temperature that the energy U plays in ordinary mechanical processes, which are always understood to be at constant entropy, because no internal changes of state are allowed. The free energy tells us how to balance the conflicting demands of a system for minimum energy and maximum entropy. The Helmholtz free energy will be a minimum for a system $\mathcal{S}$ in thermal contact with a reservoir $\mathcal{R}$, if the volume of the system is constant.

We first show that F is an extremum in equilibrium at constant τ and V. By definition, for infinitesimal reversible transfer from $\mathcal{R}$ to $\mathcal{S}$,

$$dF_{\mathcal{S}} = dU_{\mathcal{S}} - \tau d\sigma_{\mathcal{S}} \tag{36}$$

at constant temperature. But $1/\tau \equiv (\partial\sigma_{\mathcal{S}}/\partial U_{\mathcal{S}})_V$, so that $dU_{\mathcal{S}} = \tau d\sigma$ at constant volume. Therefore (36) becomes

$$dF_{\mathcal{S}} = 0 \ , \tag{37}$$

which is the condition for F to be an extremum with respect to all variations at constant volume and temperature. We like F because we can calculate it from the energy eigenvalues ε_s of the system (see p. 72).

Comment. We can show that the extremum is a minimum. The total energy is $U = U_{\Re} + U_{\delta}$. Then the total entropy is

$$\sigma = \sigma_{\Re} + \sigma_{\delta} = \sigma_{\Re}(U - U_{\delta}) + \sigma_{\delta}(U_{\delta})$$
$$\simeq \sigma_{\Re}(U) - U_{\delta}(\partial\sigma_{\Re}/\partial U_{\Re})_{V,N} + \sigma_{\delta}(U_{\delta}). \tag{38}$$

We know that

$$(\partial\sigma_{\Re}/\partial U_{\Re})_{V,N} \equiv 1/\tau \;, \tag{39}$$

so that (38) becomes

$$\sigma = \sigma_{\Re}(U) - F_{\delta}/\tau \;, \tag{40}$$

where $F_{\delta} = U_{\delta} - \tau\sigma_{\delta}$ is the free energy of the system. Now $\sigma_{\Re}(U)$ is constant; and we recall that $\sigma = \sigma_{\Re} + \sigma_{\delta}$ in equilibrium is a maximum with respect to U_{δ}. It follows from (40) that F_{δ} must be a minimum with respect to U_{δ} when the system is in the most probable configuration. The free energy of the system at constant τ, V will increase for any departure from the equilibrium configuration.

Example: Minimum property of the free energy of a paramagnetic system. Consider the model system of Chapter 1, with $N_{\uparrow}$ spins up and $N_{\downarrow}$ spins down. Let $N = N_{\uparrow} + N_{\downarrow}$; the spin excess is $2s = N_{\uparrow} - N_{\downarrow}$. The entropy in the Stirling approximation is found with the help of an approximate form of (1.31):

$$\sigma(s) \simeq -\left(\frac{1}{2}N + s\right)\log\left(\frac{1}{2} + \frac{s}{N}\right) - \left(\frac{1}{2}N - s\right)\log\left(\frac{1}{2} - \frac{s}{N}\right). \tag{41}$$

The energy in a magnetic field B is $-2smB$, where m is the magnetic moment of an elementary magnet. The free energy function (to be called the Landau function in Chapter 10) is $F_L(\tau,s,B) \equiv U(s,B) - \tau\sigma(s)$, or

$$F_L(\tau,s,B) = -2smB + \left(\frac{1}{2}N + s\right)\tau\log\left(\frac{1}{2} + \frac{s}{N}\right) + \left(\frac{1}{2}N - s\right)\tau\log\left(\frac{1}{2} - \frac{s}{N}\right). \tag{42}$$

At the minimum of $F_L(\tau,s,B)$ with respect to s, this function becomes equal to the equilibrium free energy $F(\tau,B)$. That is, $F_L(\tau,\langle s\rangle,B) = F(\tau,B)$, because $\langle s\rangle$ is a function of τ and B. The minimum of F_L with respect to the spin excess occurs when

$$(\partial F_L/\partial s)_{\tau,B} = 0 = -2mB + \tau\log\frac{N + 2s}{N - 2s}. \tag{43}$$

Thus in the magnetic field B the thermal equilibrium value of the spin excess $2s$ is given by

$$\frac{N + \langle 2s \rangle}{N - \langle 2s \rangle} = \exp(2mB/\tau); \qquad \langle 2s \rangle = N\left(\frac{\exp(2mB/\tau) - 1}{\exp(2mB/\tau) + 1}\right), \qquad (44)$$

or, on dividing numerator and denominator by $\exp(mB/\tau)$,

$$\langle 2s \rangle = N \tanh(mB/\tau). \qquad (45)$$

The magnetization M is the magnetic moment per unit volume. If n is the number of spins per unit volume, the magnetization in thermal equilibrium in the magnetic field is

$$M = \langle 2s \rangle m/V = nm \tanh(mB/\tau). \qquad (46)$$

The free energy of the system in equilibrium can be obtained by substituting (45) in (42). It is easier, however, to obtain F directly from the partition function for one magnet:

$$Z = \exp(mB/\tau) + \exp(-mB/\tau) = 2\cosh(mB/\tau). \qquad (47)$$

Now use the relation $F = -\tau \log Z$ as derived below. Multiply by N to obtain the result for N magnets. (The magnetization is derived more simply by the method of Problem 2.)

<hr>

Differential Relations

The differential of F is

$$dF = dU - \tau d\sigma - \sigma d\tau ,$$

or, with use of the thermodynamic identity (34a),

$$dF = -\sigma d\tau - p dV , \qquad (48)$$

for which

$$\left(\frac{\partial F}{\partial \tau}\right)_V = -\sigma; \qquad \left(\frac{\partial F}{\partial V}\right)_\tau = -p. \qquad (49)$$

These relations are widely used.

The free energy F in the result $p = -(\partial F/\partial V)_\tau$ acts as the effective energy for an *isothermal* change of volume; contrast this result with (26). The result may be written as

$$p = -\left(\frac{\partial U}{\partial V}\right)_\tau + \tau\left(\frac{\partial \sigma}{\partial V}\right)_\tau \,, \qquad (50)$$

by use of $F \equiv U - \tau\sigma$. The two terms on the right-hand side of (50) represent what we may call the energy pressure and the entropy pressure. The energy pressure $-(\partial U/\partial V)_\tau$ is dominant in most solids and the entropy pressure $\tau(\partial\sigma/\partial V)_\tau$ is dominant in gases and in elastic polymers such as rubber (Problem 10). The entropy contribution is testimony of the importance of the entropy: the naive feeling from simple mechanics that $-dU/dV$ must tell everything about the pressure is seriously incomplete for a process at constant temperature, because the entropy can change in response to the volume change even if the energy is independent of volume, as for an ideal gas at constant temperature.

Maxwell relation. We can now derive one of a group of useful thermodynamic relations called Maxwell relations. Form the cross-derivatives $\partial^2 F/\partial V\,\partial\tau$ and $\partial^2 F/\partial\tau\,\partial V$, which must be equal to each other. It follows from (49) that

$$(\partial\sigma/\partial V)_\tau = (\partial p/\partial\tau)_V \,, \qquad (51)$$

a relation that is not at all obvious. Other Maxwell relations will be derived later at appropriate points, by similar arguments. The methodology of obtaining thermodynamic relations is discussed by R. Gilmore, J. Chem. Phys. **75**, 5964 (1981).

Calculation of *F* from *Z*

Because $F \equiv U - \tau\sigma$ and $\sigma = -(\partial F/\partial\tau)_V$, we have the differential equation

$$F = U + \tau(\partial F/\partial\tau)_V, \qquad \text{or} \qquad -\tau^2\partial(F/\tau)/\partial\tau = U. \qquad (52)$$

We show that this equation is satisfied by

$$F/\tau = -\log Z \,, \qquad (53)$$

where Z is the partition function. On substitution,

$$\partial(F/\tau)/\partial\tau = -\partial\log Z/\partial\tau = -U/\tau^2 \qquad (54)$$

by (12). This proves that

$$\boxed{F = -\tau \log Z} \qquad (55)$$

satisfies the required differential equation (52).

It would appear possible for F/τ to contain an additive constant α such that $F = -\tau \log Z + \alpha\tau$. However, the entropy must reduce to $\log g_0$ when the temperature is so low that only the g_0 coincident states at the lowest energy ε_0 are occupied. In that limit $\log Z \to \log g_0 - \varepsilon_0/\tau$, so that $\sigma = -\partial F/\partial\tau \to \partial(\tau \log Z)/\partial\tau = \log g_0$ only if $\alpha = 0$.

We may write the result as

$$Z = \exp(-F/\tau); \qquad (56)$$

and the Boltzmann factor (11) for the occupancy probability of a quantum state s becomes

$$P(\varepsilon_s) = \frac{\exp(-\varepsilon_s/\tau)}{Z} = \exp[(F - \varepsilon_s)/\tau]. \qquad (57)$$

IDEAL GAS: A FIRST LOOK

One atom in a box. We calculate the partition function Z_1 of one atom of mass M free to move in a cubical box of volume $V = L^3$. The orbitals of the free particle wave equation $-(\hbar^2/2M)\nabla^2\psi = \varepsilon\psi$ are

$$\psi(x,y,z) = A \sin(n_x\pi x/L) \sin(n_y\pi y/L) \sin(n_z\pi z/L) , \qquad (58)$$

where n_x, n_y, n_z are any positive integers, as in Chapter 1. Negative integers do not give independent orbitals, and a zero does not give a solution. The energy values are

$$\varepsilon_n = \frac{\hbar^2}{2M}\left(\frac{\pi}{L}\right)^2 (n_x^2 + n_y^2 + n_z^2). \qquad (59)$$

We neglect the spin and all other structure of the atom, so that a state of the system is entirely specified by the values of n_x, n_y, n_z.

The partition function is the sum over the states (59):

$$Z_1 = \sum_{n_x} \sum_{n_y} \sum_{n_z} \exp[-\hbar^2\pi^2(n_x{}^2 + n_y{}^2 + n_z{}^2)/2ML^2\tau].$$ (60)

Provided the spacing of adjacent energy values is small in comparison with τ, we may replace the summations by integrations:

$$Z_1 = \int_0^\infty dn_x \int_0^\infty dn_y \int_0^\infty dn_z \exp[-\alpha^2(n_x{}^2 + n_y{}^2 + n_z{}^2)].$$ (61)

The notation $\alpha^2 \equiv \hbar^2\pi^2/2ML^2\tau$ is introduced for convenience. The exponential may be written as the product of three factors

$$\exp(-\alpha^2 n_x{}^2)\exp(-\alpha^2 n_y{}^2)\exp(-\alpha^2 n_z{}^2) ,$$

so that

$$Z_1 = \left(\int_0^\infty dn_x \exp(-\alpha^2 n_x{}^2)\right)^3 = (1/\alpha)^3\left(\int_0^\infty dx \exp(-x^2)\right)^3 = \pi^{3/2}/8\alpha^3 ,$$

whence

$$Z_1 = \frac{V}{(2\pi\hbar^2/M\tau)^{3/2}} = n_Q V = n_Q/n ,$$ (62)

in terms of the concentration $n = 1/V$.

Here

$$\boxed{n_Q \equiv (M\tau/2\pi\hbar^2)^{3/2}}$$ (63)

is called the **quantum concentration**. It is the concentration associated with one atom in a cube of side equal to the thermal average de Broglie wavelength, which is a length roughly equal to $\hbar/M\langle v\rangle \sim \hbar/(M\tau)^{1/2}$. Here $\langle v\rangle$ is a thermal average velocity. This concentration will keep turning up in the thermal physics of gases, in semiconductor theory, and in the theory of chemical reactions.

For helium at atmospheric pressure at room temperature, $n \approx 2.5 \times 10^{19}$ cm^{-3} and $n_Q \approx 0.8 \times 10^{25}$ cm^{-3}. Thus, $n/n_Q \approx 3 \times 10^{-6}$, which is very small compared to unity, so that helium is very dilute under normal conditions. Whenever $n/n_Q \ll 1$ we say that the gas is in the **classical regime**. An **ideal gas is defined as a gas of noninteracting atoms in the classical regime**.

The thermal average energy of the atom in the box is, as in (12),

$$U = \frac{\sum_n \varepsilon_n \exp(-\varepsilon_n/\tau)}{Z_1} = \tau^2 (\partial \log Z_1/\partial \tau) , \qquad (64)$$

because $Z_1^{-1} \exp(-\varepsilon_n/\tau)$ is the probability the system is in the state n. From (62),

$$\log Z_1 = -\tfrac{3}{2} \log(1/\tau) + \text{terms independent of } \tau ,$$

so that for an ideal gas of one atom

$$\boxed{U = \tfrac{3}{2}\tau.} \qquad (65)$$

If $\tau = k_B T$, where k_B is the Boltzmann constant, then $U = \tfrac{3}{2}k_B T$, the well-known result for the energy per atom of an ideal gas.

The thermal average occupancy of a free particle orbital satisfies the inequality

$$Z_1^{-1} \exp(-\varepsilon_n/\tau) < Z_1^{-1} = n/n_Q ,$$

which sets an upper limit of 4×10^{-6} for the occupancy of an orbital by a helium atom at standard concentration and temperature. For the classical regime to apply, this occupancy must be $\ll 1$. We note that ε_n as defined by (59) is always positive for a free atom.

Example: N atoms in a box. There follows now a tricky argument that we will use temporarily until we develop in Chapter 6 a powerful method to deal with the problem of many noninteracting identical atoms in a box. We first treat an ideal gas of N atoms in a box, all atoms of different species or different isotopes. This is a simple extension of the one atom result. We then discuss the major correction factor that arises when all atoms are identical, of the same isotope of the same species.

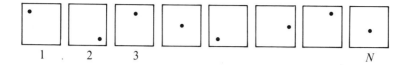

1   2   3   N

Figure 3.7 An N particle system of free particles with one particle in each of N boxes. The energy is N times that for one particle in one box.

Figure 3.8 Atoms of different species in a single box.

If we have one atom in each of N distinct boxes (Figure 3.7), the partition function is the product of the separate one atom partition functions:

$$Z_{N \text{ boxes}} = Z_1(1)\, Z_1(2) \cdots Z_1(N) \ , \tag{66}$$

because the product on the right-hand side includes every independent state of the N boxes, such as the state of energy

$$\varepsilon_\alpha(1) + \varepsilon_\beta(2) + \cdots \varepsilon_\zeta(N) \ , \tag{67}$$

where $\alpha, \beta, \ldots \zeta$ denote the orbital indices of atoms in the successive boxes. The result (66) also gives the partition function of N noninteracting atoms all of different species in a single box (Figure 3.8):

$$Z_\alpha(\bullet)\, Z_\beta(\square)\, Z_\gamma(*) \cdots Z_\zeta(\triangle) \ ,$$

this being the same problem because the energy eigenvalues are the same as for (67). If the masses of all these different atoms happened to be the same, the total partition function would be Z_1^N, where Z_1 is given by (62).

When we consider the more common problem of N identical particles in one box, we have to correct Z_1^N because it overcounts the distinct states of the N identical particle system. Particles of a single species are not distinguishable: electrons do not carry registration numbers. For two labeled particles $\bullet$ and $*$ in a single box, the state $\varepsilon_\alpha(\bullet) + \varepsilon_\beta(*)$ and the state $\varepsilon_\alpha(*) + \varepsilon_\beta(\bullet)$ are distinct states, and both combinations must be counted in the partition function. But for two identical particles the state of energy $\varepsilon_\alpha + \varepsilon_\beta$ is the identical state as $\varepsilon_\beta + \varepsilon_\alpha$, and only one entry is to be made in the state sum in the partition function.

If the orbital indices are all different, each entry will occur $N!$ times in $Z_1{}^N$, whereas the entry should occur only once if the particles are identical. Thus, $Z_1{}^N$ overcounts the states by a factor of $N!$, and the correct partition function for N identical particles is

$$Z_N = \frac{1}{N!} Z_1{}^N = \frac{1}{N!} (n_Q V)^N \qquad (68)$$

in the classical regime. Here $n_Q = (M\tau/2\pi\hbar^2)^{3/2}$ from (63).

There is a step in the argument where we assume that all N occupied orbitals are always different orbitals. It is no simple matter to evaluate directly the error introduced by this approximation, but later we will confirm by another method the validity of (68) in the classical regime $n \ll n_Q$. The $N!$ factor changes the result for the entropy of the ideal gas. The entropy is an experimentally measurable quantity, and it has been confirmed that the $N!$ factor is correct in this low concentration limit.

Energy. The energy of the ideal gas follows from the N particle partition function by use of (12):

$$U = \tau^2(\partial \log Z_N/\partial \tau) = \tfrac{3}{2}N\tau , \qquad (69)$$

consistent with (65) for one particle. The free energy is

$$F = -\tau \log Z_N = -\tau \log Z_1{}^N + \tau \log N!. \qquad (70)$$

With the earlier result $Z_1 = n_Q V = (M\tau/2\pi\hbar^2)^{3/2} V$ and the Stirling approximation $\log N! \simeq N \log N - N$, we have

$$F = -\tau N \log[(M\tau/2\pi\hbar^2)^{3/2} V] + \tau N \log N - \tau N. \qquad (71)$$

From the free energy we can calculate the entropy and the pressure of the ideal gas of N atoms. The pressure follows from (49):

$$p = -(\partial F/\partial V)_\tau = N\tau/V , \qquad (72)$$

or

$$pV = N\tau , \qquad (73)$$

which is called the **ideal gas law**. In conventional units,

$$pV = Nk_BT. \tag{74}$$

The entropy follows from (49):

$$\sigma = -(\partial F/\partial \tau)_V = N \log[(M\tau/2\pi\hbar^2)^{3/2}V] + \tfrac{3}{2}N - N \log N + N \ , \tag{75}$$

or

$$\sigma = N[\log(n_Q/n) + \tfrac{5}{2}] \ , \tag{76}$$

with the concentration $n \equiv N/V$. This result is known as the **Sackur-Tetrode equation** for the entropy of a monatomic ideal gas. It agrees with experiment. The result involves $\hbar$ through the term n_Q, so even for the classical ideal gas the entropy involves a quantum concept. We shall derive these results again in Chapter 6 by a direct method that does not explicitly involve the $N!$ or identical particle argument. The energy (69) also follows from $U = F + \tau\sigma$; with use of (71) and (76) we have $U = \tfrac{3}{2}N\tau$.

Example: Equipartition of energy. The energy $U = \tfrac{3}{2}N\tau$ from (69) is ascribed to a contribution $\tfrac{1}{2}\tau$ from each "degree of freedom" of each particle, where the number of degrees of freedom is the number of dimensions of the space in which the atoms move: 3 in this example. In the classical form of statistical mechanics, the partition function contains the kinetic energy of the particles in an integral over the momentum components p_x, p_y, p_z. For one free particle

$$Z_1 \propto \iiint \exp[-(p_x^2 + p_y^2 + p_z^2)/2M\tau]dp_x\,dp_y\,dp_z \ , \tag{77}$$

a result similar to (61). The limits of integration are $\pm\infty$ for each component. The thermal average energy may be calculated by use of (12) and is equal to $\tfrac{3}{2}\tau$.

The result is generalized in the classical theory. Whenever the hamiltonian of the system is homogeneous of degree 2 in a canonical momentum component, the classical limit of the thermal average kinetic energy associated with that momentum will be $\tfrac{1}{2}\tau$. Further, if the hamiltonian is homogeneous of degree 2 in a position coordinate component, the thermal average potential energy associated with that coordinate will also be $\tfrac{1}{2}\tau$. The result thus

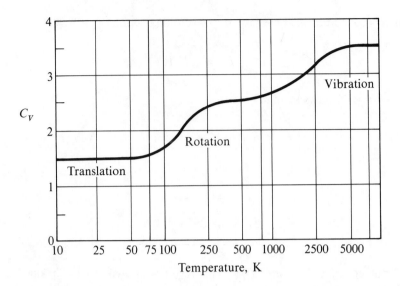

Figure 3.9   Heat capacity at constant volume of one molecule of H_2 in the gas phase. The vertical scale is in fundamental units; to obtain a value in conventional units, multiply by k_B. The contribution from the three translational degrees of freedom is $\frac{3}{2}$; the contribution at high temperatures from the two rotational degrees of freedom is 1; and the contribution from the potential and kinetic energy of the vibrational motion in the high temperature limit is 1. The classical limits are attained when $\tau \gg$ relevant energy level separations.

applies to the harmonic oscillator in the classical limit. The quantum results for the harmonic oscillator and for the diatomic rotator are derived in Problems 3 and 6, respectively. At high temperatures the classical limits are attained, as in Figure 3.9.

Example: Entropy of mixing.   In Chapter 1 we calculated the number of possible arrangements of A and B in a solid made up of $N - t$ atoms A and t atoms B. We found in (1.20) for the number of arrangements:

$$g(N,t) = \frac{N!}{(N - t)!\, t!}. \tag{78}$$

The entropy associated with these arrangements is

$$\sigma(N,t) = \log g(N,t) = \log N! - \log(N - t)! - \log t! , \tag{79}$$

and is plotted in Figure 3.10 for $N = 20$. This contribution to the total entropy of an alloy

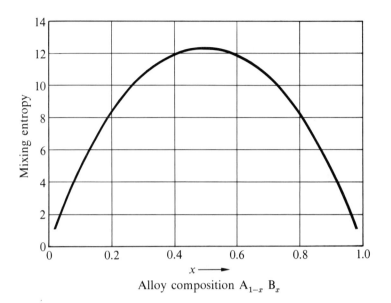

Figure 3.10   Mixing entropy of a random binary alloy as a function of the proportions of the constituent atoms A and B. The curve plotted was calculated for a total of 20 atoms. We see that this entropy is a maximum when A and B are present in equal proportions ($x = 0.5$), and the entropy is zero for pure A or pure B.

system is called the **entropy of mixing**. The result (79) may be put in a more convenient form by use of the Stirling approximation:

$$\sigma(N,t) \simeq N \log N - N - (N - t)\log(N - t) + N - t - t \log t + t$$
$$= N \log N - (N - t)\log(N - t) - t \log t$$
$$= -(N - t)\log(1 - t/N) - t \log(t/N) ,$$

or, with $x \equiv t/N$,

$$\sigma(x) = -N[(1 - x)\log(1 - x) + x \log x]. \tag{80}$$

This result gives the entropy of mixing of an alloy $A_{1-x}B_x$ treated as a random (homogeneous) solid solution. The problem is developed in detail in Chapter 11.

We ask: Is the homogeneous solid solution the equilibrium condition of a mixture of A and B atoms, or is the equilibrium a two-phase system, such as a mixture of crystallites of pure A and crystallites of pure B? The complete answer is the basis of much of the science of metallurgy: the answer will depend on the temperature and on the interatomic interaction energies U_{AA}, U_{BB}, and U_{AB}. In the special case that the interaction energies between

AA, BB, and AB neighbor pairs are all equal, the homogeneous solid solution will have a lower free energy than the corresponding mixture of crystallites of the pure elements. The free energy of the solid solution $A_{1-x}B_x$ is

$$F = F_0 - \tau\sigma(x) = F_0 + N\tau[(1-x)\log(1-x) + x\log x] , \qquad (81)$$

which we must compare with

$$F = (1-x)F_0 + xF_0 = F_0 \qquad (82)$$

for the mixture of A and B crystals in the proportion $(1-x)$ to x. The entropy of mixing is always positive—all entropies are positive—so that the solid solution has the lower free energy in this special case.

There is a tendency for at least a very small proportion of any element B to dissolve in any other element A, even if a strong repulsive energy exists between a B atom and the surrounding A atoms. Let this repulsive energy be denoted by U, a positive quantity. If a very small proportion $x \ll 1$ of B atoms is present, the total repulsive energy is xNU, where xN is the number of B atoms. The mixing entropy (80) is approximately

$$\sigma = -xN\log x \qquad (83)$$

in this limit, so that the free energy is

$$F(x) = N(xU + \tau x\log x) , \qquad (84)$$

which has a minimum when

$$\partial F/\partial x = N(U + \tau\log x + \tau) = 0 , \qquad (85)$$

or

$$x = \exp(-1)\exp(-U/\tau). \qquad (86)$$

This shows there is a natural impurity content in all crystals.

SUMMARY

1. The factor

$$P(\varepsilon_s) = \exp(-\varepsilon_s/\tau)/Z$$

is the probability of finding a system in a state s of energy ε_s when the system is in thermal contact with a large reservoir at temperature τ. The number of particles in the system is assumed constant.

2. The partition function is

$$Z \equiv \sum_s \exp(-\varepsilon_s/\tau).$$

3. The pressure is given by

$$p = -(\partial U/\partial V)_\sigma = \tau(\partial \sigma/\partial V)_U.$$

4. The Helmholtz free energy is defined as $F \equiv U - \tau\sigma$. It is a minimum in equilibrium for a system held at constant τ, V.

5. $\sigma = -(\partial F/\partial \tau)_V; \qquad p = -(\partial F/\partial V)_\tau.$

6. $F = -\tau \log Z$. This result is very useful in calculations of F and of quantities such as p and σ derived from F.

7. For an ideal monatomic gas of N atoms of spin zero,

$$Z_N = (n_Q V)^N/N! \ ,$$

if $n = N/V \ll n_Q$. The quantum concentration $n_Q \equiv (M\tau/2\pi\hbar^2)^{3/2}$. Further,

$$pV = N\tau; \qquad \sigma = N[\log(n_Q/n) + \tfrac{5}{2}]; \qquad C_V = \tfrac{3}{2}N.$$

8. A process is reversible if the system remains infinitesimally close to the equilibrium state at all times during the process.

PROBLEMS

1. Free energy of a two state system. (a) Find an expression for the free energy as a function of τ of a system with two states, one at energy 0 and one at energy ε. (b) From the free energy, find expressions for the energy and entropy of the system. The entropy is plotted in Figure 3.11.

2. Magnetic susceptibility. (a) Use the partition function to find an exact expression for the magnetization M and the susceptibility $\chi \equiv dM/dB$ as a function of temperature and magnetic field for the model system of magnetic moments in a magnetic field. The result for the magnetization is $M = nm\tanh(mB/\tau)$, as derived in (46) by another method. Here n is the particle

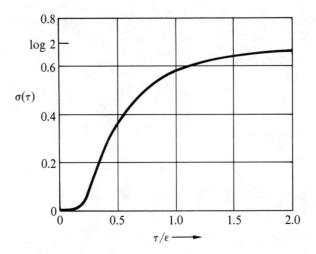

Figure 3.11   Entropy of a two-state system as a function of τ/ε. Notice that $\sigma(\tau) \to \log 2$ as $\tau \to \infty$.

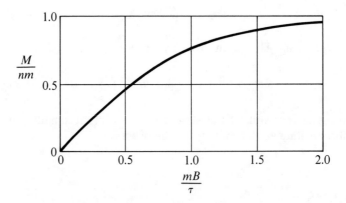

Figure 3.12   Plot of the total magnetic moment as a function of mB/τ. Notice that at low mB/τ the moment is a linear function of mB/τ, but at high mB/τ the moment tends to saturate.

concentration. The result is plotted in Figure 3.12. (b) Find the free energy and express the result as a function only of τ and the parameter $x \equiv M/nm$. (c) Show that the susceptibility is $\chi = nm^2/\tau$ in the limit $mB \ll \tau$.

3. Free energy of a harmonic oscillator.   A one-dimensional harmonic oscillator has an infinite series of equally spaced energy states, with $\varepsilon_s = sh\omega$, where

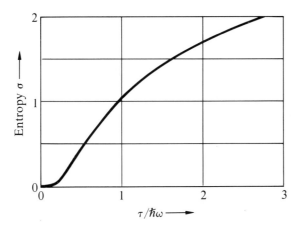

Figure 3.13   Entropy versus temperature for harmonic oscillator of frequency ω.

s is a positive integer or zero, and ω is the classical frequency of the oscillator. We have chosen the zero of energy at the state $s = 0$. (a) Show that for a harmonic oscillator the free energy is

$$F = \tau \log[1 - \exp(-\hbar\omega/\tau)]. \qquad (87)$$

Note that at high temperatures such that $\tau \gg \hbar\omega$ we may expand the argument of the logarithm to obtain $F \simeq \tau \log(\hbar\omega/\tau)$. (b) From (87) show that the entropy is

$$\sigma = \frac{\hbar\omega/\tau}{\exp(\hbar\omega/\tau) - 1} - \log[1 - \exp(-\hbar\omega/\tau)]. \qquad (88)$$

The entropy is shown in Figure 3.13 and the heat capacity in Figure 3.14.

4. Energy fluctuations.   Consider a system of fixed volume in thermal contact with a reservoir. Show that the mean square fluctuation in the energy of the system is

$$\langle(\varepsilon - \langle\varepsilon\rangle)^2\rangle = \tau^2(\partial U/\partial\tau)_V. \qquad (89)$$

Here U is the conventional symbol for $\langle\varepsilon\rangle$. *Hint*: Use the partition function Z to relate $\partial U/\partial\tau$ to the mean square fluctuation. Also, multiply out the term $(\cdots)^2$. *Note*: The temperature τ of a system is a quantity that by definition does

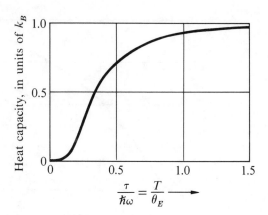

Figure 3.14 Heat capacity versus temperature for harmonic oscillator of frequency ω. The horizontal scale is in units of $\tau/\hbar\omega$, which is identical with T/θ_E, where θ_E is called the **Einstein temperature**. In the high temperature limit $C_V \rightarrow k_B$, or 1 in fundamental units. This value is known as the classical value. At low temperatures C_V decreases exponentially.

not fluctuate in value when the system is in thermal contact with a reservoir. Any other attitude would be inconsistent with our definition of the temperature of a system. The energy of such a system may fluctuate, but the temperature does not. Some workers do not adhere to a rigorous definition of temperature. Thus Landau and Lifshitz give the result

$$\langle (\Delta\tau)^2 \rangle = \tau^2/C_V \,, \tag{90}$$

but this should be viewed as just another form of (89) with $\Delta\tau$ set equal to $\Delta U/C_V$. We know that $\Delta U = C_V \Delta\tau$, whence (90) becomes $\langle (\Delta U)^2 \rangle = \tau^2 C_V$, which is our result (89).

5. Overhauser effect. Suppose that by a suitable external mechanical or electrical arrangement one can add $\alpha\varepsilon$ to the energy of the heat reservoir whenever the reservoir passes to the system the quantum of energy ε. The net increase of energy of the reservoir is $(\alpha - 1)\varepsilon$. Here α is some numerical factor, positive or negative. Show that the effective Boltzmann factor for this abnormal system is given by

$$P(\varepsilon) \propto \exp[-(1 - \alpha)\varepsilon/\tau]. \tag{91}$$

This reasoning gives the statistical basis of the Overhauser effect whereby the nuclear polarization in a magnetic field can be enhanced above the thermal equilibrium polarization. Such a condition requires the active supply of energy to the system from an external source. The system is not in equilibrium, but is said to be in a steady state. Cf. A. W. Overhauser, Phys. Rev. **92**, 411 (1953).

6. Rotation of diatomic molecules. In our first look at the ideal gas we considered only the translational energy of the particles. But molecules can rotate, with kinetic energy. The rotational motion is quantized; and the energy levels of a diatomic molecule are of the form

$$\varepsilon(j) = j(j + 1)\varepsilon_0 \qquad (92)$$

where j is any positive integer including zero: $j = 0, 1, 2, \ldots$. The multiplicity of each rotational level is $g(j) = 2j + 1$. (a) Find the partition function $Z_R(\tau)$ for the rotational states of one molecule. Remember that Z is a sum over all states, not over all levels—this makes a difference. (b) Evaluate $Z_R(\tau)$ approximately for $\tau \gg \varepsilon_0$, by converting the sum to an integral. (c) Do the same for $\tau \ll \varepsilon_0$, by truncating the sum after the second term. (d) Give expressions for the energy U and the heat capacity C, as functions of τ, in both limits. Observe that the rotational contribution to the heat capacity of a diatomic molecule approaches 1 (or, in conventional units, k_B) when $\tau \gg \varepsilon_0$. (e) Sketch the behavior of $U(\tau)$ and $C(\tau)$, showing the limiting behaviors for $\tau \to \infty$ and $\tau \to 0$.

7. Zipper problem. A zipper has N links; each link has a state in which it is closed with energy 0 and a state in which it is open with energy ε. We require, however, that the zipper can only unzip from the left end, and that the link number s can only open if all links to the left $(1,2,\ldots,s-1)$ are already open. (a) Show that the partition function can be summed in the form

$$Z = \frac{1 - \exp[-(N + 1)\varepsilon/\tau]}{1 - \exp(-\varepsilon/\tau)}. \qquad (93)$$

(b) In the limit $\varepsilon \gg \tau$, find the average number of open links. The model is a very simplified model of the unwinding of two-stranded DNA molecules—see C. Kittel, Amer. J. Physics **37**, 917 (1969).

8. Quantum concentration. Consider one particle confined to a cube of side L; the concentration in effect is $n = 1/L^3$. Find the kinetic energy of the particle when in the ground orbital. There will be a value of the concentration for which this zero-point quantum kinetic energy is equal to the temperature τ. (At this concentration the occupancy of the lowest orbital is of the order of unity; the lowest orbital always has a higher occupancy than any other orbital.) Show that the concentration n_0 thus defined is equal to the quantum concentration n_Q defined by (63), within a factor of the order of unity.

9. Partition function for two systems. Show that the partition function $Z(1 + 2)$ of two independent systems **1** and **2** in thermal contact at a common temperature τ is equal to the product of the partition functions of the separate systems:

$$Z(1 + 2) = Z(1)Z(2). \qquad (94)$$

10. Elasticity of polymers.   The thermodynamic identity for a one-dimensional system is

$$\tau d\sigma = dU - f dl \tag{95}$$

when f is the external force exerted on the line and dl is the extension of the line. By analogy with (32) we form the derivative to find

$$-\frac{f}{\tau} = \left(\frac{\partial \sigma}{\partial l}\right)_U . \tag{96}$$

The direction of the force is opposite to the conventional direction of the pressure.

We consider a polymeric chain of N links each of length ρ, with each link equally likely to be directed to the right and to the left. (a) Show that the number of arrangements that give a head-to-tail length of $l = 2|s|\rho$ is

$$g(N,-s) + g(N,s) = \frac{2N!}{(\frac{1}{2}N + s)!\,(\frac{1}{2}N - s)!} . \tag{97}$$

(b) For $|s| \ll N$ show that

$$\sigma(l) = \log[2g(N,0)] - l^2/2N\rho^2 . \tag{98}$$

(c) Show that the force at extension l is

$$f = l\tau/N\rho^2 . \tag{99}$$

The force is proportional to the temperature. The force arises because the polymer wants to curl up: the entropy is higher in a random coil than in an uncoiled configuration. Warming a rubber band makes it contract; warming a steel wire makes it expand. The theory of rubber elasticity is discussed by H. M. James and E. Guth, Journal of Chemical Physics **11**, 455 (1943); Journal of Polymer Science **4**, 153 (1949); see also L. R. G. Treloar, *Physics of rubber elasticity*, Oxford, 1958.

11. One-dimensional gas.   Consider an ideal gas of N particles, each of mass M, confined to a one-dimensional line of length L. Find the entropy at temperature τ. The particles have spin zero.

Chapter 4

Thermal Radiation and Planck Distribution

[*We consider*] *the distribution of the energy U among N oscillators of frequency v. If U is viewed as divisible without limit, then an infinite number of distributions are possible. We consider however—and this is the essential point of the whole calculation—U as made up of an entirely determined number of finite equal parts, and we make use of the natural constant* $h = 6.55 \times 10^{-27}$ *erg-sec. This constant when multiplied by the common frequency v of the oscillators gives the element of energy* ε *in ergs . . . .*

                                                                    M. Planck

PLANCK DISTRIBUTION FUNCTION

The Planck distribution describes the spectrum of the electromagnetic radiation in thermal equilibrium within a cavity. Approximately, it describes the emission spectrum of the Sun or of metal heated by a welding torch. The Planck distribution was the first application of quantum thermal physics. Thermal electromagnetic radiation is often called black body radiation. The Planck distribution also describes the thermal energy spectrum of lattice vibrations in an elastic solid.

The word "mode" characterizes a particular oscillation amplitude pattern in the cavity or in the solid. We shall always refer to $\omega = 2\pi f$ as the frequency of the radiation. The characteristic feature of the radiation problem is that a mode of oscillation of frequency ω may be excited only in units of the quantum of energy $\hbar\omega$. The energy ε_s of the state with s quanta in the mode is

$$\varepsilon_s = s\hbar\omega \ , \tag{1}$$

where s is zero or any positive integer (Figure 4.1). We omit the zero point energy $\frac{1}{2}\hbar\omega$.

These energies are the same as the energies of a quantum harmonic oscillator of frequency ω, but there is a difference between the concepts. A harmonic

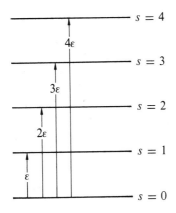

Figure 4.1   States of an oscillator that represents a mode of frequency ω of an electromagnetic field. When the oscillator is in the orbital of energy $s\hbar\omega$, the state is equivalent to s photons in the mode.

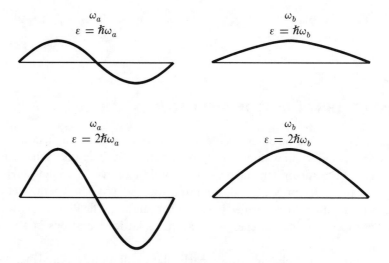

Figure 4.2 Representation in one dimension of two electromagnetic modes a and b, of frequency ω_a and ω_b. The amplitude of the electromagnetic field is suggested in the figures for one photon and two photon occupancy of each mode.

oscillator is a localized oscillator, whereas the electric and magnetic energy of an electromagnetic cavity mode is distributed throughout the interior of the cavity (Figure 4.2). For both problems the energy eigenvalues are integral multiples of $\hbar\omega$, and this is the reason for the similarity in the thermal physics of the two problems. The language used to describe an excitation is different: s for the oscillator is called the quantum number, and s for the quantized electromagnetic mode is called the number of photons in the mode.

   We first calculate the thermal average of the number of photons in a mode, when these photons are in thermal equilibrium with a reservoir at a temperature τ. The partition function (3.10) is the sum over the states (1):

$$Z = \sum_{s=0}^{\infty} \exp(-s\hbar\omega/\tau). \tag{2}$$

This sum is of the form $\sum x^s$, with $x \equiv \exp(-\hbar\omega/\tau)$. Because x is smaller than 1, the infinite series may be summed and has the value $1/(1 - x)$, whence

$$Z = \frac{1}{1 - \exp(-\hbar\omega/\tau)}. \tag{3}$$

The probability that the system is in the state s of energy $s\hbar\omega$ is given by the Boltzmann factor:

$$P(s) = \frac{\exp(-s\hbar\omega/\tau)}{Z}. \qquad (4)$$

The thermal average value of s is

$$\langle s \rangle = \sum_{s=0}^{\infty} sP(s) = Z^{-1}\sum s\exp(-s\hbar\omega/\tau). \qquad (5)$$

With $y \equiv \hbar\omega/\tau$, the summation on the right-hand side has the form:

$$\sum s\exp(-sy) = -\frac{d}{dy}\sum \exp(-sy)$$

$$= -\frac{d}{dy}\left(\frac{1}{1-\exp(-y)}\right) = \frac{\exp(-y)}{[1-\exp(-y)]^2}.$$

From (3) and (5) we find

$$\langle s \rangle = \frac{\exp(-y)}{1-\exp(-y)},$$

or

$$\boxed{\langle s \rangle = \frac{1}{\exp(\hbar\omega/\tau)-1}.} \qquad (6)$$

This is the **Planck distribution function** for the thermal average number of photons (Figure 4.3) in a single mode of frequency ω. Equally, it is the average number of phonons in the mode. The result applies to any kind of wave field with energy in the form of (1).

PLANCK LAW AND STEFAN-BOLTZMANN LAW

The thermal average energy in the mode is

$$\langle \varepsilon \rangle = \langle s \rangle \hbar\omega = \frac{\hbar\omega}{\exp(\hbar\omega/\tau)-1}. \qquad (7)$$

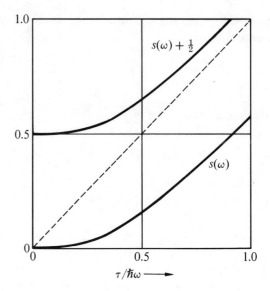

Figure 4.3 Planck distribution as a function of the reduced temperature $\tau/\hbar\omega$. Here $\langle s(\omega)\rangle$ is the thermal average of the number of photons in the mode of frequency ω. A plot of $\langle s(\omega)\rangle + \frac{1}{2}$ is also given, where $\frac{1}{2}$ is the effective zero point occupancy of the mode; the dashed line is the classical asymptote. Note that we write

$$\langle s\rangle + \tfrac{1}{2} = \tfrac{1}{2}\coth(\hbar\omega/2\tau).$$

The high temperature limit $\tau \gg \hbar\omega$ is often called the classical limit. Here $\exp(\hbar\omega/\tau)$ may be approximated as $1 + \hbar\omega/\tau + \cdots$, whence the classical average energy is

$$\langle\varepsilon\rangle \simeq \tau. \tag{8}$$

There is an infinite number of electromagnetic modes within any cavity. Each mode n has its own frequency ω_n. For radiation confined within a perfectly conducting cavity in the form of a cube of edge L, there is a set of modes of the form

$$E_x = E_{x0}\sin\omega t\cos(n_x\pi x/L)\sin(n_y\pi y/L)\sin(n_z\pi z/L)\ , \tag{9a}$$

$$E_y = E_{y0}\sin\omega t\sin(n_x\pi x/L)\cos(n_y\pi y/L)\sin(n_z\pi z/L)\ , \tag{9b}$$

$$E_z = E_{z0}\sin\omega t\sin(n_x\pi x/L)\sin(n_y\pi y/L)\cos(n_z\pi z/L). \tag{9c}$$

Here E_x, E_y and E_z are the three electric field components, and E_{x0}, E_{y0} and E_{z0} are the corresponding amplitudes. The three components are not independent, because the field must be divergence-free:

$$\operatorname{div}\mathbf{E} = \frac{\partial E_x}{\partial x} + \frac{\partial E_y}{\partial y} + \frac{\partial E_z}{\partial z} = 0. \tag{10}$$

When we insert (9) into (10) and drop all common factors, we find the condition

$$E_{x0}n_x + E_{y0}n_y + E_{z0}n_z = \mathbf{E}_0 \cdot \mathbf{n} = 0. \tag{11}$$

This states that the field vectors must be perpendicular to the vector $\mathbf{n}$ with the components n_x, n_y and n_z, so that the electromagnetic field in the cavity is a transversely polarized field. The polarization direction is defined as the direction of $\mathbf{E}_0$.

For a given triplet n_x, n_y, n_z we can choose two mutually perpendicular polarization directions, so that there are two distinct modes for each triplet n_x, n_y, n_z.

On substitution of (9) in the wave equation

$$c^2\left(\frac{\partial^2}{\partial x^2} + \frac{\partial^2}{\partial y^2} + \frac{\partial^2}{\partial z^2}\right)E_z = \frac{\partial^2 E_z}{\partial t^2}\ , \tag{12}$$

with c the velocity of light, we find

$$c^2\pi^2(n_x^2 + n_y^2 + n_z^2) = \omega^2 L^2. \tag{13}$$

This determines the frequency ω of the mode in terms of the triplet of integers n_x, n_y, n_z. If we define

$$n \equiv (n_x^2 + n_y^2 + n_z^2)^{1/2}\ , \tag{14}$$

then the frequencies are of the form

$$\omega_n = n\pi c/L. \tag{15}$$

The total energy of the photons in the cavity is, from (7),

$$U = \sum_n \langle \varepsilon_n \rangle = \sum_n \frac{\hbar\omega_n}{\exp(\hbar\omega_n/\tau) - 1}. \tag{16}$$

The sum is over the triplet of integers n_x, n_y, n_z. Positive integers alone will describe all independent modes of the form (9). We replace the sum over n_x, n_y, n_z by an integral over the volume element $dn_x\, dn_y\, dn_z$ in the space of the mode indices. That is, we set

$$\sum_n (\cdots) = \tfrac{1}{8}\int_0^\infty 4\pi n^2\, dn\, (\cdots)\ , \tag{17}$$

where the factor $\frac{1}{8} = (\frac{1}{2})^3$ arises because only the positive octant of the space is involved. We now multiply the sum or integral by a factor of 2 because there are two independent polarizations of the electromagnetic field (two independent sets of cavity modes). Thus

$$U = \pi \int_0^\infty dn\, n^2 \frac{\hbar\omega_n}{\exp(\hbar\omega_n/\tau) - 1}$$

$$= (\pi^2\hbar c/L) \int_0^\infty dn\, n^3 \frac{1}{\exp(\hbar c n\pi/L\tau) - 1}, \qquad (18)$$

with (15) for ω_n. Standard practice is to transform the definite integral to one over a dimensionless variable. We set $x \equiv \pi\hbar c n/L\tau$, and (18) becomes

$$U = (\pi^2\hbar c/L)(\tau L/\pi\hbar c)^4 \int_0^\infty dx\, \frac{x^3}{\exp x - 1}. \qquad (19)$$

The definite integral has the value $\pi^4/15$; it is found in good standard tables such as Dwight (cited in the general references). The energy per unit volume is

$$\boxed{\frac{U}{V} = \frac{\pi^2}{15\hbar^3 c^3} \tau^4,} \qquad (20)$$

with the volume $V = L^3$. The result that the radiant energy density is proportional to the fourth power of the temperature is known as the **Stefan-Boltzmann law of radiation**.

For many applications of this theory we decompose (20) into the spectral density of the radiation. The spectral density is defined as the energy per unit volume per unit frequency range, and is denoted as u_ω. We can find u_ω from (18) rewritten in terms of ω:

$$U/V = \int d\omega\, u_\omega = \frac{\hbar}{\pi^2 c^3} \int d\omega \frac{\omega^3}{\exp(\hbar\omega/\tau) - 1}, \qquad (21)$$

so that the spectral density is

$$\boxed{u_\omega = \frac{\hbar}{\pi^2 c^3} \frac{\omega^3}{\exp(\hbar\omega/\tau) - 1}.} \qquad (22)$$

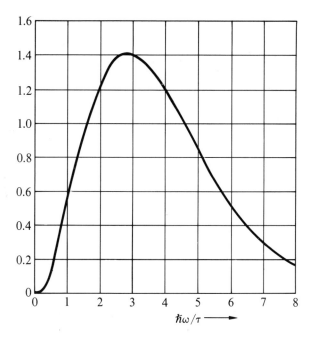

Figure 4.4 Plot of $x^3/(e^x - 1)$ with $x \equiv \hbar\omega/\tau$. This function is involved in the Planck radiation law for the spectral density u_ω. The temperature of a black body may be found from the frequency ω_{max} at which the radiant energy density is a maximum, per unit frequency range. This frequency is directly proportional to the temperature.

This result is the **Planck radiation law**; it gives the frequency distribution of thermal radiation (Figure 4.4). Quantum theory began here.

The entropy of the thermal photons can be found from the relation (3.34a) at constant volume: $d\sigma = dU/\tau$, whence from (20),

$$d\sigma = \frac{4\pi^2 V}{15\hbar^3 c^3} \tau^2 \, d\tau.$$

Thus the entropy is

$$\sigma(\tau) = (4\pi^2 V/45)(\tau/\hbar c)^3. \tag{23}$$

The constant of integration is zero, from (3.55) and the relation between F and σ.

A process carried out at constant photon entropy will have $V\tau^3 = $ constant.

The measurement of high temperatures depends on the flux of radiant energy from a small hole in the wall of a cavity maintained at the temperature of interest. Such a hole is said to radiate as a black body—which means that the radiation emission is characteristic of a thermal equilibrium distribution. The energy flux density J_U is defined as the rate of energy emission per unit area. The flux density is of the order of the energy contained in a column of unit area and length equal to the velocity of light times the unit of time. Thus,

$$J_U = [cU(\tau)/V] \times \text{(geometrical factor)}. \tag{24}$$

The geometrical factor is equal to $\frac{1}{4}$; the derivation is the subject of Problem 15. The final result for the radiant energy flux is

$$J_U = \frac{cU(\tau)}{4V} = \frac{\pi^2\tau^4}{60\hbar^3c^2} , \tag{25}$$

by use of (20) for the energy density U/V. The result is often written as

$$J_U = \sigma_B T^4; \tag{26}$$

the **Stefan-Boltzmann constant**

$$\sigma_B \equiv \pi^2 k_B{}^4/60\hbar^3c^2 \tag{26a}$$

has the value 5.670×10^{-8} W m^{-2} K^{-4} or 5.670×10^{-5} erg cm^{-2} s^{-1} K^{-4}. (Here σ_B is not the entropy.) A body that radiates at this rate is said to radiate as a black body. A small hole in a cavity whose walls are in thermal equilibrium at temperature T will radiate as a black body at the rate given in (26). The rate is independent of the physical constitution of the walls of the cavity and depends only on the temperature.

Emission and Absorption: Kirchhoff Law

The ability of a surface to emit radiation is proportional to the ability of the surface to absorb radiation. We demonstrate this relation, first for a black body or black surface and, second, for a surface with arbitrary properties. An object is defined to be black in a given frequency range if all electromagnetic radiation incident upon it in that range is absorbed. By this definition a hole in a cavity is black if the hole is small enough that radiation incident through the hole will reflect enough times from the cavity walls to be absorbed in the cavity with negligible loss back through the hole.

The radiant energy flux density J_U from a black surface at temperature τ is equal to the radiant energy flux density J_U emitted from a small hole in a cavity at the same temperature. To prove this, let us close the hole with the black surface, hereafter called the object. In thermal equilibrium the thermal average energy flux from the black object to the interior of the cavity must be equal, but opposite, to the thermal average energy flux from the cavity to the black object.

We prove the following: If a non-black object at temperature τ absorbs a fraction a of the radiation incident upon it, the radiation flux emitted by the object will be a times the radiation flux emitted by a black body at the same temperature. Let a denote the **absorptivity** and e the **emissivity**, where the emissivity is defined so that the radiation flux emitted by the object is e times the flux emitted by a black body at the same temperature. The object must emit at the same rate as it absorbs if equilibrium is to be maintained. It follows that $a = e$. This is the **Kirchhoff law**. For the special case of a perfect reflector, a is zero, whence e is zero. A perfect reflector does not radiate.

The arguments can be generalized to apply to the radiation at any frequency, as between ω and $\omega + d\omega$. We insert a filter between the object and the hole in the black body. Let the filter reflect perfectly outside this frequency range, and let it transmit perfectly within this range. The flux equality arguments now apply to the transmitted spectral band, so that $a(\omega) = e(\omega)$ for any surface in thermal equilibrium.

Estimation of Surface Temperature

One way to estimate the surface temperature of a hot body such as a star is from the frequency at which the maximum emission of radiant energy takes place (see Figure 4.4). What this frequency is depends on whether we look at the energy flux per unit frequency range or per unit wavelength range. For u_ω, the energy density per unit frequency range, the maximum is given from the Planck law, Eq. (22), as

$$\frac{d}{dx}\left(\frac{x^3}{\exp x - 1}\right) = 0 \; ,$$

or

$$3 - 3\exp(-x) = x.$$

This equation may be solved numerically. The root is

$$\hbar\omega_{max}/k_B T = x_{max} \simeq 2.82 \ , \tag{27}$$

as in Figure 4.4.

Example: Cosmic black body background radiation.   A major recent discovery is that the universe accessible to us is filled with radiation approximately like that of a black body at 2.9 K. The existence of this radiation (Figure 4.5) is important evidence for big bang cosmological models which assume that the universe is expanding and cooling with time. This radiation is left over from an early epoch when the universe was composed primarily of electrons and protons at a temperature of about 4000 K. The plasma of electrons and protons interacted strongly with electromagnetic radiation at all important frequencies, so that the matter and the black body radiation were in thermal equilibrium. By the time the universe had cooled to 3000 K, the matter was primarily in the form of atomic hydrogen. This interacts with black body radiation only at the frequencies of the hydrogen spectral lines. Most of the black body radiation energy thus was effectively decoupled from the matter. Thereafter the radiation evolved with time in a very simple way: the photon gas was cooled by expansion at constant entropy to a temperature of 2.9 K. The photon gas will remain at constant entropy if the frequency of each mode is lowered during the expansion of the universe with the number of photons in each mode kept constant. We show in (58) below that the entropy is constant if the number of photons in each mode is constant—the occupancies determine the entropy.

After the decoupling the evolution of matter into heavier atoms (which are organized into galaxies, stars, and dust clouds) was more complicated than before decoupling. Electromagnetic radiation, such as starlight, radiated by the matter since the decoupling is superimposed on the cosmic black body radiation.

ELECTRICAL NOISE

As an important example of the Planck law in one dimension, we consider the spontaneous thermal fluctuations in voltage across a resistor. These fluctuations, which are called noise, were discovered by J. B. Johnson and explained by H. Nyquist.* The characteristic property of Johnson noise is that the mean-square noise voltage is proportional to the value of the resistance R, as shown by Figure 4.6. We shall see that $\langle V^2 \rangle$ is also directly proportional to the tem-

---

* H. Nyquist, Phys. Rev. **32**, 110 (1928); a deeper discussion is given by C. Kittel, *Elementary statistical physics*, Wiley, 1958, Sections 27–30.

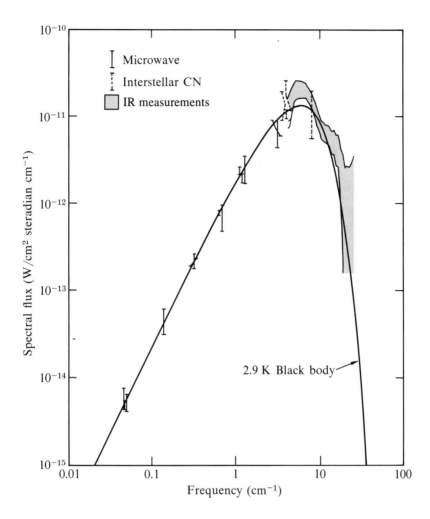

Figure 4.5 Experimental measurements of the spectrum of the cosmic black body radiation. Observations of the flux were made with microwave heterodyne receivers at frequencies below the peak, were deduced from optical measurements of the spectrum of interstellar CN molecules near the peak, and were measured with a balloon-borne infrared spectrometer at frequencies above the peak. Courtesy of P. L. Richards.

perature τ and the bandwidth Δf of the circuit. (This section presumes a knowledge of electromagnetic wave propagation at the intermediate level.)

The Nyquist theorem gives a quantitative expression for the thermal noise voltage generated by a resistor in thermal equilibrium. The theorem is therefore needed in any estimate of the limiting signal-to-noise ratio of an experimental

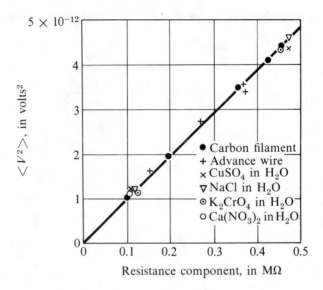

Figure 4.6   Voltage squared versus resistance for various kinds of conductors, including electrolytes. After J. B. Johnson.

apparatus. In the original form the Nyquist theorem states that the mean square voltage across a resistor of resistance R in thermal equilibrium at temperature τ is given by

$$\langle V^2 \rangle = 4R\tau \, \Delta f \; , \tag{28}$$

where Δf is the frequency* bandwidth within which the voltage fluctuations are measured; all frequency components outside the given range are ignored. We show below that the thermal noise power per unit frequency range delivered by a resistor to a matched load is τ; the factor 4 enters where it does because in the circuit of Figure 4.7, the power delivered to an arbitrary resistive load R' is

$$\langle I^2 \rangle R' = \frac{\langle V^2 \rangle R'}{(R + R')^2} \; , \tag{29}$$

which at match ($R' = R$) is $\langle V^2 \rangle / 4R$.

---

* In this section the word frequency refers to cycles per unit time, and not to radians per unit time.

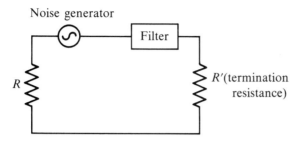

Figure 4.7   Equivalent circuit for a resistance R with a generator of thermal noise that delivers power to a load R'. The current

$$I = \frac{V}{R + R'},$$

so that the mean power dissipated in the load is

$$\mathscr{P} = \langle I^2 \rangle R' = \frac{\langle V^2 \rangle R'}{(R' + R)^2},$$

which is a maximum with respect to R' when $R' = R$. In this condition the load is said to be matched to the power supply. At match, $\mathscr{P} = \langle V^2 \rangle /4R$. The filter enables us to limit the frequency bandwidth under consideration; that is, the bandwidth to which the mean square voltage fluctuation applies.

Consider as in Figure 4.8 a lossless transmission line of length L and characteristic impedance $Z_c = R$ terminated at each end by a resistance R. Thus the line is matched at each end, in the sense that all energy traveling down the line will be absorbed without reflection in the appropriate resistance. The entire circuit is maintained at temperature τ.

A transmission line is essentially an electromagnetic system in one dimension. We follow the argument given above for the distribution of photons in thermal equilibrium, but now in a space of one dimension instead of three dimensions. The transmission line has two photon modes (one propagating in each direction) of frequency $2\pi f_n = 2n\pi/L$ from (15), so that there are two modes in the frequency range

$$\delta f = c'/L, \tag{30}$$

where c' is the propagation velocity on the line. Each mode has energy

$$\frac{\hbar\omega}{\exp(\hbar\omega/\tau) - 1} \tag{31}$$

Figure 4.8   Transmission line of length L with matched terminations, as conceived for the derivation of the Nyquist theorem. The characteristic impedance Z_c of the transmission line has the value R. According to the fundamental theorem of transmission lines, the terminal resistors are matched to the line when their resistance has the same value R.

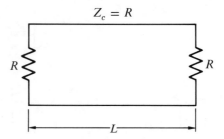

in equilibrium, according to the Planck distribution. We are usually concerned with circuits in the classical limit $\hbar\omega \ll \tau$ so that the thermal energy per mode is τ. It follows that the energy on the line in the frequency range Δf is

$$2\tau \, \Delta f/\delta f = 2\tau L \, \Delta f/c'. \tag{32}$$

The rate at which energy comes off the line in one direction is

$$\tau \, \Delta f. \tag{33}$$

The power coming off the line at one end is all absorbed in the terminal impedance R at that end; there are no reflections when the terminal impedance is matched to the line. In thermal equilibrium the load must emit energy to the line at the same rate, or else its temperature would rise. Thus the power input to the load is

$$\mathscr{P} = \langle I^2 \rangle R = \tau\Delta f \;, \tag{34}$$

but $V = 2RI$, so that (28) is obtained. The result has been used in low temperature thermometry, in temperature regions (Figure 4.9) where it is more convenient to measure $\langle V^2 \rangle$ than τ. Johnson noise is the noise across a resistor when no dc current is flowing. Additional noise (not discussed here) appears when a dc current flows.

PHONONS IN SOLIDS: DEBYE THEORY

So I decided to calculate the spectral distribution of the possible free vibrations for a continuous solid and to consider this distribution as a good enough approximation to the actual distribution. The sonic spectrum of a lattice must,

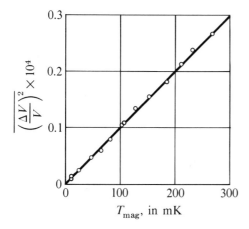

Figure 4.9 Mean square noise voltage fluctuations observed experimentally from a 3 $\mu\Omega$ resistor in the mixing chamber of a dilution refrigerator as a function of magnetic temperature indicated by a CMN powder thermometer. After R. R. Giffard, R. A. Webb, and J. C. Wheatley, J. Low Temp. Physics **6**, 533 (1972).

of course, deviate from this as soon as the wavelength becomes comparable to the distances of the atoms. . . . The only thing which had to be done was to adjust to the fact that every solid of finite dimensions contains a finite number of atoms and therefore has a finite number of free vibrations. . . . At low enough temperatures, and in perfect analogy to the radiation law of Stefan-Boltzmann . . . , the vibrational energy content of a solid will be proportional to T^4.

P. Debye

The energy of an elastic wave in a solid is quantized just as the energy of an electromagnetic wave in a cavity is quantized. The quantum of energy of an elastic wave is called a **phonon**. The thermal average number of phonons in an elastic wave of frequency ω is given by the Planck distribution function, just as for photons:

$$\langle s(\omega) \rangle = \frac{1}{\exp(\hbar\omega/\tau) - 1}. \tag{35}$$

We assume that the frequency of an elastic wave is independent of the amplitude of the elastic strain. We want to find the energy and heat capacity of the elastic waves in solids. Several of the results obtained for photons may be carried over to phonons. The results are simple if we assume that the velocities of all elastic waves are equal—independent of frequency, direction of propagation, and direction of polarization. This assumption is not very accurate, but it helps account for the general trend of the observed results in many solids, with a minimum of computation.

There are two important features of the experimental results: the heat capacity of a nonmetallic solid varies as τ^3 at low temperatures, and at high temperatures the heat capacity is independent of the temperature. In metals there is an extra contribution from the conduction electrons, treated in Chapter 7.

Number of Phonon Modes

There is no limit to the number of possible electromagnetic modes in a cavity, but the number of elastic modes in a finite solid is bounded. If the solid consists of N atoms, each with three degrees of freedom, the total number of modes is $3N$. An elastic wave has three possible polarizations, two transverse and one longitudinal, in contrast to the two possible polarizations of an electromagnetic wave. In a transverse elastic wave the displacement of the atoms is perpendicular to the propagation direction of the wave; in a longitudinal wave the displacement is parallel to the propagation direction. The sum of a quantity over all modes may be written as, including the factor 3,

$$\sum_n (\cdots) = \tfrac{3}{8} \int 4\pi n^2 \, dn (\cdots) , \tag{36}$$

by extension of (17). Here n is defined in terms of the triplet of integers n_x, n_y, n_z, exactly as for photons. We want to find n_{max} such that the total number of elastic modes is equal to $3N$:

$$\tfrac{3}{8} \int_0^{n_{max}} 4\pi n^2 \, dn = 3N. \tag{37}$$

In the photon problem there was no corresponding limitation on the total number of modes. It is customary to write n_D, after Debye, for n_{max}. Then (37) becomes

$$\tfrac{1}{2}\pi n_D{}^3 = 3N; \qquad n_D = (6N/\pi)^{1/3}. \tag{38}$$

The thermal energy of the phonons is, from (16),

$$U = \sum \langle \varepsilon_n \rangle = \sum \langle s_n \rangle \hbar \omega_n = \sum \frac{\hbar \omega_n}{\exp(\hbar \omega_n / \tau) - 1} , \tag{39}$$

or, by (36) and (38),

$$U = \frac{3\pi}{2} \int_0^{n_D} dn\, n^2 \frac{\hbar\omega_n}{\exp(\hbar\omega_n/\tau) - 1}.$$ (40)

By analogy with the evaluation of (19), with the velocity of sound v written in place of the velocity of light c,

$$U = (3\pi^2 \hbar v/2L)(\tau L/\pi\hbar v)^4 \int_0^{x_D} dx\, \frac{x^3}{\exp x - 1},$$ (41)

where $x \equiv \pi\hbar v n/L\tau$. For L^3 we write the volume V. Here, with (38), the upper limit of integration is

$$x_D = \pi\hbar v n_D/L\tau = \hbar v(6\pi^2 N/V)^{1/3}/\tau,$$ (42)

usually written as

$$x_D = \theta/T = k_B\theta/\tau,$$ (43)

where θ is called the **Debye temperature**:

$$\theta = (\hbar v/k_B)(6\pi^2 N/V)^{1/3}.$$ (44)

The result (41) for the energy is of special interest at low temperatures such that $T \ll \theta$. Here the limit x_D on the integral is much larger than unity, and x_D may be replaced by infinity. We note from Figure 4.4 that there is little contribution to the integrand out beyond $x = 10$. For the definite integral we have

$$\int_0^\infty dx\, \frac{x^3}{\exp x - 1} = \frac{\pi^4}{15},$$ (45)

as earlier. Thus the energy in the low temperature limit is

$$U(T) \simeq \frac{3\pi^4 N\tau^4}{5(k_B\theta)^3} \simeq \frac{3\pi^4 Nk_B T^4}{5\theta^3},$$ (46)

proportional to T^4. The heat capacity is, for $\tau \ll k_B\theta$ or $T \ll \theta$,

$$C_V = \left(\frac{\partial U}{\partial \tau}\right)_V = \frac{12\pi^4 N}{5}\left(\frac{\tau}{k_B\theta}\right)^3.$$ (47a)

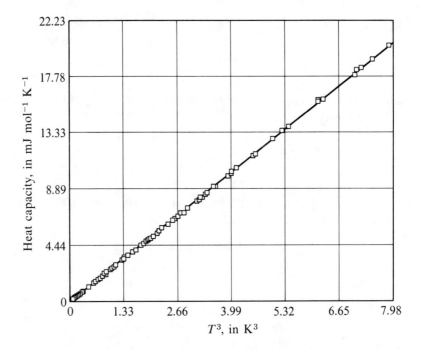

Figure 4.10   Low temperature heat capacity of solid argon, plotted against T^3 to show the excellent agreement with the Debye T^3 law. The value of θ from these data is 92 K. Courtesy of L. Finegold and N. E. Phillips.

In conventional units,

$$C_V = \left(\frac{\partial U}{\partial T}\right)_V = \frac{12\pi^4 N k_B}{5}\left(\frac{T}{\theta}\right)^3. \tag{47b}$$

This result is known as the **Debye T^3 law.*** Experimental results for argon are plotted in Figure 4.10. Representative experimental values of the Debye temperature are given in Table 4.1. The calculated variation of C_V versus T/θ is plotted in Figure 4.11. The high temperature limit $T \gg \theta$ is the subject of Problem 11. Several related thermodynamic functions for a Debye solid are given in Table 4.2 and are plotted in Figure 4.12.

---

* P. Debye, Annalen der Physik **39**, 789 (1912); M. Born and T. v. Kármán, Physikalische Zeitschrift **13**, 297 (1912); **14**, 65 (1913).

Table 4.1 Debye temperature θ_0 in K

Li 344	Be 1440											B	C 2230	N	O	F	Ne 75
Na 158	Mg 400											Al 428	Si 645	P	S	Cl	Ar 92
K 91	Ca 230	Sc 360	Ti 420	V 380	Cr 630	Mn 410	Fe 470	Co 445	Ni 450	Cu 343	Zn 327	Ga 320	Ge 374	As 282	Se 90	Br	Kr 72
Rb 56	Sr 147	Y 280	Zr 291	Nb 275	Mo 450	Tc	Ru 600	Rh 480	Pd 274	Ag 225	Cd 209	In 108	Sn w 200	Sb 211	Te 153	I	Xe 64
Cs 38	Ba 110	Laβ 142	Hf 252	Ta 240	W 400	Re 430	Os 500	Ir 420	Pt 240	Au 165	Hg 71.9	Tl 78.5	Pb 105	Bi 119	Po	At	Rn
Fr	Ra	Ac															

Ce	Pr	Nd	Pm	Sm	Eu	Gd 200	Tb	Dy 210	Ho	Er	Tm	Yb 120	Lu 210
Th 163	Pa	U 207	Np	Pu	Am	Cm	Bk	Cf	Es	Fm	Md	No	Lr

NOTE: The subscript zero on the θ denotes the low temperature limit of the experimental values.

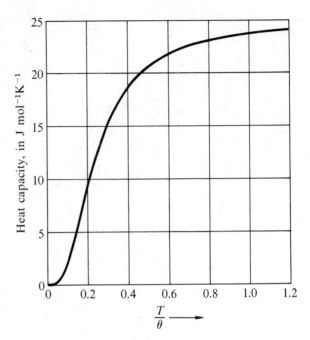

Figure 4.11 Heat capacity C_V of a solid, according to the Debye approximation. The vertical scale is in $J \, mol^{-1} \, K^{-1}$. The horizontal scale is the temperature normalized to the Debye temperature θ. The region of the T^3 law is below 0.1θ. The asymptotic value at high values of T/θ is $24.943 \, J \, mol^{-1} \, K^{-1}$.

Table 4.2 Values of C_V, S, U, and F on the Debye theory, in units $J \, mol^{-1} \, K^{-1}$

θ/T	C_V	$S = k_B\sigma$	U/θ	F/θ
0	24.943	∞	∞	
0.1	24.93	90.70	240.2	−666.8
0.2	24.89	73.43	115.6	−251
0.3	24.83	63.34	74.2	−137
0.4	24.75	56.21	53.5	−87
0.5	24.63	50.70	41.16	−60.3
0.6	24.50	46.22	32.9	−44.1
0.7	24.34	42.46	27.1	−33.5
0.8	24.16	39.22	22.8	−26.2
0.9	23.96	36.38	19.5	−20.9
1.0	23.74	33.87	16.82	−17.05
1.5	22.35	24.49	9.1	−7.23
2	20.59	18.30	5.5	−3.64
3	16.53	10.71	2.36	−1.21
4	12.55	6.51	1.13	−0.49
5	9.20	4.08	0.58	−0.23
6	6.23	2.64	0.323	−0.118
7	4.76	1.77	0.187	−0.066
8	3.45	1.22	0.114	−0.039
9	2.53	0.874	0.073	−0.025
10	1.891	0.643	0.048	−0.016
15	0.576	0.192	0.0096	−0.0032

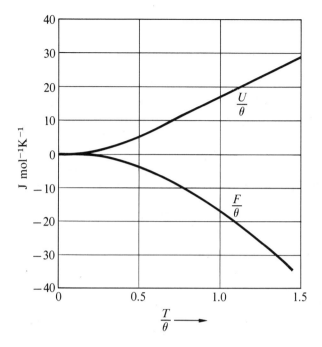

Figure 4.12 Energy U and free energy $F \equiv U - \tau\sigma$ of a solid, according to the Debye theory. The Debye temperature of the solid is θ.

SUMMARY

1. The Planck distribution function is

$$\langle s \rangle = \frac{1}{\exp(\hbar\omega/\tau) - 1} ,$$

for the thermal average number of photons in a cavity mode of frequency ω.

2. The Stefan-Boltzmann law is

$$\frac{U}{V} = \frac{\pi^2}{15\hbar^3 c^3} \tau^4 ,$$

for the radiant energy density in a cavity at temperature τ.

3. The Planck radiation law is

$$u_\omega = \frac{\hbar}{\pi^2 c^3} \frac{\omega^3}{\exp(\hbar\omega/\tau) - 1} \, ,$$

for the radiation energy per unit volume per unit range of frequency.

4. The flux density of radiant energy is $J_U = \sigma_B T^4$, where σ_B is the Stefan-Boltzmann constant $\pi^2 k_B^4 / 60\hbar^3 c^2$.

5. The Debye low temperature limit of the heat capacity of a dielectric solid is, in conventional units,

$$C_V = \frac{12\pi^4 N k_B}{5} \left(\frac{T}{\theta}\right)^3,$$

where the Debye temperature

$$\theta \equiv (\hbar v / k_B)(6\pi^2 N/V)^{1/3}.$$

PROBLEMS

1. Number of thermal photons. Show that the number of photons $\sum \langle s_n \rangle$ in equilibrium at temperature τ in a cavity of volume V is

$$N = 2.404\pi^{-2} V(\tau/\hbar c)^3. \tag{48}$$

From (23) the entropy is $\sigma = (4\pi^2 V/45)(\tau/\hbar c)^3$, whence $\sigma/N \simeq 3.602$. It is believed that the total number of photons in the universe is 10^8 larger than the total number of nucleons (protons, neutrons). Because both entropies are of the order of the respective number of particles (see Eq. 3.76), the photons provide the dominant contribution to the entropy of the universe, although the particles dominate the total energy. We believe that the entropy of the photons is essentially constant, so that the entropy of the universe is approximately constant with time.

2. Surface temperature of the Sun. The value of the total radiant energy flux density at the Earth from the Sun normal to the incident rays is called the **solar constant** of the Earth. The observed value integrated over all emission wavelengths and referred to the mean Earth-Sun distance is:

$$\text{solar constant} = 0.136 \, \text{J s}^{-1} \, \text{cm}^{-2}. \tag{49}$$

(a) Show that the total rate of energy generation of the Sun is $4 \times 10^{26} \, \text{J} \, \text{s}^{-1}$.
(b) From this result and the Stefan-Boltzmann constant $\sigma_B = 5.67 \times 10^{-12} \, \text{J} \, \text{s}^{-1} \, \text{cm}^{-2} \, \text{K}^{-4}$, show that the effective temperature of the surface of the Sun treated as a black body is $T \cong 6000$ K. Take the distance of the Earth from the Sun as 1.5×10^{13} cm and the radius of the Sun as 7×10^{10} cm.

3. Average temperature of the interior of the Sun. (a) Estimate by a dimensional argument or otherwise the order of magnitude of the gravitational self-energy of the Sun, with $M_\odot = 2 \times 10^{33}$ g and $R_\odot = 7 \times 10^{10}$ cm. The gravitational constant G is 6.6×10^{-8} dyne cm^2 g^{-2}. The self-energy will be negative referred to atoms at rest at infinite separation. (b) Assume that the total thermal kinetic energy of the atoms in the Sun is equal to $-\frac{1}{2}$ times the gravitational energy. This is the result of the virial theorem of mechanics. Estimate the average temperature of the Sun. Take the number of particles as 1×10^{57}. This estimate gives somewhat too low a temperature, because the density of the Sun is far from uniform. "The range in central temperature for different stars, excluding only those composed of degenerate matter for which the law of perfect gases does not hold (white dwarfs) and those which have excessively small average densities (giants and supergiants), is between 1.5 and 3.0×10^7 degrees." (O. Struve, B. Lynds, and H. Pillans, *Elementary astronomy*, Oxford, 1959.)

4. Age of the Sun. Suppose $4 \times 10^{26} \, \text{J} \, \text{s}^{-1}$ is the total rate at which the Sun radiates energy at the present time. (a) Find the total energy of the Sun available for radiation, on the rough assumptions that the energy source is the conversion of hydrogen (atomic weight 1.0078) to helium (atomic weight 4.0026) and that the reaction stops when 10 percent of the original hydrogen has been converted to helium. Use the Einstein relation $E = (\Delta M)c^2$. (b) Use (a) to estimate the life expectancy of the Sun. It is believed that the age of the universe is about 10×10^9 years. (A good discussion is given in the books by Peebles and by Weinberg, cited in the general references.)

5. Surface temperature of the Earth. Calculate the temperature of the surface of the Earth on the assumption that as a black body in thermal equilibrium it reradiates as much thermal radiation as it receives from the Sun. Assume also that the surface of the Earth is at a constant temperature over the day-night cycle. Use $T_\odot = 5800$ K; $R_\odot = 7 \times 10^{10}$ cm; and the Earth-Sun distance of 1.5×10^{13} cm.

6. Pressure of thermal radiation. Show for a photon gas that:

(a) $$ p = -(\partial U / \partial V)_\sigma = -\sum_j s_j \hbar (d\omega_j / dV) , \qquad (50) $$

where s_j is the number of photons in the mode j;

(b) $$d\omega_j/dV = -\omega_j/3V;$$ (51)

(c) $$p = U/3V.$$ (52)

Thus the radiation pressure is equal to $\frac{1}{3} \times$ (energy density).

(d) Compare the pressure of thermal radiation with the kinetic pressure of a gas of H atoms at a concentration of $1 \, \text{mole cm}^{-3}$ characteristic of the Sun. At what temperature (roughly) are the two pressures equal? The average temperature of the Sun is believed to be near $2 \times 10^7 \, \text{K}$. The concentration is highly nonuniform and rises to near $100 \, \text{mole cm}^{-3}$ at the center, where the kinetic pressure is considerably higher than the radiation pressure.

7. Free energy of a photon gas. (a) Show that the partition function of a photon gas is given by

$$Z = \prod_n [1 - \exp(-\hbar\omega_n/\tau)]^{-1} ,$$ (53)

where the product is over the modes n. (b) The Helmholtz free energy is found directly from (53) as

$$F = \tau \sum_n \log[1 - \exp(-\hbar\omega_n/\tau)].$$ (54)

Transform the sum to an integral; integrate by parts to find

$$F = -\pi^2 V\tau^4/45\hbar^3c^3.$$ (55)

8. Heat shields. A black (nonreflective) plane at temperature T_u is parallel to a black plane at temperature T_l. The net energy flux density in vacuum between the two planes is $J_U = \sigma_B(T_u^4 - T_l^4)$, where σ_B is the Stefan-Boltzmann constant used in (26). A third black plane is inserted between the other two and is allowed to come to a steady state temperature T_m. Find T_m in terms of T_u and T_l, and show that the net energy flux density is cut in half because of the presence of this plane. This is the principle of the heat shield and is widely used to reduce radiant heat transfer. *Comment:* The result for N independent heat shields floating in temperature between the planes T_u and T_l is that the net energy flux density is $J_U = \sigma_B(T_u^4 - T_l^4)/(N + 1)$.

9. Photon gas in one dimension. Consider a transmission line of length L on which electromagnetic waves satisfy the one-dimensional wave equation $v^2\partial^2 E/\partial x^2 = \partial^2 E/\partial t^2$, where E is an electric field component. Find the heat capacity of the photons on the line, when in thermal equilibrium at temperature

τ. The enumeration of modes proceeds in the usual way for one dimension: take the solutions as standing waves with zero amplitude at each end of the line.

10. Heat capacity of intergalactic space. Intergalactic space is believed to be occupied by hydrogen atoms in a concentration ≈ 1 atom m^{-3}. The space is also occupied by thermal radiation at 2.9 K, from the Primitive Fireball. Show that the ratio of the heat capacity of matter to that of radiation is $\sim 10^{-9}$.

11. Heat capacity of solids in high temperature limit. Show that in the limit $T \gg \theta$ the heat capacity of a solid goes towards the limit $C_V \rightarrow 3Nk_B$, in conventional units. To obtain higher accuracy when T is only moderately larger than θ, the heat capacity can be expanded as a power series in $1/T$, of the form

$$C_V = 3Nk_B \times \left[1 - \sum_n a_n/T^n \right]. \qquad (56)$$

Determine the first nonvanishing term in the sum. Check your result by inserting $T = \theta$ and comparing with Table 4.2.

12. Heat capacity of photons and phonons. Consider a dielectric solid with a Debye temperature equal to 100 K and with 10^{22} atoms cm^{-3}. Estimate the temperature at which the photon contribution to the heat capacity would be equal to the phonon contribution evaluated at 1 K.

13. Energy fluctuations in a solid at low temperatures. Consider a solid of N atoms in the temperature region in which the Debye T^3 law is valid. The solid is in thermal contact with a heat reservoir. Use the results on energy fluctuations from Chapter 3 to show that the root mean square fractional energy fluctuation $\mathscr{F}$ is given by

$$\mathscr{F}^2 = \langle (\varepsilon - \langle \varepsilon \rangle)^2 \rangle / \langle \varepsilon \rangle^2 \approx \frac{0.07}{N} \left(\frac{\theta}{T} \right)^3. \qquad (57)$$

Suppose that $T = 10^{-2}$ K; $\theta = 200$ K; and $N \approx 10^{15}$ for a particle 0.01 cm on a side; then $\mathscr{F} \approx 0.02$. At 10^{-5} K the fractional fluctuation in energy is of the order of unity for a dielectric particle of volume 1 cm^3.

14. Heat capacity of liquid ^{4}He at low temperatures. The velocity of longitudinal sound waves in liquid ^{4}He at temperatures below 0.6 K is 2.383×10^4 cm s^{-1}. There are no transverse sound waves in the liquid. The density is 0.145 g cm^{-3}. (a) Calculate the Debye temperature. (b) Calculate the heat capacity per gram on the Debye theory and compare with the experimental value $C_V = 0.0204 \times T^3$, in J g^{-1} K^{-1}. The T^3 dependence of the experimental value suggests that phonons are the most important excitations in liquid ^{4}He below 0.6 K. Note that the experimental value has been expressed per gram of liquid. The experiments are due to J. Wiebes, C. G. Niels-Hakkenberg, and H. C. Kramers, Physica **32**, 625 (1957).

15. Angular distribution of radiant energy flux. (a) Show that the spectral density of the radiant energy flux that arrives in the solid angle $d\Omega$ is $cu_\omega \cos\theta \cdot d\Omega/4\pi$, where θ is the angle the normal to the unit area makes with the incident ray, and u_ω is the energy density per unit frequency range. (b) Show that the sum of this quantity over all incident rays is $\frac{1}{4}cu_\omega$.

16. Image of a radiant object. Let a lens image the hole in a cavity of area A_H on a black object of area A_0. Use an equilibrium argument to relate the product $A_H\Omega_H$ to $A_0\Omega_0$ where Ω_H and Ω_0 are the solid angles subtended by the lens as viewed from the hole and from the object. This general property of focusing systems is easily derived from geometrical optics. It is also true when diffraction is important. Make the approximation that all rays are nearly parallel (all axial angles small).

17. Entropy and occupancy. We argued in this chapter that the entropy of the cosmic black body radiation has not changed with time because the number of photons in each mode has not changed with time, although the frequency of each mode has decreased as the wavelength has increased with the expansion of the universe. Establish the implied connection between entropy and occupancy of the modes, by showing that for one mode of frequency ω the entropy is a function of the photon occupancy $\langle s \rangle$ only:

$$\sigma = \langle s + 1 \rangle \log\langle s + 1 \rangle - \langle s \rangle \log\langle s \rangle. \tag{58}$$

It is convenient to start from the partition function.

18. Isentropic expansion of photon gas. Consider the gas of photons of the thermal equilibrium radiation in a cube of volume V at temperature τ. Let the cavity volume increase; the radiation pressure performs work during the expansion, and the temperature of the radiation will drop. From the result for the entropy we know that $\tau V^{1/3}$ is constant in such an expansion. (a) Assume that the temperature of the cosmic black-body radiation was decoupled from the temperature of the matter when both were at 3000 K. What was the radius of the universe at that time, compared to now? If the radius has increased linearly with time, at what fraction of the present age of the universe did the decoupling take place? (b) Show that the work done by the photons during the expansion is

$$W = (\pi^2/15\hbar^3c^3)V_i\tau_i^3(\tau_i - \tau_f).$$

The subscripts i and f refer to the initial and final states.

19. *Reflective heat shield and Kirchhoff's law.* Consider a plane sheet of material of absorptivity a, emissivity e, and reflectivity $r = 1 - a$. Let the sheet be suspended between and parallel with two black sheets maintained at temperatures τ_u and τ_l. Show that the net flux density of thermal radiation between the black sheets is $(1 - r)$ times the flux density when the intermediate sheet is also black as in Problem 8, which means with $a = e = 1; r = 0$. Liquid helium dewars are often insulated by many, perhaps 100, layers of an aluminized Mylar film called Superinsulation.

SUPPLEMENT: GREENHOUSE EFFECT

The Greenhouse Effect describes the warming of the surface of the Earth caused by the interposition of an infrared absorbent layer of water, as vapor and in clouds, and of carbon dioxide in the atmosphere between the Sun and the Earth. The water may contribute as much 90 percent of the warming effect.

Absent such a layer, the temperature of the surface of the Earth is determined primarily by the requirement of energy balance between the flux of solar radiation incident on the Earth and the flux of reradiation from the Earth; the reradiation flux is proportional to the fourth power of the temperature of the Earth, as in (4.26). This energy balance is the subject of Problem 4.5 and leads to the result $T_E = (R_S/2D_{SE})^{1/2}T_S$, where T_E is the temperature of the Earth and T_S is that of the Sun; here R_S is the radius of the Sun and D_{SE} is the Sun-Earth distance.

The result of that problem is $T_E = 280$ K, assuming $T_S = 5800$ K. The Sun is much hotter than the Earth, but the geometry (the small solid angle subtended by the Sun) reduces the solar flux density incident at the Earth by a factor of roughly $(1/20)^4$.

We assume as an example that the atmosphere is a perfect greenhouse, defined as an absorbent layer that transmits all of the visible radiation that falls on it from the Sun, but absorbs and re-emits all the radiation (which lies in the infrared), from the surface of the Earth. We may idealize the problem by neglecting the absorption by the layer of the infrared portion of the incident solar radiation, because the solar spectrum lies almost entirely at higher frequencies, as evident from Figure 4.4. The layer will emit energy flux I_L up and I_L down; the upward flux will balance the solar flux I_S, so that $I_L = I_S$. The net downward flux will be the sum of the solar flux I_S and the flux I_L down from the layer. The latter increases the net thermal flux incident at the surface of the Earth. Thus

$$I_{Eg} = I_S + I_L = 2I_S, \tag{59}$$

where I_{Eg} is the thermal flux from the Earth in the presence of the perfect greenhouse effect. Because the thermal flux varies as T^4, the new temperature of the surface of the Earth is

$$T_{Eg} = 2^{1/4}T_E \approx (1.19)\,280\text{ K} \approx 333\text{ K}, \tag{60}$$

so that the greenhouse warming of the Earth is $333\text{ K} - 280\text{ K} = 53\text{ K}$ for this extreme example.*

---

* For detailed discussions see *Climate change* and *Climate change 1992*, Cambridge U.P., 1990 and 1992: J. T. Houghton et al, editors.

Chapter 5

Chemical Potential and Gibbs Distribution

We considered in Chapter 2 the properties of two systems in thermal contact, and we were led naturally to the definition of the temperature. If the two systems have the same temperature, there is no net energy flow between them. If the temperatures of two systems are different, energy will flow from the system with the higher temperature to the system with the lower temperature.

Now consider systems that can exchange particles as well as energy. Such systems are said to be in diffusive (and thermal) contact: molecules can move from one system to the other by diffusion through a permeable interface. Two systems are in equilibrium with respect to particle exchange when the net particle flow is zero.

The **chemical potential** governs the flow of particles between the systems, just as the temperature governs the flow of energy. If two systems with a single chemical species are at the same temperature and have the same value of the chemical potential, there will be no net particle flow and no net energy flow between them. If the chemical potentials of the two systems are different, particles will flow from the system at the higher chemical potential to the system at the lower chemical potential. As an example, the chemical potential of electrons at one terminal of a storage battery is higher than at the other terminal. When the terminals are connected by a wire, electrons will flow in the wire from high to low chemical potential.

Consider the establishment of diffusive equilibrium between two systems $\mathcal{S}_1$ and $\mathcal{S}_2$ that are in thermal and diffusive contact. We maintain τ constant by placing both systems in thermal contact (Figure 5.1) with a large reservoir $\mathcal{R}$. We found earlier that for a single system $\mathcal{S}$ in thermal equilibrium with a reservoir $\mathcal{R}$, the Helmholtz free energy of $\mathcal{S}$ will assume the minimum value compatible with the common temperature τ and with other restraints on the system, such as the volume and the number of particles. This result applies equally to the combined $\mathcal{S}_1 + \mathcal{S}_2$ in equilibrium with $\mathcal{R}$. In diffusive equilibrium between $\mathcal{S}_1$ and $\mathcal{S}_2$, the particle distribution N_1, N_2 between the systems makes the total Helmholtz free energy

$$F = F_1 + F_2 = U_1 + U_2 - \tau(\sigma_1 + \sigma_2) \tag{1}$$

a minimum, subject to $N = N_1 + N_2 = $ constant. Because N is constant, the Helmholtz free energy of the combined system is a minimum with respect to

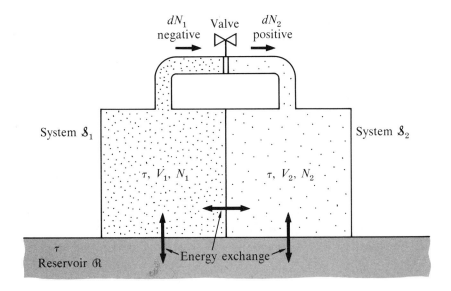

Figure 5.1   Example of two systems, $\mathcal{S}_1$ and $\mathcal{S}_2$, in thermal contact with each other and with a large reservoir $\mathcal{R}$, forming a closed total system. By opening the valve, $\mathcal{S}_1$ and $\mathcal{S}_2$ can be brought in diffusive contact while remaining at the common temperature τ. The arrows at the valve have been drawn for a net particle transfer from $\mathcal{S}_1$ to $\mathcal{S}_2$.

variations $\delta N_1 = -\delta N_2$. At the minimum,

$$dF = (\partial F_1/\partial N_1)_\tau dN_1 + (\partial F_2/\partial N_2)_\tau dN_2 = 0 \ , \tag{2}$$

with V_1, V_2, also held constant. With $dN_1 = -dN_2$, we have

$$dF = [(\partial F_1/\partial N_1)_\tau - (\partial F_2/\partial N_2)_\tau]dN_1 = 0 \ , \tag{3}$$

so that at equilibrium

$$(\partial F_1/\partial N_1)_\tau = (\partial F_2/\partial N_2)_\tau. \tag{4}$$

DEFINITION OF CHEMICAL POTENTIAL

We define the chemical potential as

$$\boxed{\mu(\tau,V,N) \equiv \left(\frac{\partial F}{\partial N}\right)_{\tau,V} \ ,} \tag{5}$$

where μ is the Greek letter mu. Then

$$\mu_1 = \mu_2$$

expresses the condition for diffusive equilibrium. If $\mu_1 > \mu_2$, we see from (3) that dF will be negative when dN_1 is negative: When particles are transferred from $\mathcal{S}_1$ to $\mathcal{S}_2$, the value of dN_1 is negative, and dN_2 is positive. Thus the free energy decreases as particles flow from $\mathcal{S}_1$ to $\mathcal{S}_2$; that is, particles flow from the system of high chemical potential to the system of low chemical potential. The strict definition of μ is in terms of a difference and not a derivative, because particles are not divisible:

$$\mu(\tau, V, N) \equiv F(\tau, V, N) - F(\tau, V, N - 1). \tag{6}$$

The chemical potential regulates the particle transfer between systems in contact, and it is fully as important as the temperature, which regulates the energy transfer. Two systems that can exchange both energy and particles are in combined thermal and diffusive equilibrium when their temperatures and chemical potentials are equal: $\tau_1 = \tau_2; \mu_1 = \mu_2$.

A difference in chemical potential acts as a driving force for the transfer of particles just as a difference in temperature acts as a driving force for the transfer of energy.

If several chemical species are present, each has its own chemical potential. For species j,

$$\mu_j = (\partial F / \partial N_j)_{\tau, V, N_1, N_2, \ldots}, \tag{7}$$

where in the differentiation the numbers of all particles are held constant except for the species j.

Example: Chemical potential of the ideal gas. In (3.70) we showed that the free energy of the monatomic ideal gas is

$$F = -\tau[N \log Z_1 - \log N!], \tag{8}$$

where

$$Z_1 = n_Q V = (M\tau/2\pi\hbar^2)^{3/2} V \tag{9}$$

is the partition function for a single particle. From (8),

$$\mu = (\partial F/\partial N)_{\tau,V} = -\tau \left[\log Z_1 - \frac{d}{dN} \log N! \right]. \tag{10}$$

If we use the Stirling approximation for $N!$ and assume that we can differentiate the factorial, we find

$$\frac{d}{dN} \log N! = \frac{d}{dN} \left[\log \sqrt{2\pi} + (N + \tfrac{1}{2}) \log N - N \right]$$

$$= \log N + (N + \tfrac{1}{2}) \cdot \frac{1}{N} - 1 = \log N + \frac{1}{2N} \, , \tag{11}$$

which approaches $\log N$ for large values of N. Hence the chemical potential of the ideal gas is

$$\mu = -\tau(\log Z_1 - \log N) = \tau \log(N/Z_1) \, ,$$

or, by (9),

$$\boxed{\mu = \tau \log(n/n_Q) \, ,} \tag{12a}$$

where $n = N/V$ is the concentration of particles and $n_Q = (M\tau/2\pi\hbar^2)^{3/2}$ is the quantum concentration defined by (3.63).

If we use $\mu = F(N) - F(N - 1)$ from (6) as the definition of μ, we do not need to use the Stirling approximation. From (8) we obtain $\mu = -\tau[\log Z_1 - \log N]$, which agrees with (12). The result depends on the concentration of particles, not on their total number or on the system volume separately. By use of the ideal gas law $p = n\tau$ we can write (12) as

$$\mu = \tau \log(p/\tau n_Q). \tag{12b}$$

The chemical potential increases as the concentration of particles increases. This is what we expect intuitively: particles flow from higher to lower chemical potential, from higher to lower concentration. Figure 5.2 shows the dependence on concentration of an ideal gas composed of electrons or of helium atoms, for two temperatures, the boiling temperature

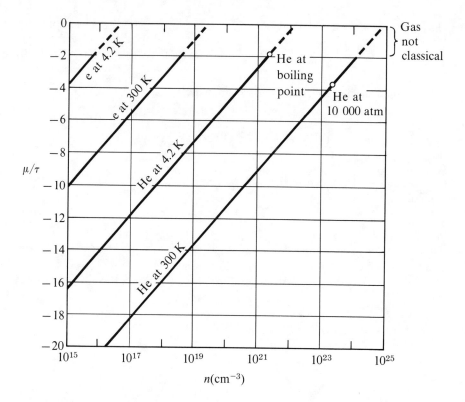

Figure 5.2 The concentration dependence of μ, in units of τ, of an ideal gas composed of electrons or helium atoms, at 4.2 K and 300 K. To be in the classical regime with $n \ll n_Q$, a gas must have a value of $-\mu$ at least τ. For electrons this is satisfied only for concentrations appreciably less than those in metals, as in the range of typical semiconductors. For gases it is always satisfied under normal conditions.

of liquid helium at atmospheric pressure, 4.2 K, and room temperature, 300 K. Atomic and molecular gases always have negative chemical potentials under physically realizable conditions: at classical concentrations such that $n/n_Q \ll 1$, we see from (12) that μ is negative.

Internal and Total Chemical Potential

The best way to understand the chemical potential is to discuss diffusive equilibrium in the presence of a potential step that acts on the particles. This

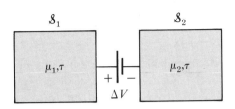

Figure 5.3 A potential step between two systems of charged particles can be established by applying a voltage between the systems. For the voltage polarity shown, the potential energy of positive particles with charge $q > 0$ in system $\mathcal{S}_1$ would be raised by $q\Delta V$ with respect to $\mathcal{S}_2$. The potential energy of negative particles would be lowered in $\mathcal{S}_1$ with respect to $\mathcal{S}_2$.

problem has wide application and includes the semiconductor $p-n$ junction discussed in Chapter 13. We again consider two systems, $\mathcal{S}_1$ and $\mathcal{S}_2$, at the same temperature and capable of exchanging particles, but not yet in diffusive equilibrium. We assume that initially $\mu_2 > \mu_1$, and we denote the initial non-equilibrium chemical potential difference by $\Delta\mu(\text{initial}) = \mu_2 - \mu_1$. Now let a difference in potential energy be established between the two systems, such that the potential energy of each particle in system $\mathcal{S}_1$ is raised by exactly $\Delta\mu(\text{initial})$ above its initial value. If the particles carry a charge q, one simple way to establish this potential step is to apply between the two systems a voltage ΔV such that

$$q\,\Delta V = q(V_2 - V_1) = \Delta\mu(\text{initial}) \,, \tag{13}$$

with the polarity shown in Figure 5.3. A difference in gravitational potential also can serve as a potential difference: when we raise a system of particles each of mass M by the height h, we establish a potential difference Mgh, where g is the gravitational acceleration.

Once a potential step is present, the potential energy of the particles produced by this step is included in the energy U and in the free energy F of the system. If in Figure 5.3 we keep the free energy of system $\mathcal{S}_2$ fixed, the step raises the free energy of $\mathcal{S}_1$ by $N_1\,\Delta\mu(\text{initial}) = N_1 q\,\Delta V$ relative to its initial value. In the language of energy states, to the energy of each state of $\mathcal{S}_1$ the potential energy $N_1\,\Delta\mu(\text{initial})$ has been added. The insertion of the potential barrier specified by (13) raises the chemical potential of $\mathcal{S}_1$ by $\Delta\mu(\text{initial})$, to make the final chemical potential of $\mathcal{S}_1$ equal to that of $\mathcal{S}_2$:

$$\mu_1(\text{final}) = \mu_1(\text{initial}) + [\mu_2(\text{initial}) - \mu_1(\text{initial})]$$
$$= \mu_2(\text{initial}) = \mu_2(\text{final}). \tag{14}$$

When the barrier was inserted, μ_2 was held fixed. Thus the barrier $q\,\Delta V = \mu_2(\text{initial}) - \mu_1(\text{initial})$ brings the two systems into diffusive equilibrium.

> The chemical potential is equivalent to a true potential energy: the difference in chemical potential between two systems is equal to the potential barrier that will bring the two systems into diffusive equilibrium.

This statement gives us a feeling for the physical effect of the chemical potential, and it forms the basis for the measurement of chemical potential differences between two systems. To measure $\mu_2 - \mu_1$, we establish a potential step between two systems that can transfer particles, and we determine the step height at which the net particle transfer vanishes.

Only differences of chemical potential have a physical meaning. The absolute value of the chemical potential depends on the zero of the potential energy scale. The ideal gas result (12) depends on the choice of the zero of energy of a free particle as equal to the zero of the kinetic energy.

When external potential steps are present, we can express the **total chemical potential** of a system as the sum of two parts:

$$\mu = \mu_{tot} = \mu_{ext} + \mu_{int}. \tag{15}$$

Here μ_{ext} is the potential energy per particle in the external potential, and μ_{int} is the **internal chemical potential*** defined as the chemical potential that would be present if the external potential were zero. The term μ_{ext} may be mechanical, electrical, magnetic, gravitational, etc. in origin. The equilibrium condition $\mu_2 = \mu_1$ can be expressed as

$$\Delta\mu_{ext} = -\Delta\mu_{int}. \tag{16}$$

Unfortunately, the distinction between external and internal chemical potential sometimes is not made in the literature. Some writers, particularly those working with charged particles in the fields of electrochemistry and of semiconductors, often mean the internal chemical potential when they use the words chemical potential without a further qualifier.

The total chemical potential may be called the electrochemical potential if the potential barriers of interest are electrostatic. Although the term electro-

---

* Gibbs called μ the potential and μ_{int} the intrinsic potential. He recognized that a voltmeter measures differences in μ.

System (2)

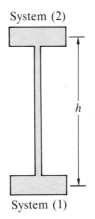

h

System (1)

Figure 5.4 A model of the variation of atmospheric pressure with altitude: two volumes of gas at different heights in a uniform gravitational field, in thermal and diffusive contact.

chemical potential is clear and unambiguous, we shall use "total chemical potential." The use of "chemical potential" without an adjective should be avoided in situations in which any confusion about its meaning could occur.

Example: Variation of barometric pressure with altitude. The simplest example of the diffusive equilibrium between systems in different external potentials is the equilibrium between layers at different heights of the Earth's atmosphere, assumed to be isothermal. The real atmosphere is in imperfect equilibrium; it is constantly upset by meteorological processes, both in the form of macroscopic air movements and of strong temperature gradients from cloud formation, and because of heat input from the ground. We may make an approximate model of the atmosphere by treating the different air layers as systems of ideal gases in thermal and diffusive equilibrium with each other, in different external potentials (Figure 5.4). If we place the zero of the potential energy at ground level, the potential energy per molecule at height h is Mgh, where M is the particle mass and g the gravitational acceleration. The internal chemical potential of the particles is given by (12). The total chemical potential is

$$\mu = \tau \log(n/n_Q) + Mgh. \tag{17}$$

In equilibrium, this must be independent of the height. Thus

$$\tau \log[n(h)/n_Q] + Mgh = \tau \log[n(0)/n_Q] \ ,$$

and the concentration $n(h)$ at height h satisfies

$$n(h) = n(0)\exp(-Mgh/\tau). \tag{18}$$

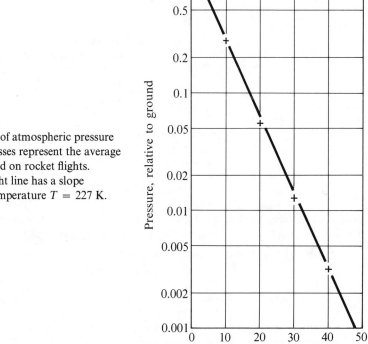

Figure 5.5 Decrease of atmospheric pressure with altitude. The crosses represent the average atmosphere as sampled on rocket flights. The connecting straight line has a slope corresponding to a temperature $T = 227$ K.

The pressure of an ideal gas is proportional to the concentration; therefore the pressure at altitude h is

$$p(h) = p(0)\exp(-Mgh/\tau) = p(0)\exp(-h/h_c). \qquad (19)$$

This is the **barometric pressure equation**. It gives the dependence of the pressure on altitude in an isothermal atmosphere of a single chemical species. At the characteristic height $h_c = \tau/Mg$ the atmospheric pressure decreases by the fraction $e^{-1} \simeq 0.37$. To estimate the characteristic height, consider an isothermal atmosphere composed of nitrogen molecules with a molecular weight of 28. The mass of an N_2 molecule is 48×10^{-24} gm. At a temperature of 290 K the value of $\tau \equiv k_B T$ is 4.0×10^{-14} erg. With $g = 980$ cm s^{-2}, the characteristic height h_c is 8.5 km, approximately 5 miles. Lighter molecules, H_2 and He, will extend farther up, but these have largely escaped from the atmosphere: see Problem 2.

Because the Earth's atmosphere is not accurately isothermal, $n(h)$ has a more complicated behavior. Figure 5.5 is a logarithmic plot of pressure data between 10 and 40 kilometers, taken on rocket flights. The data points fall near a straight line, suggesting roughly isothermal behavior. The straight line connecting the data points of Figure 5.5 spans a pressure range $p(h_2):p(h_1) = 1000:1$, over an altitude range from $h_1 = 2\,km$ to $h_2 = 48\,km$. Now, from (19),

$$\log \frac{p(h_1)}{p(h_2)} = \frac{Mg}{\tau}(h_2 - h_1)\,, \tag{20}$$

so that the slope of the line is Mg/τ, which leads to $T = \tau/k_B = 227\,K$. The non-intersection of the observed curve with the point $h = 0$, $p(h)/p(0) = 1$, is caused by the higher temperature at lower altitudes.

The atmosphere consists of more than one species of gas. In atomic percent, the composition of dry air at sea level is 78 pct N_2, 21 pct O_2, and 0.9 pct Ar; other constituents account for less than 0.1 pct each. The water vapor content of the atmosphere may be appreciable: at $T = 300\,K$ (27°C), a relative humidity of 100 pct corresponds to 3.5 pct H_2O. The carbon dioxide concentration varies about a nominal value of 0.03 pct. In an ideal static isothermal atmosphere each gas would be in equilibrium with itself. The concentration of each would fall off with a separate Boltzmann factor of the form $\exp(-Mgh/\tau)$, with M the appropriate molecular mass. Because of the differences in mass, the different constituents fall off at different rates.

Example: Chemical potential of mobile magnetic particles in a magnetic field. Consider a system of N identical particles each with a magnetic moment **m**. For simplicity suppose each moment is directed either parallel ↑ or antiparallel ↓ to an applied magnetic field **B**. Then the potential energy of a ↑ particle is $-mB$, and the potential energy of a ↓ particle is $+mB$. We may treat the particles as belonging to the two distinct chemical species labelled ↑ and ↓, one with external chemical potential $\mu_{ext}(\uparrow) = -mB$ and the other with $\mu_{ext}(\downarrow) = mB$. The particles ↑ and ↓ are as distinguishable as two different isotopes of an element or as two different elements; we speak of ↑ and ↓ as distinct species in equilibrium with each other. The internal chemical potentials of the particles viewed as ideal gases with concentrations $n_\uparrow$ and $n_\downarrow$ are

$$\mu_{int}(\uparrow) = \tau \log(n_\uparrow/n_Q); \qquad \mu_{int}(\downarrow) = \tau \log(n_\downarrow/n_Q)\,, \tag{21}$$

where $n_Q \equiv (M\tau/2\pi\hbar^2)^{3/2}$ is the same for both species.

The total chemical potentials are

$$\mu_{tot}(\uparrow) = \tau \log(n_\uparrow/n_Q) - mB; \tag{22a}$$

$$\mu_{tot}(\downarrow) = \tau \log(n_\downarrow/n_Q) + mB. \tag{22b}$$

If the magnetic field B varies in magnitude over the volume of the system, the concentration $n_\uparrow$ must vary over the volume in order to maintain a constant total chemical potential $\mu_{tot}(\uparrow)$ over the volume (Figure 5.6). (The total chemical potential of a species is constant independent of position, if there is free diffusion of particles within the volume.) Because the two

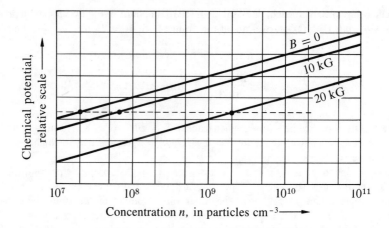

Figure 5.6 Dependence of the chemical potential of a gas of magnetic particles on the concentration, at several values of the magnetic field intensity. If $n = 2 \times 10^7$ cm^{-3} for $B = 0$, then at a point where $B = 20$ kilogauss (2 tesla) the concentration will be 2×10^9 cm^{-3}.

species in equilibrium have equal chemical potentials,

$$\mu_{tot}(\uparrow) = \text{constant} = \mu_{tot}(\downarrow). \tag{23}$$

The desired solutions of (22) and (23) are easily seen by substitution to be:

$$n_\uparrow(B) = \tfrac{1}{2}n(0)\exp(mB/\tau); \qquad n_\downarrow(B) = \tfrac{1}{2}n(0)\exp(-mB/\tau) , \tag{24}$$

where $n(0)$ is the total concentration $n_\uparrow + n_\downarrow$ at a point where the field $B = 0$. The total concentration at a point at magnetic field B is

$$n(B) = n_\uparrow(B) + n_\downarrow(B) = \tfrac{1}{2}n(0)[\exp(mB/\tau) + \exp(-mB/\tau)];$$

$$n(B) = n(0)\cosh(mB/\tau) \simeq n(0)\left(1 + \frac{m^2B^2}{2\tau^2} + \cdots\right). \tag{25}$$

The result shows the tendency of magnetic particles to concentrate in regions of high magnetic field intensity. The functional form of the result is not limited to atoms with two magnetic orientations, but is applicable to fine ferromagnetic particles in suspension in a colloidal solution. Such suspensions are used in the laboratory in the study of the magnetic flux structure of superconductors and the domain structure of ferromagnetic materials. In engineering, the suspensions are used to test for fine structural cracks in high strength steel, such as turbine blades and aircraft landing gear. When these are coated with a ferromagnetic suspension and placed in a magnetic field, the particle concentration becomes enhanced at the intense fields at the edges of the crack.

In the preceding discussion we added to μ_{ext} the internal chemical potential of the particles. If the particles were ideal gas atoms, μ_{int} would be given by (12). The logarithmic form for μ_{int} is not restricted to ideal gases, but is a consequence of the conditions that the particles do not interact and that their concentration is sufficiently low. Hence, (12) applies to macroscopic particles as well as to atoms that satisfy these assumptions. The only difference is the value of the quantum concentration n_Q. We can therefore write

$$\mu_{int} = \tau \log n + \text{constant} , \tag{26}$$

where the constant $(= -\tau \log n_Q)$ does not depend on the concentration of the particles.

Example: Batteries. One of the most vivid examples of chemical potentials and potential steps is the electrochemical battery. In the familiar lead-acid battery the negative electrode consists of metallic lead, Pb, and the positive electrode is a layer of reddish-brown lead oxide, PbO_2, on a Pb substrate. The electrodes are immersed in diluted sulfuric acid, H_2SO_4, which is partially ionized into H^+ ions (protons) and SO_4^{--} ions (Figure 5.7). It is the ions that matter.

In the discharge process both the metallic Pb of the negative electrode and the PbO_2 of the positive electrode are converted to lead sulfate, $PbSO_4$, via the two reactions:*

Negative electrode:

$$Pb + SO_4^{--} \rightarrow PbSO_4 + 2e^- ; \tag{27a}$$

Positive electrode:

$$PbO_2 + 2H^+ + H_2SO_4 + 2e^- \rightarrow PbSO_4 + 2H_2O. \tag{27b}$$

Because of (27a) the negative electrode acts as a sink for SO_4^{--} ions, keeping the internal chemical potential $\mu(SO_4^{--})$ of the sulfate ions at the surface of the negative electrode lower than inside the electrolyte (see Figure 5.7b). Similarly, because of (27b) the positive electrode acts as a sink for H^+ ions, keeping the internal chemical potential $\mu(H^+)$ of the hydrogen ions lower at the surface of the positive electrode than inside the electrolyte. The chemical potential gradients drive the ions towards the electrodes, and they drive the electrical currents during the discharge process.

If the battery terminals are not connected, electrons are depleted from the positive electrode and accumulate in the negative electrode, thereby charging both. As a result, electrochemical potential steps develop at the electrode-electrolyte interfaces, steps of exactly the correct magnitude to equalize the chemical potential steps and to stop the diffusion of ions, which stops the chemical reactions from proceeding further. If an external current is permitted to flow, the reactions resume. Electron flow directly through the electrolyte is negligible, because of a negligible electron concentration in the electrolyte.

---

* The reactions given are net reactions. The actual reaction steps are more complicated.

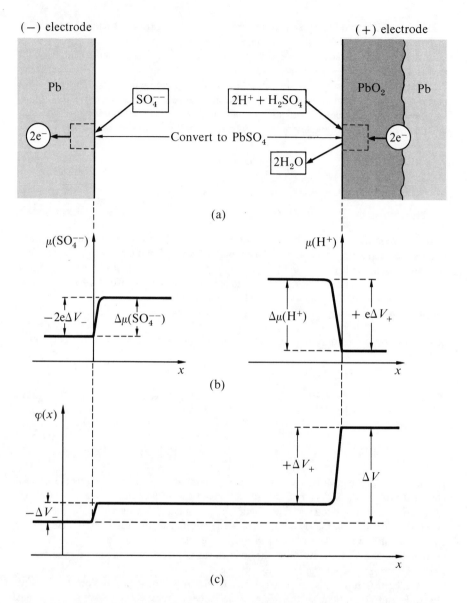

Figure 5.7   (a) The lead-acid battery consists of a Pb and a PbO_2 electrode immersed in partially ionized H_2SO_4. One SO_4^{--} ion converts one Pb atom into $PbSO_4 + 2e^-$; two H^+ ions plus one un-ionized H_2SO_4 molecule convert one PbO_2 molecule into $PbSO_4 + 2H_2O$, consuming two electrons. (b) The electrochemical potentials for SO_4^{--} and H^+ before the development of internal potential barriers that stop the diffusion and the chemical reaction. (c) The electrostatic potential $\varphi(x)$ after the formation of the barrier.

During the charging process the reactions opposite to (27a,b) take place, because now an external voltage is applied that generates electrostatic potential steps at the surface of the electrode of such magnitude as to reverse the sign of the (total) chemical potential gradients, and hence the direction of ion flow.

We denote by ΔV_- and ΔV_+ the differences in electrostatic potential of the negative and positive electrodes relative to the common electrolyte. Because the sulfate ions carry two negative charges, diffusion will stop when

$$-2q\,\Delta V_- = \Delta\mu(SO_4^{--}). \tag{28a}$$

Diffusion of the H^+ ions will stop when

$$+q\,\Delta V_+ = \Delta\mu(H^+). \tag{28b}$$

The two potentials ΔV_- and ΔV_+ are called **half-cell potentials** or half-cell EMF's (electromotive forces); their magnitudes are known:

$$\Delta V_- = -0.4\,\text{volt}; \qquad \Delta V_+ = +1.6\,\text{volt}.$$

The total electrostatic potential difference developed across one full cell of the battery, as required to stop the diffusion reaction, is

$$\Delta V = \Delta V_+ - \Delta V_- = 2.0\,\text{volt}. \tag{29}$$

This is the **open-circuit voltage** or **EMF** of the battery. It drives the electrons from the negative terminal to the positive terminal, when the two are connected.

We have ignored free electrons in the electrolyte. The potential steps tend to drive electrons from the negative electrodes into the electrolyte, and from the electrolyte into the positive electrode. Such an electron current is present, but the magnitude is so small as to be practically negligible, because the concentration of electrons in the electrolyte is many orders of magnitude less than that of the ions. The only effective electron flow path is through the external connection between the electrodes.

Chemical Potential and Entropy

In (5) we defined the chemical potential as a derivative of the Helmholtz free energy. Here we derive an alternate relation, needed later:

$$\frac{\mu(U,V,N)}{\tau} = -\left(\frac{\partial\sigma}{\partial N}\right)_{U,V}. \tag{30}$$

This expresses the ratio μ/τ as a derivative of the entropy, similar to the way $1/\tau$ was defined in Chapter 2.

To derive (30), consider the entropy as a function of the independent variables U, V, and N. The differential

$$d\sigma = \left(\frac{\partial\sigma}{\partial U}\right)_{V,N} dU + \left(\frac{\partial\sigma}{\partial V}\right)_{U,N} dV + \left(\frac{\partial\sigma}{\partial N}\right)_{U,V} dN \qquad (31)$$

gives the differential change of the entropy for arbitrary, independent differential changes dU, dV, and dN. Let $dV = 0$ for the processes under consideration. Further, select the ratios of $d\sigma$, dU, and dN in such a way that the overall temperature change $d\tau$ will be zero. If we denote these interdependent values of $d\sigma$, dU, and dN by $(\delta\sigma)_\tau$, $(\delta U)_\tau$ and $(\delta N)_\tau$, then $d\tau = 0$ when

$$(\delta\sigma)_\tau = \left(\frac{\partial\sigma}{\partial U}\right)_N (\delta U)_\tau + \left(\frac{\partial\sigma}{\partial N}\right)_U (\delta N)_\tau.$$

After division by $(\delta N)_\tau$,

$$\frac{(\delta\sigma)_\tau}{(\delta N)_\tau} = \left(\frac{\partial\sigma}{\partial U}\right)_N \frac{(\delta U)_\tau}{(\delta N)_\tau} + \left(\frac{\partial\sigma}{\partial N}\right)_U. \qquad (32)$$

The ratio $(\delta\sigma)_\tau/(\delta N)_\tau$ is $(\partial\sigma/\partial N)_\tau$, and $(\delta U)_\tau/(\delta N)_\tau$ is $(\partial U/\partial N)_\tau$, all at constant volume. With the original definition of $1/\tau$, we have

$$\tau\left(\frac{\partial\sigma}{\partial N}\right)_{\tau,V} = \left(\frac{\partial U}{\partial N}\right)_{\tau,V} + \tau\left(\frac{\partial\sigma}{\partial N}\right)_{U,V}. \qquad (33)$$

This expresses a derivative at constant U in terms of derivatives at constant τ. By the original definition of the chemical potential,

$$\mu \equiv \left(\frac{\partial F}{\partial N}\right)_{\tau,V} \equiv \left(\frac{\partial U}{\partial N}\right)_{\tau,V} - \tau\left(\frac{\partial\sigma}{\partial N}\right)_{\tau,V}, \qquad (34)$$

and on comparison with (33) we obtain

$$\boxed{\mu = -\tau(\partial\sigma/\partial N)_{U,V}.} \qquad (35)$$

The two expressions (5) and (35) represent two different ways to express the same quantity μ. The difference between them is the following. In (5), F is a

Table 5.1 Summary of relations expressing the temperature
τ, the pressure p, and the chemical potential μ in terms of
partial derivatives of the entropy σ, the energy U, and the free
energy F, with σ, U, and F given as functions of their natural
independent variables

	$\sigma(U,V,N)$	$U(\sigma,V,N)$	$F(\tau,V,N)$
τ:	$\dfrac{1}{\tau} = \left(\dfrac{\partial \sigma}{\partial U}\right)_{V,N}$	$\tau = \left(\dfrac{\partial U}{\partial \sigma}\right)_{V,N}$	τ is independent variable
p:	$\dfrac{p}{\tau} = \left(\dfrac{\partial \sigma}{\partial V}\right)_{U,N}$	$-p = \left(\dfrac{\partial U}{\partial V}\right)_{\sigma,N}$	$-p = \left(\dfrac{\partial F}{\partial V}\right)_{\tau,N}$
μ:	$-\dfrac{\mu}{\tau} = \left(\dfrac{\partial \sigma}{\partial N}\right)_{U,V}$	$\mu = \left(\dfrac{\partial U}{\partial N}\right)_{\sigma,V}$	$\mu = \left(\dfrac{\partial F}{\partial N}\right)_{\tau,V}$

function of its natural independent variables τ, V, and N, so that μ appears as
a function of the same variables. In (31) we assumed $\sigma = \sigma(U,V,N)$, so that
(35) yields μ as a function of U, V, N. The quantity μ is the same in both (5)
and (35), but expressed in terms of different variables. The object of Problem 11
is to find a third relation for μ:

$$\mu(\sigma,V,N) = (\partial U/\partial N)_{\sigma,V} , \tag{36}$$

and in Chapter 10 we derive a relation for μ as a function of τ, p, and N. Table 5.1
compiles expressions for τ, p, and μ as derivatives of σ, U, and F. All forms
have their uses.

Thermodynamic identity. We can generalize the statement of the thermodynamic identity given in (3.34a) to include systems in which the number of
particles is allowed to change. As in (31),

$$d\sigma = \left(\frac{\partial \sigma}{\partial U}\right)_{V,N} dU + \left(\frac{\partial \sigma}{\partial V}\right)_{U,N} dV + \left(\frac{\partial \sigma}{\partial N}\right)_{U,V} dN. \tag{37}$$

By use of the definition (2.26) of $1/\tau$, the relation (3.32) for p/τ, and the relation
(30) for $-\mu/\tau$, we write $d\sigma$ as

$$d\sigma = dU/\tau + pdV/\tau - \mu dN/\tau. \tag{38}$$

This may be rearranged to give

$$dU = \tau d\sigma - p dV + \mu dN \; , \tag{39}$$

which is a broader statement of the thermodynamic identity than we were able to develop in Chapter 3.

GIBBS FACTOR AND GIBBS SUM

The Boltzmann factor, derived in Chapter 3, allows us to give the ratio of the probability that a system will be in a state of energy ε_1 to the probability the system will be in a state of energy ε_2, for a system in thermal contact with a reservoir at temperature τ:

$$\frac{P(\varepsilon_1)}{P(\varepsilon_2)} = \frac{\exp(-\varepsilon_1/\tau)}{\exp(-\varepsilon_2/\tau)}. \tag{40}$$

This is perhaps the best known result of statistical mechanics. The Gibbs factor is the generalization of the Boltzmann factor to a system in thermal and diffusive contact with a reservoir at temperature τ and chemical potential μ. The argument retraces much of that presented in Chapter 3.

We consider a very large body with constant energy U_0 and constant particle number N_0. The body is composed of two parts, the very large reservoir $\mathcal{R}$ and the system $\mathcal{S}$, in thermal and diffusive contact (Figure 5.8). They may exchange particles and energy. The contact assures that the temperature and the chemical potential of the system are equal to those of the reservoir. When the system has N particles, the reservoir has $N_0 - N$ particles; when the system has energy ε, the reservoir has energy $U_0 - \varepsilon$. To obtain the statistical properties of the system, we make observations as before on identical copies of the system + reservoir, one copy for each accessible quantum state of the combination. What is the probability in a given observation that the system will be found to contain N particles and to be in a state s of energy ε_s?

The state s is a state of a system having some specified number of particles. The energy $\varepsilon_{s(N)}$ is the energy of the state s of the N-particle system; sometimes we write only ε_s, if the meaning is clear. When can we write the energy of a system having N particles in an orbital as N times the energy of one particle in the orbital? Only when interactions between the particles are neglected, so that the particles may be treated as independent of each other.

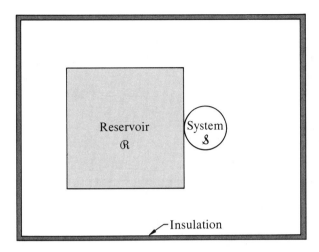

Figure 5.8 A system in thermal and diffusive contact with a large reservoir of energy and of particles. The total system $\mathcal{R} + \mathcal{S}$ is insulated from the external world, so that the total energy and the total number of particles are constant. The temperature of the system is equal to the temperature of the reservoir, and the chemical potential of the system is equal to the chemical potential of the reservoir. The system may be as small as one atom or it may be macroscopic, but the reservoir is always to be thought of as much larger than the system.

Let $P(N, \varepsilon_s)$ denote the probability that the system has N particles and is in a particular state s. This probability is proportional to the number of accessible states of the reservoir when the state of the system is exactly specified. That is, if we specify the state of $\mathcal{S}$, the number of accessible states of $\mathcal{R} + \mathcal{S}$ is just the number of accessible states of $\mathcal{R}$:

$$g(\mathcal{R} + \mathcal{S}) = g(\mathcal{R}) \times 1. \tag{41}$$

The factor 1 reminds us that we are looking at the system $\mathcal{S}$ in a single specified state. The $g(\mathcal{R})$ states of the reservoir have $N_0 - N$ particles and have energy $U_0 - \varepsilon_s$. Because the system probability $P(N, \varepsilon_s)$ is proportional to the number of accessible states of the reservoir,

$$P(N, \varepsilon_s) \propto g(N_0 - N, U_0 - \varepsilon_s). \tag{42}$$

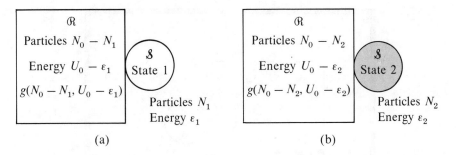

Figure 5.9 The reservoir is in thermal and diffusive contact with the system. In (a) the system is in quantum state 1, and the reservoir has $g(N_0 - N_1, U_0 - \varepsilon_1)$ states accessible to it. In (b) the system is in quantum state 2, and the reservoir has $g(N_0 - N_2, U_0 - \varepsilon_2)$ states accessible to it. Because we have specified the exact state of the system, the total number of states accessible to $\mathcal{R} + \mathcal{S}$ is just the number of states accessible to $\mathcal{R}$.

Here g refers to the reservoir alone and depends on the number of particles in the reservoir and on the energy of the reservoir.

We can express (42) as a ratio of two probabilities, one that the system is in state 1 and the other that the system is in state 2:

$$\frac{P(N_1, \varepsilon_1)}{P(N_2, \varepsilon_2)} = \frac{g(N_0 - N_1, U_0 - \varepsilon_1)}{g(N_0 - N_2, U_0 - \varepsilon_2)} ,\qquad(43)$$

where g refers to the state of the reservoir. The situation is shown in Figure 5.9. By definition of the entropy

$$g(N_0, U_0) \equiv \exp[\sigma(N_0, U_0)] ,\qquad(44)$$

so that the probability ratio in (43) may be written as

$$\frac{P(N_1, \varepsilon_1)}{P(N_2, \varepsilon_2)} = \frac{\exp[\sigma(N_0 - N_1, U_0 - \varepsilon_1)]}{\exp[\sigma(N_0 - N_2, U_0 - \varepsilon_2)]};\qquad(45)$$

or

$$\frac{P(N_1, \varepsilon_1)}{P(N_2, \varepsilon_2)} = \exp[\sigma(N_0 - N_1, U_0 - \varepsilon_1) - \sigma(N_0 - N_2, U_0 - \varepsilon_2)]$$

$$= \exp(\Delta\sigma).\qquad(46)$$

Here, $\Delta\sigma$ is the entropy difference:

$$\Delta\sigma \equiv \sigma(N_0 - N_1, U_0 - \varepsilon_1) - \sigma(N_0 - N_2, U_0 - \varepsilon_2). \tag{47}$$

The reservoir is very large in comparison with the system, and $\Delta\sigma$ may be approximated quite accurately by the first order terms in a series expansion in the two quantities N and ε that relate to the system. The entropy of the reservoir becomes

$$\sigma(N_0 - N, U_0 - \varepsilon) = \sigma(N_0, U_0) - N\left(\frac{\partial\sigma}{\partial N_0}\right)_{U_0} - \varepsilon\left(\frac{\partial\sigma}{\partial U_0}\right)_{N_0} + \cdots. \tag{48}$$

For $\Delta\sigma$ defined by (47) we have, to the first order in $N_1 - N_2$ and in $\varepsilon_1 - \varepsilon_2$,

$$\Delta\sigma = -(N_1 - N_2)\left(\frac{\partial\sigma}{\partial N_0}\right)_{U_0} - (\varepsilon_1 - \varepsilon_2)\left(\frac{\partial\sigma}{\partial U_0}\right)_{N_0}. \tag{49}$$

We know that

$$\frac{1}{\tau} \equiv \left(\frac{\partial\sigma}{\partial U_0}\right)_{N_0}, \tag{50a}$$

by our original definition of the temperature. This is written for the reservoir, but the system will have the same temperature. Also,

$$-\frac{\mu}{\tau} \equiv \left(\frac{\partial\sigma}{\partial N_0}\right)_{U_0}, \tag{50b}$$

by (30).

The entropy difference (49) is

$$\Delta\sigma = \frac{(N_1 - N_2)\mu}{\tau} - \frac{(\varepsilon_1 - \varepsilon_2)}{\tau}. \tag{51}$$

Here $\Delta\sigma$ refers to the reservoir, but $N_1, N_2, \varepsilon_1, \varepsilon_2$ refer to the system. The central result of statistical mechanics is found on combining (46) and (51):

$$\boxed{\frac{P(N_1, \varepsilon_1)}{P(N_2, \varepsilon_2)} = \frac{\exp[(N_1\mu - \varepsilon_1)/\tau]}{\exp[(N_2\mu - \varepsilon_2)/\tau]}.} \tag{52}$$

The probability is the ratio of two exponential factors, each of the form $\exp[(N\mu - \varepsilon)/\tau]$. A term of this form is called a **Gibbs factor**. The Gibbs factor is proportional to the probability that the system is in a state s of energy ε_s and number of particles N. The result was first given by J. W. Gibbs, who referred to it as the grand canonical distribution.

The sum of Gibbs factors, taken over all states of the system for all numbers of particles, is the normalizing factor that converts relative probabilities to absolute probabilities:

$$\mathcal{Z}(\mu,\tau) = \sum_{N=0}^{\infty} \sum_{s(N)} \exp[(N\mu - \varepsilon_{s(N)})/\tau] = \sum_{\text{ASN}} \exp[(N\mu - \varepsilon_{s(N)})/\tau]. \qquad (53)$$

This is called the **Gibbs sum**, or the **grand sum**, or the grand partition function. The sum is to be carried out over all states of the system for all numbers of particles: this defines the abbreviation ASN. We have written ε_s as $\varepsilon_{s(N)}$ to emphasize the dependence of the state on the number of particles N. That is, $\varepsilon_{s(N)}$ is the energy of the state $s(N)$ of the exact N-particle hamiltonian. The term $N = 0$ must be included; if we assign its energy as zero, then the first term in $\mathcal{Z}$ will be 1.

The absolute probability that the system will be found in a state N_1, ε_1 is given by the Gibbs factor divided by the Gibbs sum:

$$P(N_1,\varepsilon_1) = \frac{\exp[(N_1\mu - \varepsilon_1)/\tau]}{\mathcal{Z}}. \qquad (54)$$

This applies to a system that is at temperature τ and chemical potential μ. The ratio of any two P's is consistent with our central result (52) for the Gibbs factors. Thus (52) gives the correct relative probabilities for the states N_1, ε_1 and N_2, ε_2. The sum of the probabilities of all states for all numbers of particles of the system is unity:

$$\sum_N \sum_s P(N,\varepsilon_s) = \sum_{\text{ASN}} P(N,\varepsilon_s) = \frac{\sum_{\text{ASN}} \exp[(N\mu - \varepsilon_{s(N)})/\tau]}{\mathcal{Z}} = \frac{\mathcal{Z}}{\mathcal{Z}} = 1 \;, \qquad (55)$$

by the definition of $\mathcal{Z}$. Thus (54) gives the correct absolute probability.*

---

* Readers interested in probability theory will find Appendix C on the Poisson distribution to be particularly helpful. The method used there to derive the Poisson distribution depends on the Gibbs sum. See also Problem (6.13).

Average values over the systems in diffusive and thermal contact with a reservoir are easily found. If $X(N,s)$ is the value of X when the system has N particles and is in the quantum state s, then the thermal average of X over all N and all s is

$$\langle X \rangle = \sum_{\text{ASN}} X(N,s)P(N,\varepsilon_s) = \frac{\sum_{\text{ASN}} X(N,s)\exp[(N\mu - \varepsilon_s)/\tau]}{\mathfrak{Z}}. \tag{56}$$

We shall use this result to calculate thermal averages.

Number of particles.   The number of particles in the system can vary because the system is in diffusive contact with a reservoir. The thermal average of the number of particles in the system is

$$\langle N \rangle = \frac{\sum_{\text{ASN}} N\exp[(N\mu - \varepsilon_s)/\tau]}{\mathfrak{Z}}, \tag{57}$$

according to (56). To obtain the numerator, each term in the Gibbs sum has been multiplied by the appropriate value of N. More convenient forms of $\langle N \rangle$ can be obtained from the definition of $\mathfrak{Z}$:

$$\frac{\partial \mathfrak{Z}}{\partial \mu} = \frac{1}{\tau}\sum_{\text{ASN}} N\exp[(N\mu - \varepsilon_s)/\tau] , \tag{58}$$

whence

$$\langle N \rangle = \frac{\tau}{\mathfrak{Z}}\frac{\partial \mathfrak{Z}}{\partial \mu} = \tau\frac{\partial \log \mathfrak{Z}}{\partial \mu}. \tag{59}$$

The thermal average number of particles is easily found from the Gibbs sum $\mathfrak{Z}$ by direct use of (59). When no confusion arises, we shall write N for the thermal average $\langle N \rangle$. When we speak later of the occupancy of an orbital, f or $\langle f \rangle$ will be written interchangeably for N or $\langle N \rangle$.

We often employ the handy notation

$$\boxed{\lambda \equiv \exp(\mu/\tau) ,} \qquad \text{fugacity} \tag{60}$$

where λ is called the **absolute activity**. Here λ is the Greek letter lambda. We see from (12) that for an ideal gas λ is directly proportional to the concentration.

The Gibbs sum is written as

$$\mathcal{Z} = \sum_N \sum_s \lambda^N \exp(-\varepsilon_s/\tau) = \sum_{ASN} \lambda^N \exp(-\varepsilon_s/\tau) , \qquad (61)$$

and the ensemble average number of particles (57) is

$$\langle N \rangle = \lambda \frac{\partial}{\partial \lambda} \log \mathcal{Z} . \qquad (62)$$

This relation is useful, because in many actual problems we determine λ by finding the value that will make $\langle N \rangle$ come out equal to the given number of particles.

Energy.   The thermal average energy of the system is

$$U = \langle \varepsilon \rangle = \frac{\sum_{ASN} \varepsilon_s \exp[\beta(N\mu - \varepsilon_s)]}{\mathcal{Z}} , \qquad (63)$$

where we have temporarily introduced the notation $\beta \equiv 1/\tau$. We shall usually write U for $\langle \varepsilon \rangle$. Observe that

$$\langle N\mu - \varepsilon \rangle = \langle N \rangle \mu - U = \frac{1}{\mathcal{Z}} \frac{\partial \mathcal{Z}}{\partial \beta} = \frac{\partial}{\partial \beta} \log \mathcal{Z} , \qquad (64)$$

so that (59) and (63) may be combined to give

$$U = \left(\frac{\mu}{\beta} \frac{\partial}{\partial \mu} - \frac{\partial}{\partial \beta} \right) \log \mathcal{Z} = \left(\tau\mu \frac{\partial}{\partial \mu} - \frac{\partial}{\partial(1/\tau)} \right) \log \mathcal{Z} . \qquad (65)$$

A simpler expression that is more widely used in calculations was obtained in Chapter 3 in terms of the partition function Z.

Example: Occupancy zero or one.   A red-blooded example of a system that may be occupied by zero molecules or by one molecule is the heme group, which may be vacant or may be occupied by one O_2 molecule—and never by more than one O_2 molecule (Figure 5.10). A single heme group occurs in the protein myoglobin, which is responsible for the red color of meat. If ε is the energy of an adsorbed molecule of O_2 relative to O_2 at rest at

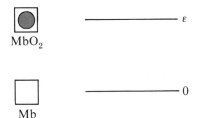

Figure 5.10 Adsorption of an O_2 by a heme, where ε is the energy of an adsorbed O_2 relative to an O_2 at infinite separation from the site. If energy must be supplied to detach the O_2 from the heme, then ε will be negative.

infinite distance, then the Gibbs sum is

$$\mathcal{Z} = 1 + \lambda \exp(-\varepsilon/\tau). \tag{66}$$

If energy must be added to remove the atom from the heme, ε will be negative. The term 1 in the sum arises from occupancy zero; the term $\lambda \exp(-\varepsilon/\tau)$ arises from single occupancy. These are the only possibilities. We have Mb + O_2 or MbO_2 present, where Mb denotes myoglobin, a protein of molecular weight 17 000.

Experimental results for the fractional occupancy versus the concentration of oxygen are shown in Figure 5.11. We compare the observed oxygen saturation curves of myoglobin and hemoglobin in Figure 5.12. Hemoglobin is the oxygen-carrying component of blood. It is made up of four molecular strands, each strand nearly identical with the single strand of myoglobin, and each capable of binding a single oxygen molecule. Historically, the classic work on the adsorption of oxygen by hemoglobin was done by Christian Bohr, the father of Niels Bohr. The oxygen saturation curve for hemoglobin (Hb) has a slower rise at low pressures, because the binding energy of a single O_2 to a molecule of Hb is lower than for Mb. At higher pressures of oxygen the Hb curve has a region that is concave upwards, because the binding energy per O_2 increases after the first O_2 is adsorbed.

The O_2 molecules on hemes are in equilibrium with the O_2 in the surrounding liquid, so that the chemical potentials of O_2 are equal on the myoglobin and in solution:

$$\mu(MbO_2) = \mu(O_2); \qquad \lambda(MbO_2) = \lambda(O_2) \tag{67}$$

where $\lambda \equiv \exp(\mu/\tau)$. From Chapter 3 we find the value of λ in terms of the gas pressure by the relation

$$\lambda = n/n_Q = p/\tau n_Q. \tag{68}$$

We assume the ideal gas result applies to O_2 in solution. At constant temperature $\lambda(O_2)$ is directly proportional to the pressure p.

The fraction f of Mb occupied by O_2 is found from (66) to be

$$f = \frac{\lambda \exp(-\varepsilon/\tau)}{1 + \lambda \exp(-\varepsilon/\tau)} = \frac{1}{\lambda^{-1} \exp(\varepsilon/\tau) + 1}, \tag{69}$$

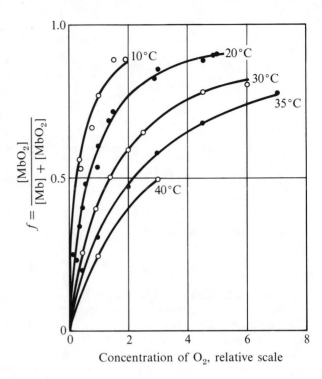

Figure 5.11 The reaction of a myoglobin (Mb) molecule with oxygen may be viewed as the adsorption of a molecule of O_2 at a site on the large myoglobin molecule. The results follow a Langmuir isotherm quite accurately. Each myoglobin molecule can adsorb one O_2 molecule. These curves show the fraction of myoglobin with adsorbed O_2 as a function of the partial pressure of O_2. The curves are for human myoglobin in solution. Myoglobin is found in muscles; it is responsible for the color of steak. After A. Rossi-Fanelli and E. Antonini, Archives of Biochemistry and Biophysics **77**, 478 (1958).

Figure 5.12 Saturation curves of O_2 bound to myoglobin (Mb) and hemoglobin (Hb) molecules in solution in water. The partial pressure of O_2 is plotted as the horizontal axis. The vertical axis gives the fraction of the molecules of Mb which has one bound O_2 molecule, or the fraction of the strands of Hb which have one bound O_2 molecule. Hemoglobin has a much larger change in oxygen content in the pressure range between the arteries and the veins. This circumstance facilitates the action of the heart, viewed as a pump. The curve for myoglobin has the predicted form for the reaction $Mb + O_2 \leftrightarrow MbO_2$. The curve for hemoglobin has a different form because of interactions between O_2 molecules bound to the four strands of the Hb molecule. The drawing is after J. S. Fruton and S. Simmonds, *General biochemistry*, Wiley, 1961.

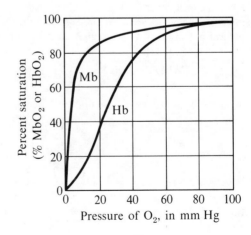

which is the same as the Fermi-Dirac distribution function derived in Chapter 7. We substitute (68) in (69) to obtain

$$f = \frac{1}{(n_Q \tau / p) \exp(\varepsilon/\tau) + 1} = \frac{p}{n_Q \tau \exp(\varepsilon/\tau) + p} , \tag{70}$$

or, with $p_0 \equiv n_Q \tau \exp(\varepsilon/\tau)$,

$$f = \frac{p}{p_0 + p} , \tag{71}$$

where p_0 is constant with respect to pressure, but depends on the temperature. The result (71) is known as the **Langmuir adsorption isotherm** when used to describe the adsorption of gases on the surfaces of solids.

Example: Impurity atom ionization in a semiconductor. Atoms of numerous chemical elements when present as impurities in a semiconductor may lose an electron by ionization to the conduction band of the semiconductor crystal. In the conduction band the electron moves about much as if it were a free particle, and the electron gas in the conduction band may often be treated as an ideal gas. The impurity atoms are small systems $\mathcal{S}$ in thermal and diffusive equilibrium with the large reservoir formed by the rest of the semiconductor; the atoms exchange electrons and energy with the semiconductor.

Let I be the ionization energy of the impurity atom. We suppose that one, but only one, electron can be bound to an impurity atom; either orientation ↑ or ↓ of the electron spin is accessible. Therefore the system $\mathcal{S}$ has three allowed states—one without an electron, one with an electron attached with spin ↑, and one with an electron attached with spin ↓. When $\mathcal{S}$ has zero electrons, the impurity atom is ionized. We choose the zero of energy of $\mathcal{S}$ as this state; the other two states therefore have the common energy $\varepsilon = -I$. The accessible states of $\mathcal{S}$ are summarized below.

State number	Description	N	ε
1	Electron detached	0	0
2	Electron attached, spin ↑	1	$-I$
3	Electron attached, spin ↓	1	$-I$

The Gibbs sum is given by

$$\mathcal{Z} = 1 + 2 \exp[(\mu + I)/\tau]. \tag{72}$$

The probability that $\mathcal{S}$ is ionized ($N = 0$) is

$$P(\text{ionized}) = P(0,0) = \frac{1}{\mathcal{Z}} = \frac{1}{1 + 2\exp[(\mu + I)/\tau]}. \tag{73}$$

The probability that $\mathcal{S}$ is neutral (un-ionized) is

$$P(\text{neutral}) = P(1\uparrow, -I) + P(1\downarrow, -I), \tag{74}$$

which is just $1 - P(0,0)$.

SUMMARY

1. The chemical potential is defined as $\mu(\tau,V,N) \equiv (\partial F/\partial N)_{\tau,V}$ and may also be found from $\mu = (\partial U/\partial N)_{\sigma,V} = -\tau(\partial\sigma/\partial N)_{U,V}$. Two systems are in diffusive equilibrium if $\mu_1 = \mu_2$.

2. The chemical potential is made up of two parts, external and internal. The external part is the potential energy of a particle in an external field of force. The internal part is of thermal origin; for an ideal monatomic gas $\mu(\text{int}) = \tau\log(n/n_Q)$, where n is the concentration and $n_Q \equiv (M\tau/2\pi\hbar^2)^{3/2}$ is the quantum concentration.

3. The Gibbs factor

$$P(N,\varepsilon_s) = \exp[(N\mu - \varepsilon_s)/\tau]/\mathcal{Z}$$

gives the probability that a system at chemical potential μ and temperature τ will have N particles and be in a quantum state s of energy ε_s.

4. The Gibbs sum

$$\mathcal{Z} \equiv \sum_{\text{ASN}} \exp[(N\mu - \varepsilon_{s(N)})/\tau]$$

is taken over all states for all numbers of particles.

5. The absolute activity λ is defined by $\lambda \equiv \exp(\mu/\tau)$.

6. The thermal average number of particles is

$$\langle N \rangle = \lambda \frac{\partial}{\partial\lambda} \log \mathcal{Z}.$$

PROBLEMS

1. *Centrifuge*. A circular cylinder of radius R rotates about the long axis with angular velocity ω. The cylinder contains an ideal gas of atoms of mass M at temperature τ. Find an expression for the dependence of the concentration $n(r)$ on the radial distance r from the axis, in terms of $n(0)$ on the axis. Take μ as for an ideal gas.

2. *Molecules in the Earth's atmosphere*. If n is the concentration of molecules at the surface of the Earth, M the mass of a molecule, and g the gravitational acceleration at the surface, show that at constant temperature the total number of molecules in the atmosphere is

$$N = 4\pi n(R)\exp(-MgR/\tau) \int_R^\infty dr\, r^2 \exp(MgR^2/r\tau) \;, \qquad (75)$$

with r measured from the center of the Earth; here R is the radius of the Earth. The integral diverges at the upper limit, so that N cannot be bounded and the atmosphere cannot be in equilibrium. Molecules, particularly light molecules, are always escaping from the atmosphere.

3. *Potential energy of gas in gravitational field*. Consider a column of atoms each of mass M at temperature τ in a uniform gravitational field g. Find the thermal average potential energy per atom. The thermal average kinetic energy density is independent of height. Find the total heat capacity per atom. The total heat capacity is the sum of contributions from the kinetic energy and from the potential energy. Take the zero of the gravitational energy at the bottom $h = 0$ of the column. Integrate from $h = 0$ to $h = \infty$.

4. *Active transport*. The concentration of potassium K^+ ions in the internal sap of a plant cell (for example, a fresh water alga) may exceed by a factor of 10^4 the concentration of K^+ ions in the pond water in which the cell is growing. The chemical potential of the K^+ ions is higher in the sap because their concentration n is higher there. Estimate the difference in chemical potential at 300 K and show that it is equivalent to a voltage of 0.24 V across the cell wall. Take μ as for an ideal gas. Because the values of the chemical potentials are different, the ions in the cell and in the pond are not in diffusive equilibrium. The plant cell membrane is highly impermeable to the passive leakage of ions through it. Important questions in cell physics include these: How is the high concentration of ions built up within the cell? How is metabolic energy applied to energize the active ion transport?

5. *Magnetic concentration*. Determine the ratio m/τ for which Figure 5.6 is drawn. If $T = 300\,K$, how many Bohr magnetons $\mu_B \equiv e\hbar/2mc$ would the particles contain to give a magnetic concentration effect of the magnitude shown?

6. Gibbs sum for a two level system.   (a) Consider a system that may be unoccupied with energy zero or occupied by one particle in either of two states, one of energy zero and one of energy ε. Show that the Gibbs sum for this system is

$$\mathcal{Z} = 1 + \lambda + \lambda \exp(-\varepsilon/\tau). \tag{76}$$

Our assumption excludes the possibility of one particle in each state at the same time. Notice that we include in the sum a term for $N = 0$ as a particular state of a system of a variable number of particles.

   (b) Show that the thermal average occupancy of the system is

$$\langle N \rangle = \frac{\lambda + \lambda \exp(-\varepsilon/\tau)}{\mathcal{Z}}. \tag{77}$$

   (c) Show that the thermal average occupancy of the state at energy ε is

$$\langle N(\varepsilon) \rangle = \lambda \exp(-\varepsilon/\tau)/\mathcal{Z}. \tag{78}$$

   (d) Find an expression for the thermal average energy of the system.

   (e) Allow the possibility that the orbital at 0 and at ε may be occupied each by one particle at the same time; show that

$$\mathcal{Z} = 1 + \lambda + \lambda \exp(-\varepsilon/\tau) + \lambda^2 \exp(-\varepsilon/\tau) = (1 + \lambda)[1 + \lambda \exp(-\varepsilon/\tau)]. \tag{79}$$

Because $\mathcal{Z}$ can be factored as shown, we have in effect two independent systems.

7. States of positive and negative ionization.   Consider a lattice of fixed hydrogen atoms; suppose that each atom can exist in four states:

State	Number of electrons	Energy
Ground	1	$-\frac{1}{2}\Delta$
Positive ion	0	$-\frac{1}{2}\delta$
Negative ion	2	$\frac{1}{2}\delta$
Excited	1	$\frac{1}{2}\Delta$

Find the condition that the average number of electrons per atom be unity. The condition will involve δ, λ, and τ.

8. Carbon monoxide poisoning.   In carbon monoxide poisoning the CO replaces the O_2 adsorbed on hemoglobin (Hb) molecules in the blood. To show the effect, consider a model for which each adsorption site on a heme may be vacant or may be occupied either with energy ε_A by one molecule O_2 or with energy ε_B by one molecule CO. Let N fixed heme sites be in equilibrium with

O_2 and CO in the gas phases at concentrations such that the activities are $\lambda(O_2) = 1 \times 10^{-5}$ and $\lambda(CO) = 1 \times 10^{-7}$, all at body temperature 37°C. Neglect any spin multiplicity factors. (a) First consider the system in the absence of CO. Evaluate ε_A such that 90 percent of the Hb sites are occupied by O_2. Express the answer in eV per O_2. (b) Now admit the CO under the specified conditions. Find ε_B such that only 10 percent of the Hb sites are occupied by O_2.

9. Adsorption of O_2 in a magnetic field. Suppose that at most one O_2 can be bound to a heme group (see Problem 8), and that when $\lambda(O_2) = 10^{-5}$ we have 90 percent of the hemes occupied by O_2. Consider O_2 as having a spin of 1 and a magnetic moment of 1 μ_B. How strong a magnetic field is needed to change the adsorption by 1 percent at $T = 300\,\text{K}$? (The Gibbs sum in the limit of zero magnetic field will differ from that of Problem 8 because there the spin multiplicity of the bound state was neglected.)

10. Concentration fluctuations. The number of particles is not constant in a system in diffusive contact with a reservoir. We have seen that

$$\langle N \rangle = \frac{\tau}{\mathfrak{Z}} \left(\frac{\partial \mathfrak{Z}}{\partial \mu} \right)_{\tau, V}, \tag{80}$$

from (59). (a) Show that

$$\langle N^2 \rangle = \frac{\tau^2}{\mathfrak{Z}} \frac{\partial^2 \mathfrak{Z}}{\partial \mu^2}. \tag{81}$$

The mean-square deviation $\langle (\Delta N)^2 \rangle$ of N from $\langle N \rangle$ is defined by

$$\langle (\Delta N)^2 \rangle = \langle (N - \langle N \rangle)^2 \rangle = \langle N^2 \rangle - 2\langle N \rangle \langle N \rangle + \langle N \rangle^2 = \langle N^2 \rangle - \langle N \rangle^2;$$

$$\langle (\Delta N)^2 \rangle = \tau^2 \left[\frac{1}{\mathfrak{Z}} \frac{\partial^2 \mathfrak{Z}}{\partial \mu^2} - \frac{1}{\mathfrak{Z}^2} \left(\frac{\partial \mathfrak{Z}}{\partial \mu} \right)^2 \right]. \tag{82}$$

(b) Show that this may be written as

$$\langle (\Delta N)^2 \rangle = \tau \partial \langle N \rangle / \partial \mu. \tag{83}$$

In Chapter 6 we apply this result to the ideal gas to find that

$$\frac{\langle (\Delta N)^2 \rangle}{\langle N \rangle^2} = \frac{1}{\langle N \rangle} \tag{84}$$

is the mean square fractional fluctuation in the population of an ideal gas in diffusive contact with a reservoir. If $\langle N \rangle$ is of the order of 10^{20} atoms, then the fractional fluctuation is exceedingly small. In such a system the number of particles is well defined even though it cannot be rigorously constant because diffusive contact is allowed with the reservoir. When $\langle N \rangle$ is low, this relation can be used in the experimental determination of the molecular weight of large molecules such as DNA of MW $10^8 - 10^{10}$; see M. Weissman, H. Schindler, and G. Feher, Proc. Nat. Acad. Sci. **73**, 2776 (1976).

11. Equivalent definition of chemical potential. The chemical potential was defined by (5) as $(\partial F/\partial N)_{\tau,V}$. An equivalent expression listed in Table 5.1 is

$$\mu = (\partial U/\partial N)_{\sigma,V}. \tag{85}$$

Prove that this relation, which was used by Gibbs to define μ, is equivalent to the definition (5) that we have adopted. It will be convenient to make use of the results (31) and (35). Our reasons for treating (5) as the definition of μ, and (85) as a mathematical consequence, are two-fold. In practice, we need the chemical potential more often as a function of the temperature τ than as a function of the entropy σ. Operationally, a process in which a particle is added to a system while the temperature of the system is kept constant is a more natural process than one in which the entropy is kept constant: Adding a particle to a system at a finite temperature tends to increase its entropy unless we can keep each system of the ensemble in a definite, although new, quantum state. There is no natural laboratory process by which this can be done. Hence the definition (5) or (6), in which the chemical potential is expressed as the change in free energy per added particle under conditions of constant temperature, is operationally the simpler. We point out that (85) will not give $U = \mu N$ on integration, because $\mu(N,\sigma,V)$ is a function of N; compare with (9.13).

12. Ascent of sap in trees. Find the maximum height to which water may rise in a tree under the assumption that the roots stand in a pool of water and the uppermost leaves are in air containing water vapor at a relative humidity $r = 0.9$. The temperature is $25°C$. If the relative humidity is r, the actual concentration of water vapor in the air at the uppermost leaves is rn_0, where n_0 is the concentration in the saturated air that stands immediately above the pool of water.

13. Isentropic expansion. (a) Show that the entropy of an ideal gas can be expressed as a function only of the orbital occupancies. (b) From this result show that $\tau V^{2/3}$ is constant in an isentropic expansion of an ideal monatomic gas.

14. Multiple binding of O_2. A hemoglobin molecule can bind four O_2 molecules. Assume that ε is the energy of each bound O_2, relative to O_2 at rest at infinite distance. Let λ denote the absolute activity $\exp(\mu/\tau)$ of the free O_2 (in solution). (a) What is the probability that one and only one O_2 is adsorbed on a hemoglobin molecule? Sketch the result qualitatively as a function of λ. (b) What is the probability that four and only four O_2 are adsorbed? Sketch this result also.

15. External chemical potential.   Consider a system at temperature τ, with N atoms of mass M in volume V. Let $\mu(0)$ denote the value of the chemical potential at the surface of the earth. (a) Prove carefully and honestly that the value of the total chemical potential for the identical system when translated to altitude h is

$$\mu(h) = \mu(0) + Mgh ,$$

where g is the acceleration of gravity. (b) Why is this result different from that applicable to the barometric equation of an isothermal atmosphere?

Chapter 6

Ideal Gas

The ideal gas is a gas of noninteracting atoms in the limit of low concentration. The limit is defined below in terms of the thermal average value of the number of particles that occupy an orbital. The thermal average occupancy is called the distribution function, usually designated as $f(\varepsilon,\tau,\mu)$, where ε is the energy of the orbital.

An **orbital** is a state of the Schrödinger equation for only one particle. This term is widely used particularly by chemists. If the interactions between particles are weak, the orbital model allows us to approximate an exact quantum state of the Schrödinger equation of a system of N particles in terms of an approximate quantum state that we construct by assigning the N particles to orbitals, with each orbital a solution of a one-particle Schrödinger equation. There are usually an infinite number of orbitals available for occupancy. The term "orbital" is used even when there is no analogy to a classical orbit or to a Bohr orbit. The orbital model gives an exact solution of the N-particle problem only if there are no interactions between the particles.

It is a fundamental result of quantum mechanics (the derivation of which would lead us astray here) that all species of particles fall into two distinct classes, fermions and bosons. Any particle with half-integral spin is a **fermion**, and any particle with zero or integral spin is a **boson**. There are no intermediate classes. Composite particles follow the same rule: an atom of ^{3}He is composed of an odd number of particles—2 electrons, 2 protons, 1 neutron— each of spin $\frac{1}{2}$, so that ^{3}He must have half-integral spin and must be a fermion. An atom of ^{4}He has one more neutron, so there are an even number of particles of spin $\frac{1}{2}$, and ^{4}He must be a boson.

The fermion or boson nature of the particle species that make up a many-body system has a profound and important effect on the states of the system. The results of quantum theory as applied to the orbital model of noninteracting particles appear as occupancy rules:

1. An orbital can be occupied by any integral number of bosons of the same species, including zero.
2. An orbital can be occupied by 0 or 1 fermion of the same species.

The second rule is a statement of the **Pauli exclusion principle**. Thermal averages of occupancies need not be integral or half-integral, but the orbital occupancies of any individual system must conform to one or the other rule.

The two different occupancy rules give rise to two different Gibbs sums for each orbital: there is a boson sum over all integral values of the orbital occupancy N, and there is a fermion sum in which $N = 0$ or $N = 1$ only. Different Gibbs sums lead to different quantum distribution functions $f(\varepsilon,\tau,\mu)$ for the thermal average occupancy. If conditions are such that $f \ll 1$, it will not matter whether the occupancies $N = 2, 3, \ldots$ are excluded or are allowed. Thus when $f \ll 1$ the fermion and boson distribution functions must be similar. This limit in which the orbital occupancy is small in comparison with unity is called the **classical regime**.

We now treat the Fermi-Dirac distribution function for the thermal average occupancy of an orbital by fermions and the Bose-Einstein distribution function for the thermal average occupancy of an orbital by bosons. We show the equivalence of the two functions in the limit of low occupancy, and we go on to treat the properties of a gas in this limit. In Chapter 7 we treat the properties of fermion and boson gases in the opposite limit, where the nature of the particles is absolutely crucial for the properties of the gas.

FERMI-DIRAC DISTRIBUTION FUNCTION

We consider a system composed of a single orbital that may be occupied by a fermion. The system is placed in thermal and diffusive contact with a reservoir, as in Figures 6.1 and 6.2. A real system may consist of a large number N_0 of fermions, but it is very helpful to focus on one orbital and call it the system. All other orbitals of the real system are thought of as the reservoir. Our problem is to find the thermal average occupancy of the orbital thus singled out. An orbital can be occupied by zero or by one fermion. No other occupancy is allowed by the Pauli exclusion principle. The energy of the system will be taken to be zero if the orbital is unoccupied. The energy is ε if the orbital is occupied by one fermion.

The Gibbs sum now is simple: from the definition in Chapter 5 we have

$$\mathfrak{Z} = 1 + \lambda \exp(-\varepsilon/\tau). \tag{1}$$

The term 1 comes from the configuration with occupancy $N = 0$ and energy $\varepsilon = 0$. The term $\lambda \exp(-\varepsilon/\tau)$ comes when the orbital is occupied by one fermion, so that $N = 1$ and the energy is ε. The thermal average value of the occupancy of the orbital is the ratio of the term in the Gibbs sum with $N = 1$ to the entire Gibbs sum:

$$\langle N(\varepsilon) \rangle = \frac{\lambda \exp(-\varepsilon/\tau)}{1 + \lambda \exp(-\varepsilon/\tau)} = \frac{1}{\lambda^{-1} \exp(\varepsilon/\tau) + 1}. \tag{2}$$

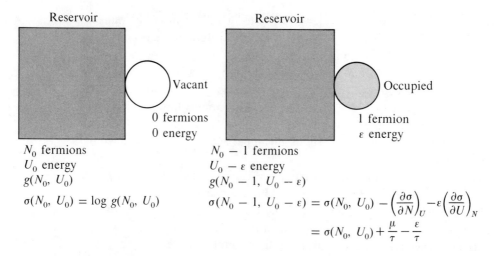

Figure 6.1 We consider as the system a single orbital that may be occupied at most by one fermion. The system is in thermal and diffusive contact with the reservoir at temperature τ. The energy ε of the occupied orbital might be the kinetic energy of a free electron of a definite spin orientation and confined to a fixed volume. Other allowed quantum states may be considered as forming the reservoir. The reservoir will contain N_0 fermions if the system is unoccupied and $N_0 - 1$ fermions if the system is occupied by one fermion.

We introduce for the average occupancy the conventional symbol $f(\varepsilon)$ that denotes the thermal average number of particles in an orbital of energy ε:

$$f(\varepsilon) \equiv \langle N(\varepsilon) \rangle. \tag{3}$$

Recall from Chapter 5 that $\lambda \equiv \exp(\mu/\tau)$, where μ is the chemical potential. We may write (2) in the standard form

$$f(\varepsilon) = \frac{1}{\exp[(\varepsilon - \mu)/\tau] + 1}. \tag{4}$$

This result is known as the Fermi-Dirac distribution function.* Equation (4) gives the average number of fermions in a single orbital of energy ε. The value

---

* This distribution function was discovered independently by E. Fermi, Zeitschrift für Physik **36** 902 (1926), and P. A. M. Dirac, Proceedings of the Royal Society of London **A112**, 661 (1926). Both workers drew on Pauli's paper of the preceding year in which the exclusion principle was discovered. The paper by Dirac is concerned with the new quantum mechanics and contains a general statement of the form assumed by the Pauli principle on this theory.

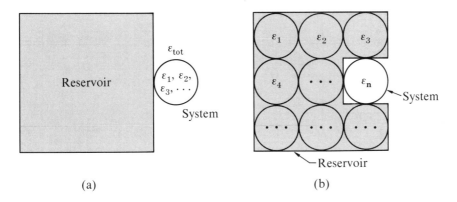

(a)   (b)

Figure 6.2 (a) The obvious method of viewing a system of noninteracting particles is shown here. The energy levels each refer to an orbital that is a solution of a single-particle Schrödinger equation. The total energy of the system is

$$\varepsilon_{tot} = \sum N_n \varepsilon_n \; ,$$

where N_n is the number of particles in the orbital **n** of energy ε_n. For fermions $N_n = 0$ or 1. (b) It is much simpler than (a), and equally valid, to treat a single orbital as the system. The system in this scheme may be the orbital **n** of energy ε_n. All other orbitals are viewed as the reservoir. The total energy of this one-orbital system is $N_n \varepsilon_n$, where N_n is the number of particles in the orbital. This device of using one orbital as the system works because the particles are supposed to interact only weakly with each other. If we think of the fermion system associated with the orbital **n**, these are two possibilities: either the system has 0 particles and energy 0, or the system has 1 particle and energy ε_n. Thus, the Gibbs sum consists of only two terms:

$$\mathfrak{Z} = 1 + \lambda \exp(-\varepsilon_n/\tau).$$

The first term arises from the orbital occupancy $N_n = 0$, and the second term arises from $N_n = 1$.

of f always lies between zero and one. The Fermi-Dirac distribution function is plotted in Figure 6.3.

In the field of solid state physics the chemical potential μ is often called the **Fermi level**. The chemical potential usually depends on the temperature. The value of μ at zero temperature is often written as ε_F; that is,

$$\mu(\tau = 0) \equiv \mu(0) = \varepsilon_F. \tag{5}$$

We call ε_F the **Fermi energy**, not to be confused[*] with the Fermi level which

---

[*] In the semiconductor literature the symbol ε_F is often used for μ at any temperature, and ε_F is then called the Fermi level.

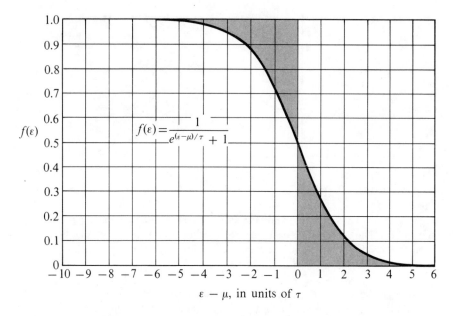

Figure 6.3 Plot of the Fermi-Dirac distribution function $f(\varepsilon)$ versus $\varepsilon - \mu$ in units of the temperature τ. The value of $f(\varepsilon)$ gives the fraction of orbitals at a given energy which are occupied when the system is in thermal equilibrium. When the system is heated from absolute zero, fermions are transferred from the shaded region at $\varepsilon/\mu < 1$ to the shaded region at $\varepsilon/\mu > 1$. For conduction electrons in a metal, μ might correspond to 50 000 K.

is the temperature dependent $\mu(\tau)$. Consider a system of many independent orbitals, as in Figure 6.4. At the temperature $\tau = 0$, all orbitals of energy below the Fermi energy are occupied by exactly one fermion each, and all orbitals of higher energy are unoccupied. At nonzero temperatures the value of the chemical potential μ departs from the Fermi energy, as we will see in Chapter 7.

If there is an orbital of energy equal to the chemical potential ($\varepsilon = \mu$), the orbital is exactly half-filled, in the sense of a thermal average:

$$f(\varepsilon = \mu) = \frac{1}{1 + 1}. \tag{6}$$

Orbitals of lower energy are more than half-filled, and orbitals of higher energy are less than half-filled.

We shall discuss the physical consequences of the Fermi-Dirac distribution in Chapter 7. Right now we go on to discuss the distribution function of non-

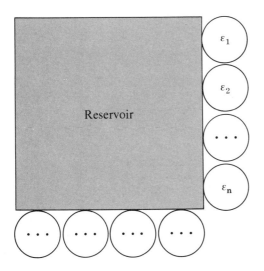

Figure 6.4 A convenient pictorial way to think of a system composed of independent orbitals that do not interact with each other, but interact with a common reservoir.

interacting bosons, and then we establish the ideal gas law for both fermions and bosons in the appropriate limit.

BOSE-EINSTEIN DISTRIBUTION FUNCTION

A boson is a particle with an integral value of the spin. The occupancy rule for bosons is that an orbital can be occupied by any number of bosons, so that bosons have an essentially different quality than fermions. Systems of bosons can have rather different physical properties than systems of fermions. Atoms of ^{4}He are bosons; atoms of ^{3}He are fermions. The remarkable superfluid properties of the low temperature ($T < 2.17\,\text{K}$) phase of liquid helium can be attributed to the properties of a boson gas. There is a sudden increase in the fluidity and in the heat conductivity of liquid ^{4}He below this temperature. In experiments by Kapitza the flow viscosity of ^{4}He below 2.17 K was found to be less than 10^{-7} of the viscosity of the liquid above 2.17 K.

Photons (the quanta of the electromagnetic field) and phonons (the quanta of elastic waves in solids) can be considered to be bosons whose number is not conserved, but it is simpler to think of photons and phonons as excitations of an oscillator, as we did in Chapter 4.

We consider the distribution function for a system of noninteracting bosons in thermal and diffusive contact with a reservoir. We assume the bosons are all of the same species. Let ε denote the energy of a single orbital when occupied by one particle; when there are N particles in the orbital, the energy is $N\varepsilon$, as in Figure 6.5. We treat one orbital as the system and view all other orbitals

Figure 6.5 Energy-level scheme for non-interacting bosons. Here ε is the energy of an orbital when occupied by one particle; $N\varepsilon$ is the energy of the same orbital when occupied by N particles. Any number of bosons can occupy the same orbital. The lowest level of this orbital contributes a term 1 to the grand sum; the next highest level contributes $\lambda \exp(-\varepsilon/\tau)$; and the subsequent contributions are $\lambda^2 \exp(-2\varepsilon/\tau)$; $\lambda^3 \exp(-3\varepsilon/\tau)$; $\lambda^4 \exp(-4\varepsilon/\tau)$; and so on. The Gibbs sum is $\mathcal{Z} = 1 + \lambda \exp(-\varepsilon/\tau) + \lambda^2 \exp(-2\varepsilon/\tau) + \cdots$.

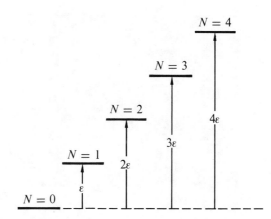

as part of the reservoir. Any arbitrary number of particles may be in the orbital. The Gibbs sum taken for the orbital is

$$\mathcal{Z} = \sum_{N=0}^{\infty} \lambda^N \exp(-N\varepsilon/\tau) = \sum_{N=0}^{\infty} [\lambda \exp(-\varepsilon/\tau)]^N. \tag{7}$$

The upper limit on N should be the total number of particles in the combined system and reservoir. However, the reservoir may be arbitrarily large, so that N may run from zero to infinity. The series (7) may be summed in closed form. Let $x \equiv \lambda \exp(-\varepsilon/\tau)$; then

$$\mathcal{Z} = \sum_{N=0}^{\infty} x^N = \frac{1}{1-x} = \frac{1}{1 - \lambda \exp(-\varepsilon/\tau)}, \tag{8}$$

provided that $\lambda \exp(-\varepsilon/\tau) < 1$. In all applications, $\lambda \exp(-\varepsilon/\tau)$ will satisfy this inequality; otherwise the number of bosons in the system would not be bounded.

The thermal average of the number of particles in the orbital is found from the Gibbs sum by use of (5.62):

$$f(\varepsilon) = \lambda \frac{\partial}{\partial \lambda} \log \mathcal{Z} = -x \frac{d}{dx} \log(1-x) = \frac{x}{1-x} = \frac{1}{\lambda^{-1} \exp(\varepsilon/\tau) - 1} \tag{9}$$

or

$$f(\varepsilon) = \frac{1}{\exp[(\varepsilon - \mu)/\tau] - 1}. \tag{10}$$

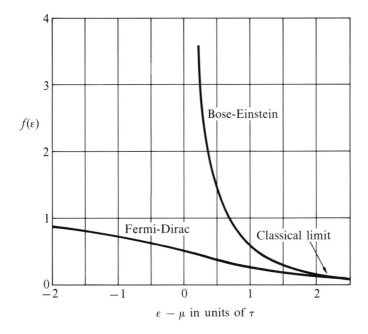

$f(\varepsilon)$

Bose-Einstein

Fermi-Dirac

Classical limit

$\varepsilon - \mu$ in units of τ

Figure 6.6 Comparison of Bose-Einstein and Fermi-Dirac distribution functions. The classical regime is attained for $(\varepsilon - \mu) \gg \tau$, where the two distributions become nearly identical. We shall see in Chapter 7 that in the degenerate regime at low temperature the chemical potential μ for a FD distribution is positive, and changes to negative at high temperature.

This defines the **Bose-Einstein distribution function**. It differs mathematically from the Fermi-Dirac distribution function only by having -1 instead of $+1$ in the denominator. The change can have very significant physical consequences, as we shall see in Chapter 7. The two distribution functions are compared in Figure 6.6. The ideal gas represents the limit $\varepsilon - \mu \gg \tau$ in which the two distribution functions are approximately equal, as discussed below. The choice of the zero of the energy ε is always arbitrary. The particular choice made in any problem will affect the value of the chemical potential μ, but the value of the difference $\varepsilon - \mu$ has to be independent of the choice of the zero of ε. This point is discussed further in (20) below.

A gas is in the **classical regime** when the average number of atoms in each orbital is much less than one. The average orbital occupancy for a gas at room temperature and atmospheric pressure is of the order of only 10^{-6}, safely in the classical regime. Differences between fermions (half-integral spin) and bosons

Table 6.1 Comparison of the orbital occupancies in the classical and the quantum regimes

Regime	Class of particle	Thermal average occupancy of any orbital
Classical	Fermion	Always much less than one.
	Boson	Always much less than one.
Quantum	Fermion	Close to but less than one.
	Boson	Orbital of lowest energy has an occupancy much greater than one.

arise only for occupancies of the order of one or more, so that in the classical regime their equilibrium properties are identical. The quantum regime is the opposite of the classical regime. These characteristic features are summarized in Table 6.1.

CLASSICAL LIMIT

An ideal gas is defined as a system of free noninteracting particles in the classical regime. "Free" means confined in a box with no restrictions or external forces acting within the box. We develop the properties of an ideal gas with the use of the powerful method of the Gibbs sum. In Chapter 3 we treated the ideal gas by use of the partition function, but the identical particle problem encountered there was resolved by a method whose validity was not perfectly clear.

The Fermi-Dirac and Bose-Einstein distribution functions in the classical limit lead to the identical result for the average number of atoms in an orbital. Write $f(\varepsilon)$ for the average occupancy of an orbital at energy ε. Here ε is the energy of an orbital occupied by one particle; it is not the energy of a system of N particles. The Fermi-Dirac (FD) and Bose-Einstein (BE) distribution functions are

$$f(\varepsilon) = \frac{1}{\exp[(\varepsilon - \mu)/\tau] \pm 1} \; , \tag{11}$$

where the plus sign is for the FD distribution and the minus sign for the BE distribution. In order that $f(\varepsilon)$ be much smaller than unity for all orbitals, we must have in this classical regime

$$\exp[(\varepsilon - \mu)/\tau] \gg 1 \; , \tag{12}$$

for all ε. When this inequality is satisfied we may neglect the term ± 1 in the denominator of (11). Then for either fermions or bosons, the average occupancy of an orbital of energy ε is

$$f(\varepsilon) \simeq \exp[(\mu - \varepsilon)/\tau] = \lambda \exp(-\varepsilon/\tau) , \tag{13}$$

(grand canonical)

with $\lambda \equiv \exp(\mu/\tau)$. The limiting result (13) is called the **classical distribution function**. It is the limit of the Fermi-Dirac and Bose-Einstein distribution functions when the average occupancy $f(\varepsilon)$ is very small in comparison with unity. Equation (13), although called classical, is still a result for particles described by quantum mechanics: we shall find that the expression for λ or μ always involves the quantum constant $\hbar$. Any theory which contains $\hbar$ cannot be a classical theory.

We use the classical distribution function $f(\varepsilon) = \lambda \exp(-\varepsilon/\tau)$ to study the thermal properties of the ideal gas. There are many topics of importance: the entropy, chemical potential, heat capacity, the pressure-volume-temperature relation, and the distribution of atomic velocities. To obtain results from the classical distribution function, we need first to find the chemical potential in terms of the concentration of atoms.

Chemical Potential

The chemical potential is found from the condition that the thermal average of the total number of atoms equals the number of atoms known to be present. This number must be the sum over all orbitals of the distribution function $f(\varepsilon_s)$:

$$N = \langle N \rangle = \sum_s f(\varepsilon_s) , \tag{14}$$

where s is the index of an orbital of energy ε_s. We start with a monatomic gas of N identical atoms of zero spin, and later we include spin and molecular modes of motion. The total number of atoms is the sum of the average number of atoms in each orbital. We use (13) in (14) to obtain

$$N = \lambda \sum_s \exp(-\varepsilon_s/\tau). \tag{15}$$

To evaluate this sum, observe that the summation over free particle orbitals is just the partition function Z_1 for a single free atom in volume V, whence $N = \lambda Z_1$.

In Chapter 3 it was shown that $Z_1 = n_Q V$, where $n_Q \equiv (M\tau/2\pi\hbar^2)^{3/2}$ is the quantum concentration. Thus

$$N = \lambda Z_1 = \lambda n_Q V; \qquad \lambda = N/n_Q V = n/n_Q , \qquad (16)$$

in terms of the number density $n = N/V$. Finally,

$$\lambda = \exp(\mu/\tau) = n/n_Q , \qquad (17)$$

which is equal to the number of atoms in the quantum volume $1/n_Q$. In the classical regime n/n_Q is $\ll 1$. The chemical potential of the ideal monatomic gas is

$$\mu = \tau \log(n/n_Q) , \qquad (18)$$

in agreement with (5.12a) obtained in another way. The result may be written out to give

$$\mu = \tau[\log N - \log V - \tfrac{3}{2}\log\tau + \tfrac{3}{2}\log(2\pi\hbar^2/M)]. \qquad (19)$$

We see that the chemical potential increases as the concentration increases and decreases as the temperature increases.

Comment: The simple expression (18) for the chemical potential can be subject to several modifications. We mention four examples.

(a)   If the zero of the energy scale is shifted by an energy Δ so that the zero of the kinetic energy of an orbital falls at $\varepsilon_0 = \Delta$ instead of at $\varepsilon_0 = 0$, then

$$\mu = \Delta + \tau \log(n/n_Q). \qquad (20)$$

(b)   If the atoms have spin S, the number of orbitals in the sum in (15) is multiplied by the spin multiplicity $2S + 1$. For spin $\tfrac{1}{2}$ it is doubled; the value of the partition function Z_1 is doubled; n_Q will be replaced everywhere by $2n_Q$, and the right-hand side of (18) will have an added term $-\tau \log 2$. The effect of the spin on the entropy is treated below.

(c)   If the gas is not monatomic, the internal energy states associated with rotational and vibrational motion will enter the partition function, and the chemical potential will have an added term $-\tau \log Z_{int}$, per (48) below, where Z_{int} is the partition function of the internal degrees of freedom of one molecule.

(d) If the gas is nonideal, the result for μ may be considerably more complicated; see Chapter 10 for the relatively simple van der Waals approximation to a gas of interacting atoms.

Free Energy

The chemical potential is related to the free energy by

$$(\partial F/\partial N)_{\tau,V} = \mu \; , \tag{21}$$

according to Chapter 5. From this,

$$F(N,\tau,V) = \int_0^N dN \, \mu(N,\tau,V) = \tau \int_0^N dN \, [\log N + \cdots] \; , \tag{22}$$

where the integrand is found in brackets in (19). Now $\int dx \log x = x \log x - x$, so that

$$F = N\tau[\log N - 1 - \log V - \tfrac{3}{2}\log \tau + \tfrac{3}{2}\log(2\pi\hbar^2/M)] \; , \tag{23}$$

or

$$F = N\tau[\log(n/n_Q) - 1]. \tag{24}$$

The free energy increases with concentration and decreases with temperature.

Comment: The integral in (22) should strictly be a sum, because N is a discrete variable. Thus, from (5.6),

$$F(N,\tau,V) = \sum_{N=1}^{N} \mu(N) \; , \tag{25}$$

which differs from the integral only in the term in $\log N$ in (19), for

$$\sum_{N=1}^{N} \log N = \log(1 \times 2 \times 3 \times \cdots \times N) = \log N! \; , \tag{26}$$

where the integral gave $N \log N - N$ in (23). But for large N the Stirling approximation

$$\log N! \simeq N \log N - N \ , \tag{27}$$

may be used, and now (25) is the same as (23).

Pressure

The pressure is related to the free energy by (3.49):

$$p = -(\partial F / \partial V)_{\tau,N}. \tag{28}$$

With (23) for F we have

$$\boxed{p = N\tau/V; \qquad pV = N\tau \ ,} \tag{29}$$

which is the ideal gas law, as derived in Chapter 3.

Energy

The thermal energy U is found from $F \equiv U - \tau\sigma$, or

$$U = F + \tau\sigma = F - \tau(\partial F/\partial\tau)_{V,N} = -\tau^2\left(\frac{\partial}{\partial\tau}\frac{F}{\tau}\right)_{V,N} \tag{30}$$

With (23) for F we have

$$\left(\frac{\partial}{\partial\tau}\frac{F}{\tau}\right)_{V,N} = -\frac{3N}{2\tau} \ , \tag{31}$$

so that for an ideal gas

$$\boxed{U = \tfrac{3}{2}N\tau.} \tag{32}$$

The factor $\frac{3}{2}$ arises from the exponent of τ in n_Q because the gas is in three dimensions; if n_Q were in one or two dimensions, the factor would be $\frac{1}{2}$ or 1, respectively. The average kinetic energy of translational motion in the classical limit is equal to $\frac{1}{2}\tau$ or $\frac{1}{2}k_B T$ per translational degree of freedom of an atom. The principle of equipartition of energy among degrees of freedom was discussed in Chapter 3.

A polyatomic molecule has rotational degrees of freedom, and the average energy of each rotational degree of freedom is $\frac{1}{2}\tau$ when the temperature is high in comparison with the energy differences between the rotational energy levels of the molecule. The rotational energy is kinetic. A linear molecule has two degrees of rotational freedom which can be excited; a nonlinear molecule has three degrees of rotational freedom.

Entropy

The entropy is related to the free energy by

$$\sigma = -(\partial F/\partial \tau)_{V,N}. \tag{33}$$

From (23) for F we have the entropy of an ideal gas:

$$\boxed{\sigma = N[\log(n_Q/n) + \tfrac{5}{2}].} \tag{34}$$

This is identical with our earlier result (3.76). In the classical regime n/n_Q is $\ll 1$, so that $\log(n_Q/n)$ is positive. The result (34) is known as the **Sackur-Tetrode equation** for the absolute entropy of a monatomic ideal gas. It is important historically and is essential in the thermodynamics of chemical reactions. Even though the equation contains $\hbar$, the result was inferred from experiments on vapor pressure and on equilibrium in chemical reactions long before the quantum-mechanical basis was fully understood. It was a great challenge to theoretical physicists to explain the Sackur-Tetrode equation, and many unsuccessful attempts to do so were made in the early years of this century. We shall encounter applications of the result in later chapters.

The entropy of the ideal gas is directly proportional to the number of particles N if their concentration n is constant, as we see from (34). When two identical gases at identical conditions are placed side by side, each system having entropy σ_1, the total entropy is $2\sigma_1$ because N is doubled. If a valve that connects the systems is opened, the entropy is unchanged. We see that the entropy scales as the size of the system: the entropy is linear in the number of particles, at constant concentration. If the gases are not identical, the entropy increases when the valve is opened (Problem 6).

Heat Capacity

The heat capacity at constant volume is defined in Chapter 3 as

$$C_V \equiv \tau(\partial \sigma/\partial \tau)_V. \tag{35}$$

We can calculate the derivative directly from the entropy (34) of an ideal gas when the expression for n_Q is written out:

$$\left(\frac{\partial \sigma}{\partial \tau}\right)_{V,N} = \frac{\partial}{\partial \tau}\left(\frac{3}{2} N \log \tau + \cdots \right) = \frac{3N}{2\tau}.$$

From this, for an ideal gas

$$C_V = \tfrac{3}{2}N \, , \tag{36}$$

or $C_V = \tfrac{3}{2}Nk_B$ in conventional units.

The heat capacity at constant pressure is larger than C_V because additional heat must be added to perform the work needed to expand the volume of the gas against the constant pressure p, as discussed in detail in Chapter 8. We use the thermodynamic identity $\tau d\sigma = dU + pdV$ to obtain

$$C_p = \tau \left(\frac{\partial \sigma}{\partial \tau}\right)_p = \left(\frac{\partial U}{\partial \tau}\right)_p + p\left(\frac{\partial V}{\partial \tau}\right)_p. \tag{37}$$

The energy of an ideal gas depends only on the temperature, so that $(\partial U/\partial \tau)_p$ will have the same value as $(\partial U/\partial \tau)_V$, which is just C_V by the argument of (3.17b). By the ideal gas law $V = N\tau/p$, so that the term $p(\partial V/\partial \tau)_p = N$. Thus (37) becomes

$$C_p = C_V + N \tag{38a}$$

in fundamental units, or

$$C_p = C_V + Nk_B \tag{38b}$$

in conventional units. We notice again the different dimensions that heat capacities have in the two systems of units. For one mole, Nk_B is usually written as R, called the **gas constant**.

The results (38a,b) are written for an ideal gas without spin or other internal degrees of freedom of a molecule. For an atom $C_V = \tfrac{3}{2}N$, so that

$$C_p = \tfrac{3}{2}N + N = \tfrac{5}{2}N \tag{38c}$$

in fundamental units, or

$$C_p = \tfrac{5}{2}Nk_B \tag{38d}$$

in conventional units. The ratio C_p/C_V is written as γ, the Greek letter gamma.

Example: Experimental tests of the Sackur-Tetrode equation. Experimental values of the entropy are often found from experimental values of C_p by numerical integration of (37) to give at constant pressure

$$\sigma(\tau) - \sigma(0) = \int_0^\tau (C_p/\tau)d\tau. \tag{39}$$

Here $\sigma(0)$ denotes the entropy at the lowest temperature attained in the measurements of C_p. The third law of thermodynamics suggests that $\sigma(0)$ may be set equal to zero unless there are multiplicities not removed at the lowest temperature attained.

We can calculate the entropy of a monatomic ideal gas by use of the Sackur-Tetrode equation (34). The value thus calculated at a selected temperature and pressure may be compared with the experimental value of the entropy of the gas. The experimental value is found by summing the following contributions:

1. Entropy increase on heating solid from absolute zero to the melting point.
2. Entropy increase in the solid-to-liquid transformation (discussed in Chapter 10).
3. Entropy increase on heating liquid from melting point to the boiling point.
4. Entropy increase in the liquid-to-gas transformation.
5. Entropy change on heating gas from the boiling point to the selected temperature and pressure.

There may further be a slight correction to (34) for the nonideality of the gas. Comparisons of experimental and theoretical values have now been carried out for many gases, and very satisfactory agreement is found between the two sets of values.*

We give details of the comparison for neon, after the measurements of Clusius. The entropy is given in terms of the conventional entropy $S = k_B\sigma$.

1. The heat capacity of the solid was measured from 12.3 K to the melting point 24.55 K under one atmosphere of pressure. The heat capacity of the solid below 12.3 K was estimated by a Debye law (Chapter 4) extrapolation to absolute zero of the measurements above 12.3 K. The entropy of the solid at the melting point is found by numerical integration of $\int dT(C_p/T)$ to be

$$S_{\text{solid}} = 14.29 \, \text{J} \, \text{mol}^{-1} \, \text{K}^{-1}.$$

---

* A classic study is "The heat capacity of oxygen from 12 K to its boiling point and its heat of vaporization. The entropy from spectroscopic data," W. F. Giauque and H. L. Johnston, Journal of the American Chemical Society **51**, 2300 (1929).

Table 6.2   Comparison of experimental and calculated values of the entropy at the boiling point under one atmosphere

| | | Entropy in $\mathrm{J\,mol^{-1}\,K^{-1}}$ | |
Gas	$T_{\text{b.p.}}$, in K	Experimental	Calculated
Ne	27.2	96.40	96.45
Ar	87.29	129.75	129.24
Kr	119.93	144.56	145.06

SOURCE: From *Landolt Börnstein* tables, 6th ed., Vol. 2, Part 4, pp. 394–399.

2. The heat input required to melt the solid at 24.55 K is observed to be 335 J mol^{-1}. The associated entropy of melting is

$$\Delta S_{\text{melting}} = \frac{335\,\text{J mol}^{-1}}{24.55\,\text{K}} = 13.64\,\text{J mol}^{-1}\,\text{K}^{-1}.$$

3. The heat capacity of the liquid was measured from the melting point to the boiling point of 27.2 K under one atmosphere of pressure. The entropy increase was found to be

$$\Delta S_{\text{liquid}} = 3.85\,\text{J mol}^{-1}\,\text{K}^{-1}.$$

4. The heat input required to vaporize the liquid at 27.2 K was observed to be 1761 J mol^{-1}. The associated entropy of vaporization is

$$\Delta S_{\text{vaporization}} = \frac{1761\,\text{J mol}^{-1}}{27.2\,\text{K}} = 64.62\,\text{J mol}^{-1}\,\text{K}^{-1}.$$

The experimental value of the entropy of neon gas at 27.2 K at a pressure of one atmosphere adds up to

$$S_{\text{gas}} = S_{\text{solid}} + \Delta S_{\text{melting}} + \Delta S_{\text{liquid}} + \Delta S_{\text{vaporization}} = 96.40\,\text{J mol}^{-1}\,\text{K}^{-1}.$$

The calculated value of the entropy of neon under the same conditions is

$$S_{\text{gas}} = 96.45\,\text{J mol}^{-1}\,K^{-1},$$

from the Sackur-Tetrode equation. The excellent agreement with the experimental value gives us confidence in the basis of the entire theoretical apparatus that led to the Sackur-Tetrode equation. The result (34) could hardly have been guessed; to find it verified by observation is a real experience. Results for argon and krypton are given in Table 6.2.

Chemical Potential of Ideal Gas with Internal Degrees of Freedom

We consider now an ideal gas of identical polyatomic molecules. Each molecule has rotational and vibrational degrees of freedom in addition to the translational degrees of freedom. The total energy ε of the molecule is the sum of two independent parts,

$$\varepsilon = \varepsilon_\mathbf{n} + \varepsilon_{int} , \tag{40}$$

where ε_{int} refers to the rotational and vibrational degrees of freedom and $\varepsilon_\mathbf{n}$ to the translational motion of the center of mass of the molecule. The vibrational energy problem is the harmonic oscillator problem treated earlier. The rotational energy was the subject of Problem 3.6.

In the classical regime the Gibbs sum for the orbital $\mathbf{n}$ is

$$\mathcal{Z} = 1 + \lambda \exp(-\varepsilon_\mathbf{n}/\tau) , \tag{41}$$

where terms in higher powers of λ are omitted because the average occupancy of the orbital $\mathbf{n}$ is assumed to be $\ll 1$. That is, we neglect the terms in $\mathcal{Z}$ which correspond to occupancies greater than unity. In the presence of internal energy states the Gibbs sum associated with the orbital $\mathbf{n}$ becomes

$$\mathcal{Z} = 1 + \lambda \sum_{int} \exp[-(\varepsilon_\mathbf{n} + \varepsilon_{int})/\tau] , \tag{42}$$

or

$$\mathcal{Z} = 1 + \lambda \exp(-\varepsilon_\mathbf{n}/\tau) \sum_{int} \exp(-\varepsilon_{int}/\tau). \tag{43}$$

The summation is just the partition function of the internal states:

$$Z_{int} = \sum_{int} \exp(-\varepsilon_{int}/\tau) , \tag{44}$$

which is related to the internal free energy of the one molecule by $F_{int} = -\tau \log Z_{int}$. From (43) the Gibbs sum is

$$\boxed{\mathcal{Z} = 1 + \lambda Z_{int} \exp(-\varepsilon_\mathbf{n}/\tau).} \tag{45}$$

The probability that the translational orbital **n** is occupied, irrespective of the state of internal motion of the molecule, is given by the ratio of the term in λ to the Gibbs sum $\mathcal{Z}$:

$$f(\varepsilon_\mathbf{n}) = \frac{\lambda Z_{\mathrm{int}} \exp(-\varepsilon_\mathbf{n}/\tau)}{1 + \lambda Z_{\mathrm{int}} \exp(-\varepsilon_\mathbf{n}/\tau)} \simeq \lambda Z_{\mathrm{int}} \exp(-\varepsilon_\mathbf{n}/\tau). \tag{46}$$

The classical regime was defined earlier as $f(\varepsilon_\mathbf{n}) \ll 1$. The result (46) is entirely analogous to (13) for the monatomic case, but λZ_{int} now plays the role of λ.

Several of the results derived for the monatomic ideal gas are different for the polyatomic ideal gas:

(a)   Equation (17) for λ is replaced by

$$\lambda = n/(n_Q Z_{\mathrm{int}}) \ , \tag{47}$$

with n_Q defined exactly as before. (We shall always use n_Q as defined for the monatomic ideal gas of atoms with zero spin.) Because $\lambda \equiv \exp(\mu/\tau)$ we have

$$\mu = \tau[\log(n/n_Q) - \log Z_{\mathrm{int}}]. \tag{48}$$

(b)   The free energy is increased by, for N molecules,

$$F_{\mathrm{int}} = -N\tau \log Z_{\mathrm{int}}. \tag{49}$$

(c)   The entropy is increased by

$$\sigma_{\mathrm{int}} = -(\partial F_{\mathrm{int}}/\partial\tau)_V. \tag{50}$$

The former result $U = \frac{3}{2}N\tau$ applies to the translational energy alone.

Example: Spin entropy in zero magnetic field.   Consider an atom of spin I, where I may represent both electronic and nuclear spins. The internal partition function associated with the spin alone is

$$Z_{int} = (2I + 1) , \tag{51}$$

this being the number of independent spin states. The spin contribution to the free energy is

$$F_{int} = -\tau \log(2I + 1) , \tag{52}$$

and the spin entropy is

$$\sigma_{int} = \log(2I + 1) , \tag{53}$$

by (50). The effect of the spin entropy on the chemical potential is found with the help of (48):

$$\mu = \tau[\log(n/n_Q) - \log(2I + 1)]. \tag{54}$$

Reversible Isothermal Expansion

Consider as a model example 1×10^{22} atoms of ^{4}He at an initial volume of 10^3 cm^3 at 300 K. Let the gas expand slowly at constant temperature until the volume is 2×10^3 cm^3. The temperature is maintained constant by thermal contact with a large reservoir. In a reversible expansion the system at any instant is in its most probable configuration.

What is the pressure after expansion?

The final volume is twice the initial volume; the final temperature is equal to the initial temperature. From $pV = N\tau$ we see that the final pressure is one-half the initial pressure.

What is the increase of entropy on expansion?

The entropy of an ideal gas at constant temperature depends on volume as

$$\sigma(V) = N \log V + \text{constant} , \tag{55}$$

whence

$$\sigma_2 - \sigma_1 = N \log(V_2/V_1) = N \log 2 = (1 \times 10^{22})(0.693) = 0.069 \times 10^{23}. \tag{56}$$

Notice that the entropy is larger at the larger volume, because the system has more accessible states in the larger volume than in the smaller volume at the same temperature.

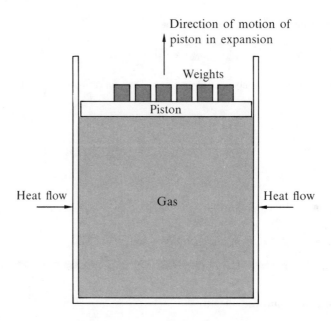

Figure 6.7   Work is done by the gas in an isothermal expansion. Here the gas does work by raising the weights. Under isothermal conditions pV is constant for an ideal gas, so that the pressure must be reduced to allow the volume to expand. The pressure is reduced by removing the load of weights a little at a time.

How much work is done by the gas in the expansion?

When the gas expands isothermally, it does work against a piston, as in Figure 6.7. The work done on the piston when the volume is doubled is

$$\int_{V_1}^{V_2} pdV = \int_{V_1}^{V_2}(N\tau/V)dV = N\tau \log(V_2/V_1) = N\tau \log 2. \qquad (57)$$

We evaluate $N\tau$ directly as $4.14 \times 10^8 \, \text{erg} = 41.4 \, \text{J}$. Thus the work done on the piston is, from (57),

$$N\tau \log 2 = (41.4 \, \text{J})(0.693) = 28.7 \, \text{J}. \qquad (58)$$

The assumption that the process is reversible enters in (57) when we assume that a knowledge of V at every stage determines p at every stage of the expansion.

We define W as the work done *on* the gas by external agencies. This is the negative of the work done by the gas on the piston. From (58),

$$W = -\int pdV = -28.7 \, \text{J}. \qquad (59)$$

What is the change of energy in the expansion?

The energy of an ideal monatomic gas is $U = \frac{3}{2}N\tau$ and does not change in an expansion at constant temperature. However, the Helmholtz free energy decreases by $N\tau \log 2$, which is the work done. The connection is discussed in Chapter 8.

How much heat flowed into the gas from the reservoir?

We have seen that the energy of the ideal gas remained constant when the gas did work on the piston. By conservation of energy it is necessary that a flow of energy in the form of heat into the gas occur from the reservoir through the walls of the container. The quantity Q of heat added to the gas must be equal, but be opposite in sign, to the work done by the piston, because $Q + W = 0$. Thus

$$Q = 28.7 \text{ J} , \tag{60}$$

from the result (59).

Reversible Expansion at Constant Entropy

We considered above an expansion at constant temperature. Suppose instead that the gas expands reversibly from $1 \times 10^3 \text{ cm}^3$ to $2 \times 10^3 \text{ cm}^3$ in an insulated container. No heat flow to or from the gas is permitted, so that $Q = 0$. The entropy is constant in a system isolated from the reservoir if the expansion process is carried out reversibly (slowly). A process without a change of entropy is called an **isentropic process** or an adiabatic process. The term "adiabatic" has the specific meaning that there is no heat transfer in the process. For simplicity, we shall stick with "isentropic."

What is the temperature of the gas after expansion?

The entropy of an ideal monatomic gas depends on the volume and the temperature as

$$\sigma(\tau, V) = N(\log \tau^{3/2} + \log V + \text{constant}) , \tag{61}$$

so that the entropy remains constant if

$$\log \tau^{3/2} V = \text{constant}; \qquad \tau^{3/2} V = \text{constant}. \tag{62}$$

In an expansion at constant entropy from V_1 to V_2 we have

$$\tau_1^{3/2} V_1 = \tau_2^{3/2} V_2 \tag{63}$$

for an ideal monatomic gas.

We use the ideal gas law $pV = N\tau$ to obtain two alternate forms. We insert $V = N\tau/p$ into (63) and cancel N on both sides to obtain

$$\frac{\tau_1^{5/2}}{p_1} = \frac{\tau_2^{5/2}}{p_2}. \tag{64}$$

Similarly, we insert $\tau = pV/N$ in (63) to obtain

$$p_1^{3/2} V_1^{5/2} = p_2^{3/2} V_2^{5/2}; \quad \text{or} \quad p_1 V_1^{5/3} = p_2 V_2^{5/3}. \tag{65}$$

Both (64) and (65) hold only for a monatomic gas.

It is the subject of Problem 10 to generalize these results for an ideal gas of molecules with internal degrees of motion (rotations, vibrations). We obtain for an isentropic process

$$\boxed{\begin{aligned} \tau_1 V_1^{\gamma-1} &= \tau_2 V_2^{\gamma-1} && (66) \\ \tau_1^{\gamma/(1-\gamma)} p_1 &= \tau_2^{\gamma/(1-\gamma)} p_2 && (67) \\ p_1 V_1^{\gamma} &= p_2 V_2^{\gamma}. && (68) \end{aligned}}$$

Here $\gamma \equiv C_p/C_V$ is the ratio of the heat capacities at constant pressure and constant volume.

With $T_1 = 300\,\mathrm{K}$ and $V_1/V_2 = \frac{1}{2}$ we find from (63):

$$T_2 = (\tfrac{1}{2})^{2/3}(300\mathrm{K}) = 189\,\mathrm{K}. \tag{69}$$

This is the final temperature after the expansion at constant entropy. The gas is cooled in the expansion process by

$$T_1 - T_2 = 300\,\mathrm{K} - 189\,\mathrm{K} = 111\,\mathrm{K}. \tag{70}$$

Expansion at constant entropy is an important method of refrigeration.

What is the change in energy in the expansion?

The energy change is calculated from the temperature change (70). For an ideal monatomic gas

$$U_2 - U_1 = C_V(\tau_2 - \tau_1) = \tfrac{3}{2}N(\tau_2 - \tau_1) , \tag{71}$$

or, in conventional units,

$$U_2 - U_1 = \tfrac{3}{2}Nk_B(T_2 - T_1)$$
$$= \tfrac{3}{2}(1 \times 10^{22})(1.38 \times 10^{-16}\,\mathrm{erg\,K^{-1}})(-111\,\mathrm{K})$$
$$= -2.3 \times 10^8\,\mathrm{erg} = -23\,\mathrm{J}. \tag{72}$$

The energy decreases in an expansion at constant entropy. The work done by the gas is equal to the decrease in energy of the gas, which is $U_1 - U_2 = 23\,\mathrm{J}$.

Sudden Expansion into a Vacuum

Let the gas expand suddenly into a vacuum from an initial volume of 1 liter to a final volume of 2 liters. This is an excellent example of an irreversible process. When a hole is opened in the partition to permit the expansion, the first atoms rush through the hole and strike the opposite wall. If no heat flow through the walls is permitted, there is no way for the atoms to lose their kinetic energy. The subsequent flow may be turbulent (irreversible), with different parts of the gas at different values of the energy density. Irreversible energy flow between regions will eventually equalize conditions throughout the gas. We assume the whole process occurs rapidly enough so that no heat flows in through the walls.

How much work is done in the expansion?

No means of doing external work is provided, so that the work done is zero. Zero work is not necessarily a characteristic of all irreversible processes, but the work is zero for expansion into a vacuum.

What is the temperature after expansion?

No work is done and no heat is added in the expansion: $W = 0$, $Q = 0$, and $U_2 - U_1 = 0$. Because the energy is unchanged, the temperature of the ideal gas is unchanged. The energy of a real gas may change in the process because the atoms are moved farther apart, which affects their interaction energy.

What is the change of entropy in the expansion?

The increase of entropy when the volume is doubled at constant temperature is given by (56):

$$\Delta\sigma = \sigma_2 - \sigma_1 = N\log 2 = 0.069 \times 10^{23}. \tag{73}$$

For the expansion into a vacuum $Q = 0$.

Expansion into a vacuum is not a reversible process: the system is not in the most probable (equilibrium) configuration at every stage of the expansion. Only

Table 6.3   Summary of ideal monatomic gas expansion experiments

	$U_2 - U_1$	$\sigma_2 - \sigma_1$	W	Q
Reversible isothermal expansion	0	$N \log \dfrac{V_2}{V_1}$	$-N\tau \log \dfrac{V_2}{V_1}$	$N\tau \log \dfrac{V_2}{V_1}$
Reversible isentropic expansion	$-\tfrac{3}{2}N\tau_1\left[1 - \left(\dfrac{V_1}{V_2}\right)^{2/3}\right]$	0	$-\tfrac{3}{2}N\tau_1\left[1 - \left(\dfrac{V_1}{V_2}\right)^{2/3}\right]$	0
Irreversible expansion into vacuum	0	$N \log \dfrac{V_2}{V_1}$	0	0

the initial configuration before removal of the partition and the final configuration after equilibration are most probable configurations. At intermediate stages the distribution in concentration and kinetic energy of atoms between the two regions into which the system is divided does not correspond to an equilibrium distribution. The central results of these calculations are summarized in Table 6.3.

SUMMARY: STEPS LEADING TO THE IDEAL GAS LAW FOR SPINLESS MONATOMIC GAS

(a) $f(\varepsilon) = \lambda \exp(-\varepsilon/\tau)$ Occupancy of an orbital in the classical limit of $f(\varepsilon) \ll 1$.

(b) $\lambda = \dfrac{N}{\sum \exp(-\varepsilon_\mathbf{n}/\tau)}$ Given N, this equation determines λ in the classical limit.

(c) $\varepsilon_\mathbf{n} = \dfrac{\hbar^2}{2M}\left(\dfrac{\pi n}{V^{1/3}}\right)^2$ Energy of a free particle orbital of quantum number $\mathbf{n}$ in a cube of volume V.

(d) $\sum\limits_\mathbf{n} \exp(-\varepsilon_\mathbf{n}/\tau) = \tfrac{1}{2}\pi \int dn\, n^2 \exp(-\varepsilon/\tau)$ Transformation of the sum to an integral.

(e) $\lambda = N/n_Q V$ Result of the integration (d) after subsitution in (b).

(f) $n_Q = (M\tau/2\pi\hbar^2)^{3/2}$ Definition of the quantum concentration.

(g) $\mu = \tau \log(n/n_Q)$

(h) $F = \int dN\, \mu(N,\tau,V) = N\tau[\log(n/n_Q) - 1]$ $3\tau =$

(i) $p = -(\partial F/\partial V)_{\tau,N} = N\tau/V$

PROBLEMS

1. Derivative of Fermi-Dirac function. Show that $-\partial f/\partial \varepsilon$ evaluated at the Fermi level $\varepsilon = \mu$ has the value $(4\tau)^{-1}$. Thus the lower the temperature, the steeper the slope of the Fermi-Dirac function.

2. Symmetry of filled and vacant orbitals. Let $\varepsilon = \mu + \delta$, so that $f(\varepsilon)$ appears as $f(\mu + \delta)$. Show that

$$f(\mu + \delta) = 1 - f(\mu - \delta). \qquad (74)$$

Thus the probability that an orbital δ above the Fermi level is occupied is equal to the probability an orbital δ below the Fermi level is vacant. A vacant orbital is sometimes known as a **hole**.

3. Distribution function for double occupancy statistics. Let us imagine a new mechanics in which the allowed occupancies of an orbital are 0, 1, and 2. The values of the energy associated with these occupancies are assumed to be 0, ε, and 2ε, respectively.

(a) Derive an expression for the ensemble average occupancy $\langle N \rangle$, when the system composed of this orbital is in thermal and diffusive contact with a reservoir at temperature τ and chemical potential μ.

(b) Return now to the usual quantum mechanics, and derive an expression for the ensemble average occupancy of an energy level which is doubly degenerate; that is, two orbitals have the identical energy ε. If both orbitals are occupied the total energy is 2ε.

4. Energy of gas of extreme relativistic particles. Extreme relativistic particles have momenta p such that $pc \gg Mc^2$, where M is the rest mass of the particle. The de Broglie relation $\lambda = h/p$ for the quantum wavelength continues to apply. Show that the mean energy per particle of an extreme relativistic ideal gas is 3τ if $\varepsilon \cong pc$, in contrast to $\frac{3}{2}\tau$ for the nonrelativistic problem. (An interesting variety of relativistic problems are discussed by E. Fermi in *Notes on Thermodynamics and Statistics*, University of Chicago Press, 1966, paperback.)

5. Integration of the thermodynamic identity for an ideal gas. From the thermodynamic identity at constant number of particles we have

$$d\sigma = \frac{dU}{\tau} + \frac{pdV}{\tau} = \frac{1}{\tau}\left(\frac{\partial U}{\partial \tau}\right)_V d\tau + \frac{1}{\tau}\left(\frac{\partial U}{\partial V}\right)_\tau dV + \frac{pdV}{\tau}. \qquad (75)$$

Show by integration that for an ideal gas the entropy is

$$\sigma = C_V \log \tau + N \log V + \sigma_1 \ , \tag{76}$$

where σ_1 is a constant, independent of τ and V.

6. Entropy of mixing. Suppose that a system of N atoms of type A is placed in diffusive contact with a system of N atoms of type B at the same temperature and volume. Show that after diffusive equilibrium is reached the total entropy is increased by $2N \log 2$. The entropy increase $2N \log 2$ is known as the entropy of mixing. If the atoms are identical (A $\equiv$ B), show that there is no increase in entropy when diffusive contact is established. The difference in the results has been called the Gibbs paradox.

7. Relation of pressure and energy density. (a) Show that the average pressure in a system in thermal contact with a heat reservoir is given by

$$p = -\frac{\sum_s (\partial \varepsilon_s / \partial V)_N \exp(-\varepsilon_s / \tau)}{Z} \ , \tag{77}$$

where the sum is over all states of the system. (b) Show for a gas of free particles that

$$\left(\frac{\partial \varepsilon_s}{\partial V}\right)_N = -\frac{2}{3} \frac{\varepsilon_s}{V} \ , \tag{78}$$

as a result of the boundary conditions of the problem. The result holds equally whether ε_s refers to a state of N noninteracting particles or to an orbital. (c) Show that for a gas of free nonrelativistic particles

$$p = 2U/3V \ , \tag{79}$$

where U is the thermal average energy of the system. This result is not limited to the classical regime; it holds equally for fermion and boson particles, as long as they are nonrelativistic.

8. Time for a large fluctuation. We quoted Boltzmann to the effect that two gases in a 0.1 liter container will unmix only in a time enormously long compared to $10^{(10^{10})}$ years. We shall investigate a related problem: we let a gas of atoms of ^{4}He occupy a container of volume of 0.1 liter at 300 K and a pressure of 1 atm, and we ask how long it will be before the atoms assume a configuration in which all are in one-half of the container.

(a) Estimate the number of states accessible to the system in this initial condition.

(b) The gas is compressed isothermally to a volume of 0.05 liter. How many states are accessible now?

(c) For the system in the 0.1 liter container, estimate the value of the ratio

$$\frac{\text{number of states for which all atoms are in one-half of the volume}}{\text{number of states for which the atoms are anywhere in the volume}}.$$

(d) If the collision rate of an atom is $\approx 10^{10}\,\text{s}^{-1}$, what is the total number of collisions of all atoms in the system in a year? We use this as a crude estimate of the frequency with which the state of the system changes.

(e) Estimate the number of years you would expect to wait before all atoms are in one-half of the volume, starting from the equilibrium configuration.

9. Gas of atoms with internal degree of freedom. Consider an ideal monatomic gas, but one for which the atom has two internal energy states, one an energy Δ above the other. There are N atoms in volume V at temperature τ. Find the (a) chemical potential; (b) free energy; (c) entropy; (d) pressure; (e) heat capacity at constant pressure.

10. Isentropic relations of ideal gas. (a) Show that the differential changes for an ideal gas in an isentropic process satisfy

$$\frac{dp}{p} + \gamma \frac{dV}{V} = 0; \qquad \frac{d\tau}{\tau} + (\gamma - 1)\frac{dV}{V} = 0; \qquad \frac{dp}{p} + \frac{\gamma}{1 - \gamma}\frac{d\tau}{\tau} = 0. \quad (80)$$

where $\gamma = C_p/C_V$; these relations apply even if the molecules have internal degrees of freedom. (b) The isentropic and isothermal bulk moduli are defined as

$$B_\sigma = -V(\partial p/\partial V)_\sigma; \qquad B_\tau = -V(\partial p/\partial V)_\tau. \quad (81)$$

Show that for an ideal gas $B_\sigma = \gamma p$; $B_\tau = p$. The velocity of sound in a gas is given by $c = (B_\sigma/\rho)^{1/2}$; there is very little heat transfer in a sound wave. For an ideal gas of molecules of mass M we have $p = \rho\tau/M$, so that $c = (\gamma\tau/M)^{1/2}$. Here ρ is the mass density.

11. Convective isentropic equilibrium of the atmosphere. The lower 10–15 km of the atmosphere—the troposphere—is often in a convective steady state at constant entropy, not constant temperature. In such equilibrium pV^γ is independent of altitude, where $\gamma = C_p/C_V$. Use the condition of mechanical equilibrium in a uniform gravitational field to: (a) Show that $dT/dz = $ constant, where z is the altitude. This quantity, important in meteorology, is called the dry adiabatic lapse rate. (Do not use the barometric pressure relation that was derived in Chapter 5 for an isothermal atmosphere.) (b) Estimate dT/dz, in °C per km. Take $\gamma = 7/5$. (c) Show that $p \propto \rho^\gamma$, where ρ is the mass density.

If the actual temperature gradient is greater than the isentropic gradient, the atmosphere may be unstable with respect to convection.

12. Ideal gas in two dimensions. (a) Find the chemical potential of an ideal monatomic gas in two dimensions, with N atoms confined to a square of area $A = L^2$. The spin is zero. (b) Find an expression for the energy U of the gas. (c) Find an expression for the entropy σ. The temperature is τ.

13. Gibbs sum for ideal gas. (a) With the help of $Z_N = (n_Q V)^N/N!$ from Chapter 3, show that the Gibbs sum for an ideal gas of identical atoms is $\mathcal{Z} = \exp(\lambda n_Q V)$. (b) Show that the probability there are N atoms in the gas in volume V in diffusive contact with a reservoir is

$$P(N) = \langle N \rangle^N \exp(-\langle N \rangle)/N! \; , \tag{82}$$

which is just the Poisson distribution function (Appendix C). Here $\langle N \rangle$ is the thermal average number of atoms in the volume, which we have evaluated previously as $\langle N \rangle = \lambda V n_Q$. (c) Confirm that $P(N)$ above satisfies

$$\sum_N P(N) = 1 \quad \text{and} \quad \sum_N N P(N) = \langle N \rangle.$$

14. Ideal gas calculations. Consider one mole of an ideal monatomic gas at 300 K and 1 atm. First, let the gas expand isothermally and reversibly to twice the initial volume; second, let this be followed by an isentropic expansion from twice to four times the initial volume. (a) How much heat (in joules) is added to the gas in each of these two processes? (b) What is the temperature at the end of the second process? Suppose the first process is replaced by an irreversible expansion into a vacuum, to a total volume twice the initial volume. (c) What is the increase of entropy in the irreversible expansion, in joules per kelvin?

15. Diesel engine compression. A diesel engine is an internal combustion engine in which fuel is sprayed into the cylinders after the air charge has been so highly compressed that it has attained a temperature sufficient to ignite the fuel. Assume that the air in the cylinders is compressed isentropically from an initial temperature of 27°C (300 K). If the compression ratio is 15, what is the maximum temperature in °C to which the air is heated by the compression? Take $\gamma = 1.4$.

Fermi and Bose Gases

It is a fundamental result of quantum theory that all particles, including atoms and molecules, are either fermions or bosons. They behave alike in the classical regime in which the concentration is small in comparison with the quantum concentration,

$$n \ll n_Q \equiv (M\tau/2\pi\hbar^2)^{3/2}. \tag{1}$$

Whenever $n \geq n_Q$ the gas is said to be in the quantum regime and is called a **quantum gas**. The difference in physical properties between a quantum gas of fermions and one of bosons is dramatic, and both are unlike a gas in the classical regime. A Fermi gas or liquid has a high kinetic energy, low heat capacity, low magnetic susceptibility, low interparticle collision rate, and exerts a high pressure on the container, even at absolute zero. A Bose gas or liquid has a high concentration of particles in the ground orbital, and these particles—called the Bose condensate—may act as a superfluid, with practically zero viscosity.

For many systems the concentration n is fixed, and the temperature is the important variable. The quantum regime obtains when the temperature is below

$$\tau_0 \equiv (2\pi\hbar^2/M)n^{2/3}, \tag{2}$$

defined by the condition $n = n_Q$. A gas in the quantum regime with $\tau \ll \tau_0$ is often said to be a **degenerate gas***.

It was realized by Nernst that the entropy of a classical gas diverges as $\log \tau$ as $\tau \to 0$. Quantum theory removes the difficulty: both fermion and boson gases approach a unique ground state as $\tau \to 0$, so that the entropy goes to zero. We say that the entropy is squeezed out on cooling a quantum gas (see Problems 3 and 8).

In the classical regime (Chapter 6) the thermal average number of particles in an orbital of energy ε is given by

$$f(\varepsilon) \simeq \exp[(\mu - \varepsilon)/\tau]. \tag{3}$$

---

* Here we have the second distinct usage of the word "degenerate" in statistical physics. The first usage was introduced in Chapter 1, where we called an energy level degenerate if more than one state has the same energy.

With the result for μ appropriate to this regime,

$$f(\varepsilon) \simeq (n/n_Q) \exp(-\varepsilon/\tau) \ , \tag{4}$$

with the usual choice of the origin of ε at zero for the energy of the lowest orbital. The form (4) assures us that the average occupancy of any orbital is always $\leq n/n_Q$, which is $\ll 1$, consistent with our original picture of the classical regime.

A fermion is any particle—elementary or composite—with a half-integral spin. A fermion is limited by the Pauli exclusion principle to an orbital occupancy of 0 or 1, with an average occupancy anywhere between these limits. At low temperatures it is clear that many low-lying orbitals will have one fermion in each orbital. At absolute zero all orbitals with $0 < \varepsilon < \varepsilon_F$ will be occupied with $f = 1$. Here ε_F is the energy below which there are just enough orbitals to hold the number of particles assigned to the system. This energy is called the **Fermi energy**. Above ε_F all orbitals will have $f = 0$ at $\tau = 0$. As τ increases the distribution function will develop a high energy tail, as in Figure 7.3.

Bosons have integral or zero spin. They may be elementary or composite; if composite, they must be made up of an even number of elementary particles if these have spin $\frac{1}{2}$, for there is no way to arrive at an integer from an odd number of half-integers. The Pauli principle does not apply to bosons, so there is no limit on the occupancy of any orbital. At absolute zero the ground orbital—the orbital of lowest energy—is occupied by all the particles in the system. As the temperature is increased the lowest single orbital loses its population only slowly, and each excited orbital—any orbital of higher energy—will contain a relatively small number of particles. We shall discuss this point carefully. Above $\tau = \tau_0$ the ground orbital loses its special feature, and its occupancy becomes much like that of any low-lying excited orbital.

FERMI GAS

A Fermi gas is called degenerate when the temperature is low in comparison with the Fermi energy. When the inequality $\tau \ll \varepsilon_F$ is satisfied the orbitals of energy lower than the Fermi energy ε_F will be almost entirely occupied, and the orbitals of higher energy will be almost entirely vacant. An orbital is occupied fully when it contains one fermion. A Fermi gas is said to be nondegenerate when the temperature is high compared with the Fermi energy, as in the classical regime treated in Chapter 6.

The important applications of the theory of degenerate Fermi gases include conduction electrons in metals; the white dwarf stars; liquid ^3He; and nuclear matter. The most striking property of a fermion gas is the high kinetic energy

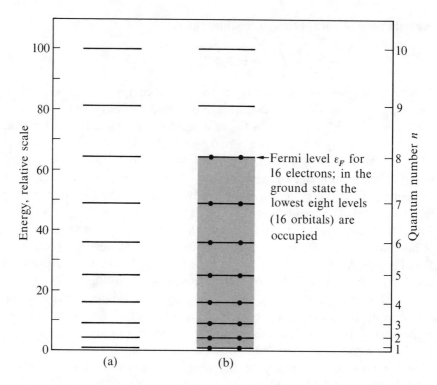

Figure 7.1 (a) The energies of the orbitals $n = 1, 2, \ldots, 10$ for an electron confined to a line of length L. Each level corresponds to two orbitals, one for spin up and one for spin down. (b) The ground state of a system of 16 electrons. Orbitals above the shaded region are vacant in the ground state.

of the ground state of the system at absolute zero. Suppose that it is necessary to accommodate N noninteracting electrons in a length L in one dimension. What orbitals will be occupied in the ground state of the N electron system? In a one-dimensional crystal the quantum number of a free electron orbital of form $\sin(n\pi x/L)$ is a positive integer n, supplemented by the spin quantum number $m_s = \pm\frac{1}{2}$ for spin up or spin down.

If the system has 8 electrons, then in the ground state the orbitals with $n = 1, 2, 3, 4$ and with $m_s = \pm\frac{1}{2}$ are filled, and the orbitals of higher n are empty. Any other arrangement gives a higher energy. To construct the ground state we fill the orbitals starting from $n = 1$ at the bottom, and we continue filling higher orbitals with electrons until all N electrons are accommodated. The orbitals that are filled in the ground state of a system of 16 electrons are shown in Figure 7.1.

Ground State of Fermi Gas in Three Dimensions

Let the system be a cube of side L and volume $V = L^3$. The orbitals have the form of (3.58) and their energy is given by (3.59). The Fermi energy ε_F is the energy of the highest filled orbital at absolute zero; it is determined by the requirement that the system in the ground state hold N electrons, with each orbital filled with one electron up to the energy

$$\varepsilon_F = \frac{\hbar^2}{2m}\left(\frac{\pi n_F}{L}\right)^2. \tag{5}$$

Here n_F is the radius of a sphere (in the space of the integers n_x, n_y, n_z) that separates filled and empty orbitals. For the system to hold N electrons the orbitals must be filled up to n_F determined by

$$N = 2 \times \tfrac{1}{8} \times \frac{4\pi}{3} n_F{}^3 = \frac{\pi}{3} n_F{}^3; \qquad n_F = (3N/\pi)^{1/3}. \tag{6}$$

The factor 2 arises because an electron has two possible spin orientations. The factor $\frac{1}{8}$ arises because only triplets n_x, n_y, n_z in the positive octant of the sphere in n space are to be counted. The volume of the sphere is $4\pi n_F{}^3/3$. We may then write (5) as

$$\varepsilon_F = \frac{\hbar^2}{2m}\left(\frac{3\pi^2 N}{V}\right)^{2/3} = \frac{\hbar^2}{2m}(3\pi^2 n)^{2/3} \equiv \tau_F. \tag{7}$$

This relates the Fermi energy to the electron concentration $N/V \equiv n$. The so-called "Fermi temperature" τ_F is defined as $\tau_F \equiv \varepsilon_F$.

The total energy of the system in the ground state is

$$U_0 = 2 \sum_{n \le n_F} \varepsilon_\mathbf{n} = 2 \times \tfrac{1}{8} \times 4\pi \int_0^{n_F} dn\, n^2 \varepsilon_\mathbf{n} = \frac{\pi^3}{2m}\left(\frac{\hbar}{L}\right)^2 \int_0^{n_F} dn\, n^4, \tag{8}$$

with $\varepsilon_\mathbf{n} = (\hbar^2/2m)(\pi n/L)^2$. In (8) and (9), n is an integer and is not N/V. Consistent with (6), we have let

$$2 \sum_\mathbf{n} (\cdots) \to 2(\tfrac{1}{8})(4\pi) \int dn\, n^2 (\cdots) \tag{9}$$

in the conversion of the sum into an integral. Integration of (8) gives the total ground state kinetic energy:

$$U_0 = \frac{\pi^3}{10m}\left(\frac{\hbar}{L}\right)^2 n_F{}^5 = \frac{3\hbar^2}{10m}\left(\frac{\pi n_F}{L}\right)^2 N = \tfrac{3}{5}N\varepsilon_F, \tag{10}$$

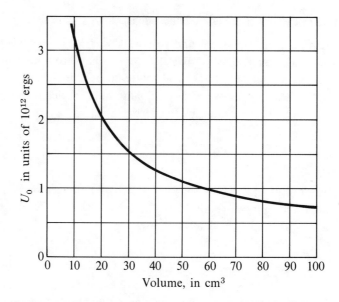

Figure 7.2   Total ground state energy U_0 of one mole of electrons, versus volume.

using (5) and (6). The average kinetic energy per particle is U_0/N and is $\frac{3}{5}$ of the Fermi energy ε_F. At constant N the energy increases as the volume decreases (Figure 7.2), so that the Fermi energy gives a repulsive contribution to the binding of any material; in most metals and in white dwarf and neutron stars it is the most important repulsive interaction. That is, the Fermi energy tends to increase the volume. It is balanced in metals by the Coulomb attraction between electrons and ions and in the stars by gravitational attraction.

Density of States

Thermal averages for independent particle problems have the form

$$\langle X \rangle = \sum_{\mathbf{n}} f(\varepsilon_{\mathbf{n}}, \tau, \mu) X_{\mathbf{n}} \; , \tag{11}$$

where $\mathbf{n}$ denotes the quantum orbital; $X_{\mathbf{n}}$ is the value of the quantity X in the orbital $\mathbf{n}$; and $f(\varepsilon_{\mathbf{n}}, \tau, \mu)$ is the thermal average occupancy, called the distribution function, of the orbital $\mathbf{n}$. We often express $\langle X \rangle$ as an integral over the orbital energy ε. Then (11) becomes

$$\langle X \rangle = \int d\varepsilon \; \mathfrak{D}(\varepsilon) f(\varepsilon, \tau, \mu) X(\varepsilon) \; , \tag{12}$$

where the sum over orbitals has been transformed to an integral by the substitution

$$\sum_{\mathbf{n}} (\cdots) \rightarrow \int d\varepsilon \; \mathfrak{D}(\varepsilon)(\cdots). \tag{13}$$

Here $\mathfrak{D}(\varepsilon)d\varepsilon$ is the number of orbitals of energy between ε and $\varepsilon + d\varepsilon$. The quantity $\mathfrak{D}(\varepsilon)$ is nearly always called the **density of states**, although it is more accurate to call it the density of orbitals because it refers to the solutions of a one particle problem and not to the states of the N particle system.

Consider an example of the calculation of $\mathfrak{D}(\varepsilon)$. We see from (7) that the number N of free electron orbitals of energy less than or equal to some ε is

$$N(\varepsilon) = (V/3\pi^2)(2m/\hbar^2)^{3/2}\varepsilon^{3/2} \; , \tag{14}$$

for volume V. Take the logarithm of both sides:

$$\log N = \tfrac{3}{2}\log \varepsilon + \text{constant}, \tag{15}$$

and take differentials of $\log N$ and $\log \varepsilon$:

$$\frac{dN}{N} = \frac{3}{2}\frac{d\varepsilon}{\varepsilon}. \tag{16}$$

The quantity $dN = (3N/2\varepsilon)d\varepsilon$ is the number of orbitals of energy between ε and $\varepsilon + d\varepsilon$, so that

$$\mathfrak{D}(\varepsilon) \equiv dN/d\varepsilon = 3N(\varepsilon)/2\varepsilon \tag{17}$$

is the density of orbitals. The two spin orientations of an electron have been counted throughout this derivation because they were counted in (6). We can write $\mathfrak{D}(\varepsilon)$ as a function of ε alone because

$$N(\varepsilon)/\varepsilon = (V/3\pi^2)(2m/\hbar^2)^{3/2}\varepsilon^{1/2} \; , \tag{18}$$

from (14). Then (17) becomes

$$\boxed{\mathfrak{D}(\varepsilon) = \frac{V}{2\pi^2}\left(\frac{2m}{\hbar^2}\right)^{3/2}\varepsilon^{1/2}.} \tag{19}$$

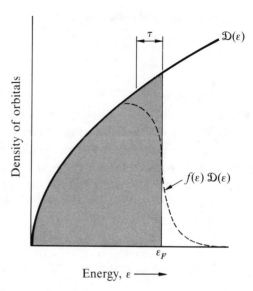

Figure 7.3   Density of orbitals as a function of energy, for a free electron gas in three dimensions. The dashed curve represents the density $f(\varepsilon)\mathfrak{D}(\varepsilon)$ of occupied orbitals at a finite temperature, but such that τ is small in comparison with ε_F. The shaded area represents the occupied orbitals at absolute zero.

When multiplied by the distribution function (Figure 6.3), the density of orbitals $\mathfrak{D}(\varepsilon)$ becomes $\mathfrak{D}(\varepsilon)f(\varepsilon)$, the density of occupied orbitals (Figure 7.3). The total number of electrons in a system may now be written as

$$N = \int_0^\infty d\varepsilon\, \mathfrak{D}(\varepsilon)f(\varepsilon,\tau,\mu) \ , \tag{20}$$

where $f(\varepsilon)$ is the Fermi-Dirac distribution function described in Chapter 6. In problems where we know the total number of particles, we determine μ by requiring that the total number of particles calculated from (20) be equal to the correct value. The total kinetic energy of the electrons is

$$U = \int_0^\infty d\varepsilon\, \varepsilon\mathfrak{D}(\varepsilon)f(\varepsilon,\tau,\mu). \tag{21}$$

If the system is in the ground state, all orbitals are filled up to the energy ε_F, above which they are vacant. The number of electrons is equal to

$$N = \int_0^{\varepsilon_F} d\varepsilon\, \mathfrak{D}(\varepsilon) \ , \tag{22}$$

and the energy is

$$U_0 = \int_0^{\varepsilon_F} d\varepsilon\, \varepsilon\mathfrak{D}(\varepsilon). \tag{23}$$

Heat Capacity of Electron Gas

We derive a quantitative expression for the heat capacity of a degenerate Fermi gas of electrons in three dimensions. The calculation is perhaps the most impressive accomplishment of the theory of the degenerate Fermi gas. For an ideal monatomic gas the heat capacity is $\frac{3}{2}N$, but for electrons in a metal very much lower values are found. The calculation that follows gives excellent agreement with the experimental results. The increase in the total energy of a system of N electrons when heated from 0 to τ is denoted by $\Delta U \equiv U(\tau) - U(0)$, whence

$$\Delta U = \int_0^\infty d\varepsilon \, \varepsilon \mathfrak{D}(\varepsilon) f(\varepsilon) - \int_0^{\varepsilon_F} d\varepsilon \, \varepsilon \mathfrak{D}(\varepsilon). \tag{24}$$

Here $f(\varepsilon)$ is the Fermi-Dirac function, and $\mathfrak{D}(\varepsilon)$ is the number of orbitals per unit energy range. We multiply the identity

$$N = \int_0^\infty d\varepsilon \, f(\varepsilon) \mathfrak{D}(\varepsilon) = \int_0^{\varepsilon_F} d\varepsilon \, \mathfrak{D}(\varepsilon) \tag{25}$$

by ε_F to obtain

$$\left(\int_0^{\varepsilon_F} + \int_{\varepsilon_F}^\infty \right) d\varepsilon \, \varepsilon_F f(\varepsilon) \mathfrak{D}(\varepsilon) = \int_0^{\varepsilon_F} d\varepsilon \, \varepsilon_F \mathfrak{D}(\varepsilon). \tag{26}$$

We use (26) to rewrite (24) as

$$\Delta U = \int_{\varepsilon_F}^\infty d\varepsilon \, (\varepsilon - \varepsilon_F) f(\varepsilon) \mathfrak{D}(\varepsilon) + \int_0^{\varepsilon_F} d\varepsilon \, (\varepsilon_F - \varepsilon) \left[1 - f(\varepsilon) \right] \mathfrak{D}(\varepsilon). \tag{27}$$

The first integral on the right-hand side of (27) gives the energy needed to take electrons from ε_F to the orbitals of energy $\varepsilon > \varepsilon_F$, and the second integral gives the energy needed to bring the electrons to ε_F from orbitals below ε_F. Both contributions to the energy are positive. The product $f(\varepsilon) \mathfrak{D}(\varepsilon) d\varepsilon$ in the first integral is the number of electrons elevated to orbitals in the energy range $d\varepsilon$ at an energy ε. The factor $\left[1 - f(\varepsilon) \right]$ in the second integral is the probability that an electron has been removed from an orbital ε. The function ΔU is plotted in Figure 7.4. In Figure 7.5 we plot the Fermi-Dirac distribution function versus ε, for six values of the temperature. The electron concentration of the Fermi gas was taken such that $\varepsilon_F / k_B = 50\,000 \, K$, characteristic of the conduction electrons in a metal.

The heat capacity of the electron gas is found on differentiating ΔU with respect to τ. The only temperature-dependent term in (27) is $f(\varepsilon)$, whence we

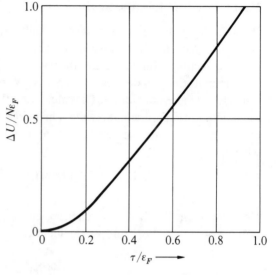

Figure 7.4 Temperature dependence of the energy of a noninteracting fermion gas in three dimensions. The energy is plotted in normalized form as $\Delta U/N\varepsilon_F$, where N is the number of electrons. The temperature is plotted as τ/ε_F.

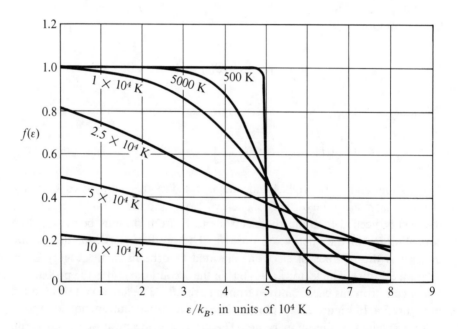

Figure 7.5 Fermi-Dirac distribution function at various temperatures, for $T_F \equiv \varepsilon_F/k_B = 50\,000$ K. The results apply to a gas in three dimensions. The total number of particles is constant, independent of temperature. The chemical potential at each temperature was calculated with the help of Eq. (20) and may be read off the graph as the energy at which $f = \frac{1}{2}$. Courtesy of B. Feldman.

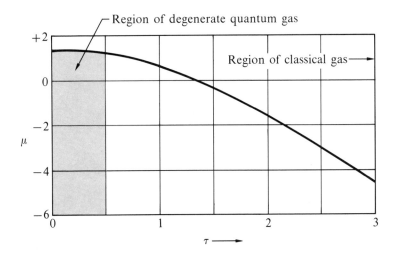

Figure 7.6 Plot of the chemical potential μ versus temperature τ for a gas of noninteracting fermions in three dimensions. For convenience in plotting, the units of μ and τ are $0.763\varepsilon_F$.

can group terms to obtain

$$C_{el} = \frac{dU}{d\tau} = \int_0^\infty d\varepsilon (\varepsilon - \varepsilon_F) \frac{df}{d\tau} \mathfrak{D}(\varepsilon). \tag{28}$$

At the temperatures of interest in metals $\tau/\varepsilon_F < 0.01$, and we see from Figure 7.5 that the derivative $df/d\tau$ is large only at energies near ε_F. It is a good approximation to evaluate the density of orbitals $\mathfrak{D}(\varepsilon)$ at ε_F and take it outside of the integral:

$$C_{el} \cong \mathfrak{D}(\varepsilon_F) \int_0^\infty d\varepsilon (\varepsilon - \varepsilon_F) \frac{df}{d\tau}. \tag{29}$$

Examination of the graphs in Figures 7.6 and 7.7 of the variation of μ with τ suggests that when $\tau \ll \varepsilon_F$ we ignore the temperature dependence of the chemical potential μ in the Fermi-Dirac distribution function and replace μ by the constant ε_F. We have then:

$$\frac{df}{d\tau} = \frac{\varepsilon - \varepsilon_F}{\tau^2} \cdot \frac{\exp[(\varepsilon - \varepsilon_F)/\tau]}{\{\exp[(\varepsilon - \varepsilon_F)/\tau] + 1\}^2}. \tag{30}$$

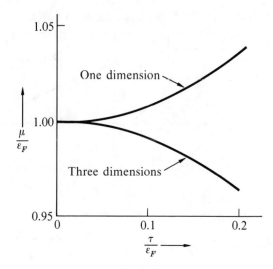

Figure 7.7 Variation with temperature of the chemical potential μ, for free electron Fermi gases in one and three dimensions. In common metals $\tau/\varepsilon_F \approx 0.01$ at room temperature, so that μ is closely equal to ε_F. These curves were calculated from series expansions of the integral for the number of particles in the system.

We set

$$x \equiv (\varepsilon - \varepsilon_F)/\tau \ , \tag{31}$$

and it follows from (29) and (30) that

$$C_{\mathrm{el}} = \tau \mathbf{D}(\varepsilon_F) \int_{-\varepsilon_F/\tau}^{\infty} dx\, x^2 \frac{e^x}{(e^x + 1)^2}. \tag{32}$$

We may safely replace the lower limit by $-\infty$ because the factor e^x in the integrand is already negligible at $x = -\varepsilon_F/\tau$ if we are concerned with low temperatures such that $\varepsilon_F/\tau \sim 100$ or more. The integral* becomes

$$\int_{-\infty}^{\infty} dx\, x^2 \frac{e^x}{(e^x + 1)^2} = \frac{\pi^2}{3} \ , \tag{33}$$

---

* The integral is not elementary, but may be evaluated from the more familiar result

$$\int_0^{\infty} dx\, \frac{x}{e^{ax} + 1} = \frac{\pi^2}{12a^2}$$

on differentiation of both sides with respect to the parameter a.

whence we have for the heat capacity of an electron gas, when $\tau \ll \tau_F$,

$$C_{el} = \tfrac{1}{3}\pi^2 \mathfrak{D}(\varepsilon_F)\tau. \tag{34}$$

In conventional units,

$$C_{el} = \tfrac{1}{3}\pi^2 \mathfrak{D}(\varepsilon_F)k_B{}^2 T. \tag{35}$$

We found that the density of orbitals at the Fermi energy is

$$\mathfrak{D}(\varepsilon_F) = 3N/2\varepsilon_F = 3N/2\tau_F \tag{36}$$

for a free electron gas, with $\tau_F \equiv \varepsilon_F$. Do not be deceived by the notation τ_F: it is *not* the temperature of the Fermi gas, but only a convenient reference point. For $\tau \ll \tau_F$ the gas is degenerate; for $\tau \gg \tau_F$ the gas is in the classical regime. Thus (34) becomes

$$C_{el} = \tfrac{1}{2}\pi^2 N\tau/\tau_F. \tag{37}$$

In conventional units there is an extra factor k_B, so that

$$C_{el} = \tfrac{1}{2}\pi^2 Nk_B T/T_F , \tag{38}$$

where $k_B T_F \equiv \varepsilon_F$. Again, T_F is not an actual temperature, but only a reference point.

We can give a physical explanation of the form of the result (37). When the specimen is heated from absolute zero, chiefly those electrons in states within an energy range τ of the Fermi level are excited thermally, because the FD distribution function is affected over a region of the order of τ in width, illustrated by Figures 7.3 and 7.5. Thus the number of excited electrons is of the order of $N\tau/\varepsilon_F$, and each of these has its energy increased approximately by τ. The total electronic thermal energy is therefore of the order of $U_{el} \approx N\tau^2/\varepsilon_F$. Thus the electronic contribution to the heat capacity is given by

$$C_{el} = dU_{el}/d\tau \approx N\tau/\varepsilon_F \approx N\tau/\tau_F , \tag{39}$$

which is directly proportional to τ, in agreement with the exact result (34) and with the experimental results.

Table 7.1   Calculated Fermi energy parameters for free electrons

	Conduction electron concentration N/V, in cm^{-3}	Velocity v_F, in cm s^{-1}	Fermi energy ε_F, in eV	Fermi temperature $T_F = \varepsilon_F/k_B$, in K
Li	4.6×10^{22}	1.3×10^8	4.7	5.5×10^4
Na	2.5	1.1	3.1	3.7
K	1.34	0.85	2.1	2.4
Rb	1.08	0.79	1.8	2.1
Cs	0.86	0.73	1.5	1.8
Cu	8.50	1.56	7.0	8.2
Ag	5.76	1.38	5.5	6.4
Au	5.90	1.39	5.5	6.4

Fermi Gas in Metals

The alkali metals and copper, silver, and gold have one valence electron per atom, and the valence electron becomes the conduction electron in the metal. Thus the concentration of conduction electrons is equal to the concentration of atoms, which may be evaluated either from the density and the atomic weight or from the crystal lattice dimensions.

If the conduction electrons act as a free fermion gas, the value of the Fermi energy ε_F may be calculated from (7):

$$\varepsilon_F = (\hbar^2/2m)(3\pi^2 n)^{2/3}. \tag{40}$$

Values of n and of ε_F are given in Table 7.1 and in Figure 7.8. The electron velocity v_F at the Fermi surface is also given in the table; it is defined so that the kinetic energy is equal to ε_F:

$$\tfrac{1}{2}mv_F{}^2 = \varepsilon_F , \tag{41}$$

where m is the mass of the electron. The values of the Fermi temperature $T_F \equiv \varepsilon_F/k_B$ for ordinary metals are of the order of 5×10^4 K, so that the assumption $T \ll T_F$ used in the derivation of (35) is an excellent approximation at room temperature and below.

The heat capacity of many metals at constant volume may be written as the sum of an electronic contribution and a lattice vibration contribution. At low temperatures the sum has the form

$$C_V = \gamma\tau + A\tau^3 , \tag{42}$$

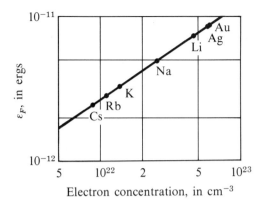

Figure 7.8 Fermi energy ε_F of a free electron gas as a function of the concentration. Calculated values are shown for several monovalent metals. The straight line is drawn for $\varepsilon_F = 5.835 \times 10^{-27}\, n^{2/3}$ ergs, with n in cm^{-3}.

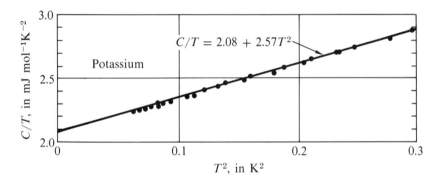

Figure 7.9 Experimental heat capacity values for potassium, plotted as C/T versus T^2. After W. H. Lien and N. E. Phillips, Phys. Rev. **133**, A1370 (1964).

where γ and A are constants characteristic of the material. Here $\gamma \equiv \frac{1}{2}\pi^2 N/\tau_F$ from (37), and the lattice vibration term $A\tau^3$ was discussed in Chapter 4. The electronic term is linear in τ and is dominant at sufficiently low temperatures. It is helpful to display the experimental values of the heat capacity for a given material as a plot of C_V/τ versus τ^2:

$$C_V/\tau = \gamma + A\tau^2 , \qquad (43)$$

for then the points should lie on a straight line. The intercept at $\tau = 0$ gives the value of γ. Such a plot is shown for potassium in Figure 7.9. Observed values of γ are given in Tables 7.2 and 7.3.

Table 7.2   Experimental and free electron electronic heat capacities of monovalent metals

Metal	γ (exp), mJ mol^{-1} K^{-2}	γ_0 (free electron), mJ mol^{-1} K^{-2}	γ/γ_0
Li	1.63	0.75	2.17
Na	1.38	1.14	1.21
K	2.08	1.69	1.23
Rb	2.41	1.97	1.22
Cs	3.20	2.36	1.35
Cu	0.695	0.50	1.39
Ag	0.646	0.65	1.00
Au	0.729	0.65	1.13

NOTE: The values of γ and γ_0 are in mJ mol^{-1} K^{-2}.
SOURCE: Courtesy of N. E. Phillips.

Table 7.3   Experimental values of electronic heat capacity constant γ of metals

Li 1.63	Be 0.17											B	C	N
Na 1.38	Mg 1.3											Al 1.35	Si	P
K 2.08	Ca 2.9	Sc 10.7	Ti 3.35	V 9.26	Cr 1.40	Mnγ 9.20	Fe 4.98	Co 4.73	Ni 7.02	Cu 0.695	Zn 0.64	Ga 0.596	Ge	As 0.19
Rb 2.41	Sr 3.6	Y 10.2	Zr 2.80	Nb 7.79	Mo 2.0	Tc —	Ru 3.3	Rh 4.9	Pd 9.42	Ag 0.646	Cd 0.688	In 1.69	Sn 1.78	Sb 0.11
Cs 3.20	Ba 2.7	La 10.	Hf 2.16	Ta 5.9	W 1.3	Re 2.3	Os 2.4	Ir 3.1	Pt 6.8	Au 0.729	Hg 1.79	Tl 1.47	Pb 2.98	Bi 0.008

NOTE: The value of γ is in mJ mol^{-1} K^{-2}.
SOURCE: From compilations furnished by N. E. Phillips and N. Pearlman.

White Dwarf Stars

White dwarf stars have masses comparable to that of the Sun. The mass and radius of the Sun are

$$M_\odot = 2.0 \times 10^{33} \text{ g}; \qquad R_\odot = 7.0 \times 10^{10} \text{ cm}. \tag{44}$$

The radii of white dwarfs are very small, perhaps 0.01 that of the Sun. The density of the Sun, which is a normal star, is of the order of 1 g cm^{-3}, like that of water on the Earth. The densities of white dwarfs are exceedingly high, of the order of 10^4 to 10^7 g cm^{-3}. Atoms under the densities prevalent in white dwarfs are entirely ionized into nuclei and free electrons, and the electron gas is a degenerate gas, as will be shown below.

The companion of Sirius was the first white dwarf to be discovered. In 1844 Bessel observed that the path of the star Sirius oscillated slightly about a straight line, as if it had an invisible companion. The companion, Sirius B, was discovered near its predicted position by Clark in 1862. The mass of Sirius B was determined to be 2.0×10^{33} g by measurements on the orbits. The radius of Sirius B is estimated as 2×10^9 cm by a comparison of the surface temperature and the radiant energy flux, using the properties of thermal radiant energy developed in Chapter 4.

The mass and radius of Sirius B lead to the mean density

$$\rho = \frac{M}{V} = \frac{2 \times 10^{33} \text{ g}}{\frac{4}{3}\pi(2 \times 10^9 \text{ cm})^3} \approx 0.7 \times 10^5 \text{ g cm}^{-3}. \tag{45}$$

This extraordinarily high density was appraised by Eddington in 1926 in the following words: "Apart from the incredibility of the result, there was no particular reason to view the calculation with suspicion." Other white dwarfs have higher densities: that named Van Maanen No. 2 has a mean density 100 times higher.

Hydrogen atoms at a density of 10^6 g cm^{-3} have a volume per atom equal to

$$V_A \approx \frac{1}{(10^6 \text{ mol cm}^{-3})(6 \times 10^{23} \text{ atoms mol}^{-1})} \approx 2 \times 10^{-30} \text{ cm}^3 \text{ per atom}$$

or 2×10^{-6} Å^3 per atom. The average nearest-neighbor separation is then of the order of 0.01 Å, as compared with the internuclear separation of 0.74 Å in a molecule of hydrogen. Under conditions of such high density the atomic electrons are no longer attached to individual nuclei. The electrons are ionized and form an electron gas. The matter in the white dwarfs is held together by gravitational attraction, which is the binding force in all stars.

In the interior of white dwarf stars* the electron gas is degenerate; the temperature is much less than the Fermi energy ε_F. The Fermi energy of an electron gas at a concentration of 1×10^{30} electrons cm^{-3} is given by

$$\varepsilon_F = (\hbar^2/2m)(3\pi^2 n)^{2/3} \approx 0.5 \times 10^{-6} \text{ erg} \approx 3 \times 10^5 \text{ eV} , \tag{46}$$

---

* A good discussion of white dwarf stars is given by W. K. Rose, *Astrophysics*, Holt, Rinehart, and Winston, 1973.

Table 7.4   Fermi energy of degenerate fermion gases (characteristic values)

Phase of matter	Particles	T_F, in K
Liquid ^{3}He	atoms	0.3
Metal	electrons	5×10^4
White dwarf stars	electrons	3×10^9
Nuclear matter	nucleons	3×10^{11}
Neutron stars	neutrons	3×10^{12}

about 10^5 higher than in a typical metal. The Fermi temperature ε_F/k_B of the electrons is $\approx 3 \times 10^9$ K, as in Table 7.4. The actual temperature in the interior of a white dwarf is believed to be of the order of 10^7 K. The electron gas in the interior of a white dwarf is highly degenerate because the thermal energy is much lower than the Fermi energy.

Are the electron energies in the relativistic regime? This question arises because our theory of the Fermi gas has used the nonrelativistic expression $p^2/2m$ for the kinetic energy of an electron of momentum p. The energy equivalence of the rest mass of an electron is

$$\varepsilon_0 = mc^2 \approx (1 \times 10^{-27}\,\text{g})(3 \times 10^{10}\,\text{cm s}^{-1})^2 \approx 1 \times 10^{-6}\,\text{erg}. \quad (47)$$

This energy is of the same order as the Fermi energy (46). Thus relativistic effects will be significant, but not dominant. At higher densities the Fermi gas is relativistic.

Nuclear Matter

We consider the state of matter within nuclei. The neutrons and protons of which nuclear matter is composed form a degenerate fermion gas, at least qualitatively. We estimate here the Fermi energy of the nucleon gas: The radius of a nucleus that contains A nucleons is given by the empirical relation

$$R \cong (1.3 \times 10^{-13}\,\text{cm}) \times A^{1/3}. \quad (48)$$

According to this relation the average volume per particle is constant, for the volume goes as R^3, which is proportional to A. The concentration of nucleons in nuclear matter is

$$n \cong \frac{A}{\frac{4}{3}\pi(1.3 \times 10^{-13}\,\text{cm})^3 A} \cong 0.11 \times 10^{39}\,\text{cm}^{-3}, \quad (49)$$

about 10^8 times higher than the concentration of nucleons in a white dwarf star. Neutrons and protons are not identical particles. The Fermi energy of the neutrons need not equal the Fermi energy of the protons. The concentration of one or the other, but not both, enters the familiar relation

$$\varepsilon_F = \frac{\hbar^2}{2M} (3\pi^2 n)^{2/3}. \tag{50}$$

For simplicity let us suppose that the number of protons is equal to the number of neutrons. Then

$$n_{\text{protons}} \approx n_{\text{neutrons}} \approx 0.05 \times 10^{39} \text{ cm}^{-3} , \tag{51}$$

as obtained from (49) on dividing by 2. The Fermi energy is

$$\varepsilon_F \cong (3.17 \times 10^{-30}) n^{2/3} \approx 0.43 \times 10^{-4} \text{ erg} \approx 27 \text{ Mev}. \tag{52}$$

The average kinetic energy of a particle in a degenerate Fermi gas is $\frac{3}{5}$ of the Fermi energy, so that in nuclear matter the average kinetic energy is 16 Mev per nucleon.

BOSON GAS AND EINSTEIN CONDENSATION

A very remarkable effect occurs in a gas of noninteracting bosons at a certain transition temperature, below which a substantial fraction of the total number of particles in the system will occupy the single orbital of lowest energy, called the ground orbital. Any other orbital, including the orbital of second lowest energy, at the same temperature will be occupied by a relatively negligible number of particles. The total occupancy of all orbitals will always be equal to the specified number of particles in the system. The ground-orbital effect is called the **Einstein condensation**.

There would be nothing surprising to us in this result for the ground state occupancy if it were valid only below 10^{-14} K. This temperature is comparable with the energy spacing between the lowest and next lowest orbitals in a system of volume 1 cm^3, as we show below. But the Einstein condensation temperature for a gas of fictitious noninteracting helium atoms at the observed density of liquid helium is very much higher, about 3 K. Helium is the most familiar example of Einstein condensation in action.

Chemical Potential Near Absolute Zero

The key to the Einstein condensation is the behavior of the chemical potential of a boson system at low temperatures. The chemical potential is responsible for the apparent stabilization of a large population of particles in the ground orbital. We consider a system composed of a large number N of noninteracting bosons. When the system is at absolute zero all particles occupy the lowest-energy orbital and the system is in the state of minimum energy. It is certainly not surprising that at $\tau = 0$ all particles should be in the orbital of lowest energy. We can show that a substantial fraction remains in the ground orbital at low, although experimentally obtainable, temperatures.

If we put the energy of the ground orbital at zero on our energy scale, then from the Bose-Einstein distribution function

$$f(\varepsilon,\tau) = \frac{1}{\exp[(\varepsilon - \mu)/\tau] - 1} \tag{53}$$

we obtain the occupancy of the ground orbital at $\varepsilon = 0$ as

$$f(0,\tau) = \frac{1}{\exp(-\mu/\tau) - 1}. \tag{54}$$

When $\tau \to 0$ the occupancy of the ground orbital becomes equal to the total number of particles in the system, so that

$$\lim_{\tau \to 0} f(0,\tau) = N \approx \lim_{\tau \to 0} \frac{1}{\exp(-\mu/\tau) - 1} \approx \frac{1}{1 - (\mu/\tau) - 1} = -\frac{\tau}{\mu}.$$

Here we have made use of the series expansion $\exp(-x) = 1 - x + \cdots$. We know that x, which is μ/τ, must be small in comparison with unity, for otherwise the total number of particles N could not be large. From this result we find

$$\boxed{N = -\tau/\mu; \qquad \mu = -\tau/N,} \tag{55}$$

as $\tau \to 0$. For $N = 10^{22}$ at $T = 1$ K, we have $\mu \cong -1.4 \times 10^{-38}$ erg. We note from (55) that

$$\lambda \equiv \exp(\mu/\tau) \cong 1 - \frac{1}{N}, \tag{56}$$

as $\tau \to 0$. The chemical potential in a boson system must always be lower in energy than the ground state orbital, in order that the occupancy of every orbital be non-negative.

Example: Spacing of lowest and second lowest orbitals of free atoms. The energy of an orbital of an atom free to move in a cube of volume $V = L^3$ is

$$\varepsilon = \frac{\hbar^2}{2M} \left(\frac{\pi}{L}\right)^2 (n_x^2 + n_y^2 + n_z^2) , \qquad (57)$$

where n_x, n_y, n_z are positive integers. The energy $\varepsilon(111)$ of the lowest orbital is

$$\varepsilon(111) = \frac{\hbar^2}{2M} \left(\frac{\pi}{L}\right)^2 (1 + 1 + 1) , \qquad (58)$$

and the energy $\varepsilon(211)$ of one of the set of next lowest orbitals is

$$\varepsilon(211) = \frac{\hbar^2}{2M} \left(\frac{\pi}{L}\right)^2 (4 + 1 + 1). \qquad (59)$$

The lowest excitation energy of the atom is

$$\Delta\varepsilon = \varepsilon(211) - \varepsilon(111) = 3 \times \frac{\hbar^2}{2M} \left(\frac{\pi}{L}\right)^2. \qquad (60)$$

If $M(^4\text{He}) = 6.6 \times 10^{-24}\,$g and $L = 1\,$cm,

$$\Delta\varepsilon = (3)(8.4 \times 10^{-32})(9.86) = 2.48 \times 10^{-30}\,\text{erg}. \qquad (61)$$

In temperature units, $\Delta\varepsilon/k_B = 1.80 \times 10^{-14}\,$K.

This splitting is extremely small, and it is difficult to conceive that it can play an important part in a physical problem even at the lowest reasonably accessible temperatures such as 1 mK, which is 10^{-3} K. However, at the 1 mK temperature (55) gives $\mu \simeq -1.4 \times 10^{-41}$ erg for $N = 10^{22}$ atoms, referred to the orbital (58) as the zero of energy. Thus μ is much closer to the ground orbital than is the next lowest orbital (59), and $\exp\{[\varepsilon(111) - \mu]/\tau\}$ is much closer to 1 than is $\exp\{[\varepsilon(211) - \mu]/\tau\}$, so that $\varepsilon(111)$ dominates the distribution function.

The Boltzmann factor $\exp(-\Delta\varepsilon/\tau)$ at 1 mK is

$$\exp(-1.8 \times 10^{-11}) \cong 1 - 1.8 \times 10^{-11} , \qquad (62)$$

which is essentially unity. By (4) we would expect that even if $n \approx n_Q$ the occupancy of the first excited orbital would only be of the order of 1. However, the Bose-Einstein distribution gives an entirely different value for the occupancy of the first excited orbital:

$$f(\Delta\varepsilon, \tau) = \frac{1}{\exp[(\Delta\varepsilon - \mu)/\tau] - 1} \cong \frac{1}{\exp(\Delta\varepsilon/\tau) - 1} , \qquad (63)$$

because $\Delta\varepsilon \gg \mu$. Thus the occupation of the first excited orbital at 1 mK is

$$f \cong \frac{1}{1.8 \times 10^{-11}} \cong 5 \times 10^{10} , \qquad (64)$$

so that the fraction of the N particles that are in this orbital is $f/N \simeq 5 \times 10^{10}/10^{22} \simeq 5 \times 10^{-12}$, which is very small. We see that the occupancy of the first excited orbital at low temperatures is relatively very much lower than would be expected at first sight from the simple Boltzmann factor (62). The Bose-Einstein distribution is quite strange; it favors a situation in which the greatest part of the population is left in the ground orbital at sufficiently low temperatures. The particles in the ground orbital, as long as their number is $\gg 1$, are called the **Bose-Einstein condensate**. The atoms in the condensate act quite differently from the atoms in excited states.

How do we understand the existence of the condensate? Suppose the atoms were governed by the Planck distribution (Chapter 4), which makes no provision for holding constant the total number of particles; instead, the thermal average number of photons increases with temperature at τ^3, as found in Problem 4.1. If the laws of nature restricted the total number of photons to a value N, we would say that the ground orbital of the photon gas contained the difference $N_0 = N - N(\tau)$ between the number allotted and the number thermally excited. The N_0 nonexcited photons would be described as condensed into the ground orbital, but N_0 becomes essentially zero at a temperature τ_c such that all N photons are excited. There is no actual constraint on the total number of photons; however, there is a constraint on the total number N of material bosons, such as ^4He atoms, in a system. This constraint is the origin of the condensation into the ground orbital. The difference between the Planck distribution and the Bose-Einstein distribution is that the latter will conserve the total number of particles, independent of temperature, so that nonexcited atoms are really in the ground state condensate.

Orbital Occupancy Versus Temperature

We saw in (19) that the number of free particle orbitals per unit energy range is

$$\mathfrak{D}(\varepsilon) = \frac{V}{4\pi^2}\left(\frac{2M}{\hbar^2}\right)^{3/2}\varepsilon^{1/2} , \qquad (65)$$

for a particle of spin zero. The total number of atoms of helium-4 in the ground and excited orbitals is given by the sum of the occupancies of all orbitals:

$$N = \sum_n f_n = N_0(\tau) + N_e(\tau) = N_0(\tau) + \int_0^\infty d\varepsilon\,\mathfrak{D}(\varepsilon)f(\varepsilon,\tau). \qquad (66)$$

We have separated the sum over n into two parts. Here $N_0(\tau)$ has been written for $f(0,\tau)$, the number of atoms in the ground orbital at temperature τ. The integral in (66) gives the number of atoms $N_e(\tau)$ in all excited orbitals, with

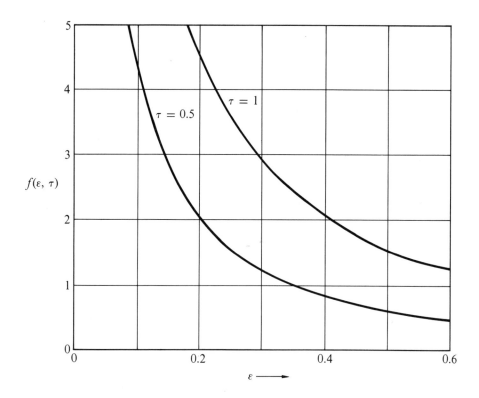

Figure 7.10　Plot of the boson distribution function for two temperatures, with sufficient particles present to ensure $\lambda \simeq 1$. The integral of the distribution times the density of states gives the number N_e of particles in excited orbitals; the rest of the particles present are condensed into the ground state orbital. The value of N_0 is too large to be shown on the plot.

$f(\varepsilon,\tau)$ as the Bose-Einstein distribution function. The integral gives only the number of atoms in excited orbitals and excludes the atoms in the ground orbital, because the function $\mathcal{D}(\varepsilon)$ is zero at $\varepsilon = 0$. To count the atoms correctly we must count separately the occupancy N_0 of the orbital with $\varepsilon = 0$. Although only a single orbital is involved, the value of N_0 may be very large in a gas of bosons. We shall call N_0 the number of atoms in the **condensed phase** and N_e the number of atoms in the **normal phase**. The whole secret of the result which follows is that at low temperatures the chemical potential μ is very much closer in energy to the ground state orbital than the first excited orbital is to the ground state orbital. This closeness of μ to the ground orbital loads most of the population of the system into the ground orbital (Figure 7.10).

The Bose-Einstein distribution function when written for the orbital at $\varepsilon = 0$ is

$$N_0(\tau) = \frac{1}{\lambda^{-1} - 1}, \tag{67}$$

as in (54), where λ will depend on the temperature τ. The number of particles in all excited orbitals increases as $\tau^{3/2}$:

$$\lambda = e^{\mu/\tau}$$

$$N_e(\tau) = \frac{V}{4\pi^2} \left(\frac{2M}{\hbar^2} \right)^{3/2} \int_0^\infty d\varepsilon \, \frac{\varepsilon^{1/2}}{\lambda^{-1} \exp(\varepsilon/\tau) - 1}.$$

or, with $x \equiv \varepsilon/\tau$,

$$N_e(\tau) = \frac{V}{4\pi^2} \left(\frac{2M}{\hbar^2} \right)^{3/2} \tau^{3/2} \int_0^\infty dx \, \frac{x^{1/2}}{\lambda^{-1} e^x - 1}. \tag{68}$$

Notice the factor $\tau^{3/2}$ which gives the temperature dependence of N_e.

At sufficiently low temperatures the number of particles in the ground state will be a very large number. Equation (67) tells us that λ must be very close to unity whenever N_0 is $\gg 1$. Then λ is very accurately constant, because a macroscopic value of N_0 forces λ to be close to unity. The condition for the validity of the calculation is that $N_0 \gg 1$, and it is not required that $N_e \ll N$. When $\varepsilon \approx \tau$ in the integrand, the value of the integrand is insensitive to small deviations of λ from 1, so that we can set $\lambda = 1$ in (68), although not in (67).

The value of the integral* in (68) is, when $\lambda = 1$,

$$\int_0^\infty dx \, \frac{x^{1/2}}{e^x - 1} = 1.306\pi^{1/2}. \tag{69}$$

---

* To evaluate the integral we write

$$\int_0^\infty dx \, \frac{x^{1/2}}{e^x - 1} = \int_0^\infty dx \, \frac{x^{1/2} e^{-x}}{1 - e^{-x}} = \sum_{s=1}^\infty \int_0^\infty dx \, x^{1/2} e^{-sx}$$

$$= \left(\sum_{s=1}^\infty s^{-3/2} \right) \int_0^\infty dy \, y^{1/2} e^{-y}.$$

The infinite sum is easily evaluated numerically to be 2.612. The integral may be transformed with $y = u^2$ to give

$$2 \int_0^\infty du \, u^2 \exp(-u^2) = \tfrac{1}{2}\sqrt{\pi}.$$

Thus the number of atoms in excited states is

$$N_e = \frac{1.306V}{4}\left(\frac{2M\tau}{\pi\hbar^2}\right)^{3/2} = 2.612n_Q V , \tag{70}$$

where $n_Q \equiv (M\tau/2\pi\hbar^2)^{3/2}$ is again the quantum concentration. We divide N_e by N to obtain the fraction of atoms in excited orbitals:

$$N_e/N \simeq 2.612n_Q V/N = 2.612n_Q/n. \tag{71}$$

The value $\lambda \simeq 1$ or $1 - 1/N$ which led to (71) is valid as long as a large number of atoms are in the ground state. All particles have to be in some orbital, either in an excited orbital or in the ground orbital. The number in excited orbitals is relatively insensitive to small changes in λ, but the rest of the particles have to be in the ground orbital. To assure this we must take λ very close to 1 as long as N_0 is a large number. Even 10^3 is a large number for the occupancy of an orbital. Yet within $\Delta\tau/\tau_E = 10^{-6}$ of the transition, where τ_E is defined by (72) below, the occupancy of the ground orbital is $> 10^{15}$ atoms cm^{-3} at the concentration of liquid ^{4}He. Thus our argument is highly accurate at $\Delta\tau/\tau_E = 10^{-6}$.

Einstein Condensation Temperature

We define the **Einstein condensation temperature*** τ_E as the temperature for which the number of atoms in excited states is equal to the total number of atoms. That is, $N_e(\tau_E) = N$. Above τ_E the occupancy of the ground orbital is not a macroscopic number; below τ_E the occupancy is macroscopic. From (70) with N for N_e we find for the condensation temperature

$$\tau_E \equiv \frac{2\pi\hbar^2}{M}\left(\frac{N}{2.612V}\right)^{2/3}. \tag{72}$$

Now (71) may be written as

$$N_e/N \simeq (\tau/\tau_E)^{3/2} , \tag{73}$$

where N is the total number of atoms. The number of atoms in excited orbitals varies as $\tau^{3/2}$ at temperatures below τ_E, as shown in Figure 7.11. The calculated value of T_E for atoms of ^{4}He is ≈ 3 K.

---

* A. Einstein, Akademie der Wissenschaften, Berlin, Sitzungsberichte **1924**, 261; **1925**, 3.

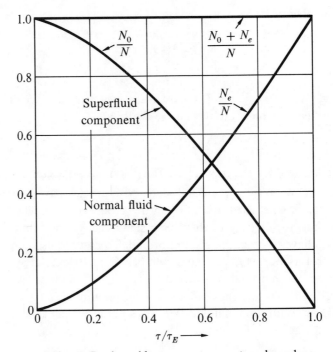

Figure 7.11 Condensed boson gas: temperature dependence of the proportion N_0/N of atoms in the ground orbital and of the proportion N_e/N of atoms in all excited orbitals. We have labeled the two components as normal and superfluid to agree with the customary description of liquid helium. The slopes of all three curves are intended to be zero at $\tau = 0$.

The number of particles in the ground orbital is found from (73):

$$N_0 = N - N_e = N[1 - (\tau/\tau_E)^{3/2}]. \tag{74}$$

We note that N may be of the order of 10^{22}. For τ even slightly less than τ_E a large number of particles will be in the ground orbital, as we see in Figure 7.11. We have said that the particles in the ground orbital below τ form the condensed phase or the superfluid phase.

The condensation temperature in kelvin is given by the numerical relation

$$T_E(\text{in K}) = (115/V_M{}^{2/3}M) , \tag{75}$$

where V_M is the molar volume in $\text{cm}^3 \, \text{mol}^{-1}$ and M is the molecular weight. For liquid helium $V_M = 27.6 \, \text{cm}^3 \, \text{mol}^{-1}$ and $M = 4$; thus $T_E = 3.1 \, \text{K}$.

Liquid ⁴He

The calculated temperature of 3 K is suggestively close to the actual temperature of 2.17 K at which a transition to a new state of matter is observed to take place in liquid helium (Figure 7.12). We believe that in liquid ⁴He below 2.17 K there is a condensation of a substantial fraction of the atoms of ⁴He into the ground orbital of the system. This is different from the condensation in coordinate space that occurs in the condensation of a gas to a liquid. Evidently the interatomic forces that lead to the liquefaction of ⁴He at 4.2 K under a pressure of one atmosphere are too weak to destroy the major effects of the boson condensation at 2.17 K. In this respect the liquid behaves as a gas. The condensation into the ground orbital is certainly connected with the properties of bosons.

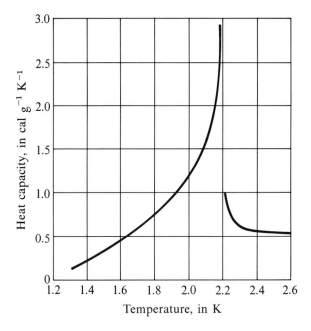

Figure 7.12   Heat capacity of liquid ⁴He. The sharp peak near 2.17 K is evidence of an important transition in the nature of the liquid. The viscosity of the liquid above the transition temperature is typical of normal liquids, whereas the viscosity below the transition as determined by rate of flow through narrow slits is vanishingly small, at least 10⁶ times smaller than the viscosity above the transition. The transition is often called a lambda transition merely because of the shape of the graph. After Keesom et al.

The condensation is normally not permitted for fermions, but pairs of fermions may act as bosons, as in the superconductivity of electron pairs (Cooper pairs) in metals. A different type of transition to complex phases with superfluid properties has been observed in liquid ^{3}He below 3 mK. Atoms of ^{3}He have spin $\frac{1}{2}$ and are fermions, but pairs of ^{3}He atoms act as bosons.

We can give several arguments in support of our view of liquid helium as a gas of noninteracting particles. At first sight this is a drastic oversimplification of the problem, but there are some important features of liquid helium for which the view is correct.

(a) The molar volume of liquid ^{4}He at absolute zero is 3.1 times the volume that we calculate from the known interactions of helium atoms. The interaction forces between pairs of helium atoms are well known experimentally and theoretically, and from these forces by standard elementary methods of solid state physics we can calculate the equilibrium volume of a *static* lattice of helium atoms. In a typical calculation we find the molar volume to be 9 cm^3 mol^{-1}, as compared with the observed 27.6 cm^3 mol^{-1}. Thus the kinetic motion of the helium atoms has a large effect on the liquid state and leads to an expanded structure in which the atoms to a certain extent can move freely over appreciable distances. We can say that the quantum zero-point motion is responsible for the expansion of the molar volume.

(b) The transport properties of liquid helium in the normal state are not very different from those of a normal classical gas. In particular, the ratio of the thermal conductivity K to the product of the viscosity η times the heat capacity per unit mass has the values

$$\frac{K}{\eta C_V} = \begin{cases} 2.6, & \text{at} \quad 2.8 \text{ K} \\ 3.2, & \text{at} \quad 4.0 \text{ K} \end{cases}$$

These values are quite close to those observed for normal gases at room temperature—see Table 14.3. The values of the transport coefficients themselves in the liquid are within an order of magnitude of those calculated for the gas at the same density. Normal liquids act quite differently.

(c) The forces in the liquid are relatively weak, and the liquid does not exist above the critical temperature of 5.2 K, which is the maximum boiling point observed. The binding energy would be perhaps ten times stronger in the equilibrium configuration of a static lattice, but the expansion of the molar volume by the quantum zero-point motion of the atoms is responsible for the reduction in the binding energy to the observed value. The value of the critical temperature is directly proportional to the binding energy.

(d) The liquid is stable at absolute zero at pressures under 25 atm; above 25 atm the solid is more stable.

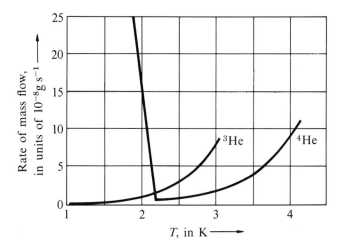

Figure 7.13   Comparison of rates of flow of liquid ^{3}He and
liquid ^{4}He under gravity through a fine hole. Notice the sudden
onset of high fluidity or superfluidity in ^{4}He. After D. W.
Osborne, B. Weinstock, and B. M. Abraham, Phys. Rev. **75**,
988 (1949).

The new state of matter into which liquid ^{4}He enters when cooled below
2.17 K has quite astonishing properties. The viscosity as measured in a flow
experiment* is essentially zero (Figure 7.13), and the thermal conductivity is
very high. We say that liquid ^{4}He below the transition temperature is a superfluid. More precisely, we denote liquid ^{4}He below the transition temperature
as liquid He II, and we say that liquid He II is a mixture of normal fluid and
superfluid components. The normal fluid component consists of the helium
atoms in thermally excited orbitals, and the superfluid component consists of
the helium atoms condensed into the ground orbital. It is known that the
radioactive boson ^{6}He in solution in liquid ^{4}He does not take part in the
superflow of the latter; neither, of course, does the fermion ^{3}He in solution in
^{4}He take part in the superflow.

We speak of liquid ^{4}He above the transition temperature as liquid He I.
There is no superfluid component in liquid He I, for here the ground orbital
occupancy is negligible, being of the same order of magnitude as the occupancy

---

* In other arrangements there may be an effective viscosity: this is true of a disk oscillating in liquid
^{4}He at any finite temperature below the condensation temperature. For a combination of two
fluids of different viscosities, some experiments measure the average viscosity, and other experiments
measure the average of $1/\eta$, or the average fluidity.

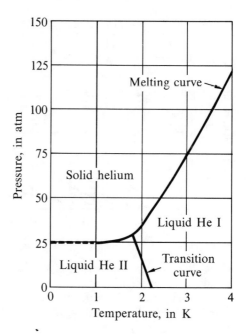

Figure 7.14   The melting curve of liquid and solid helium (^{4}He), and the transition curve between the two forms of liquid helium, He I and He II. The liquid He II form exhibits superflow properties as a consequence of the condensation of atoms into the ground orbital of the system. Note that helium is a liquid at absolute zero at pressures below 25 atm. The liquid-vapor boiling curve is not included in this graph as it would merge with the zero pressure line. After C. A. Swenson, Phys. Rev. **79**, 626 (1950).

of any other low-lying orbital, as we have seen. The regions of pressure and temperature in which liquid He I and II exist are shown in Figure 7.14.

The development of superfluid properties is not an automatic consequence of the Einstein condensation of atoms into the ground orbital. Advanced calculations show that it is the existence of some form (almost any form) of interaction among atoms that leads to the development of superfluid properties in the atoms condensed in the ground orbital.

Phase Relations of Helium

The phase diagram of ^{4}He was shown in Figure 7.14. The liquid-vapor curve can be followed from the critical point of 5.2 K down to absolute zero without any appearance of the solid. At the transition temperature the normal liquid, called He I, makes a transition to the form with superfluid properties, called He II. A temperature called the λ point is the triple point at which liquid He I, liquid He II, and vapor coexist. Keesom, who first solidified helium, found that the solid* did not exist below a pressure of 25 atm. Another triple point exists

---

* An interesting discussion of solid helium is given by B. Bertram and R. A. Guyer, Scientific American, August 1967, pp. 85–95. Solid ^{4}He exists in three crystal structures according to the conditions of temperature and pressure.

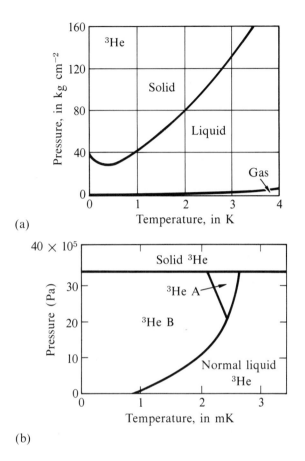

(a)

(b)

Figure 7.15 Phase diagrams for liquid ^{3}He, (a) in kelvin and (b) in millikelvin. In the region of negative slope shown in (a) on the phase boundary the solid has a higher entropy than the liquid, and we have to add heat to the liquid to solidify it. Superfluid properties appear in (b) in the A and B phases of liquid ^{3}He. The A phase is double—in a magnetic field the phase divides into two components with opposite nuclear magnetic moments.

at 1.743 K: here the solid is in equilibrium with the two liquid modifications, He I and He II. The two triple points are connected by a line that separates the regions of existence of He II and He I.

The phase diagram of ^{3}He differs in a remarkable way from the phase diagram of ^{4}He. Figure 7.15 exhibits the importance of the fermion nature of ^{3}He. Note the negative slope of the coexistence curve at low temperatures. As explained in Chapter 10, the negative slope means that the entropy of the liquid phase is lower than the entropy of the solid phase.

Quasiparticles and Superfluidity, ⁴He

For many purposes the superfluid component of liquid helium II behaves as if it were a vacuum, as if it were not there at all. The N_0 atoms of the superfluid are condensed into the ground orbital and have no excitation energy, for the ground orbital by definition has no excitation energy. The superfluid has energy only when the center of mass of the superfluid is given a velocity relative to the laboratory reference frame—as when the superfluid is set into flow relative to the laboratory.

The condensed component of N_0 atoms will flow with zero viscosity so long as the flow does not create excitations in the superfluid—that is, so long as no atoms make transitions between the ground orbital and the excited orbitals. Such transitions might be caused by collisions of helium atoms with irregularities in the wall of the tube through which the helium atoms are flowing. The transitions, if they occur, are a cause of energy loss and of momentum loss from the moving fluid, and the flow is not resistanceless if such collisions can occur.

The criterion for superfluidity involves the energy and momentum relationship of the excitations in liquid He II. If the excited orbitals were really like the orbitals of free atoms, with a free particle relation

$$\varepsilon = \tfrac{1}{2}Mv^2 = \frac{1}{2M}\,(\hbar k)^2 \tag{76}$$

between the energy ε and the momentum Mv or $\hbar \mathbf{k}$ of an atom, then we can show that superfluidity would not be expected. Here $k = 2\pi/\text{wavelength}$. But because of the existence of interactions between the atoms the low energy excitations do not resemble free particle excitations, but are longitudinal sound waves, longitudinal phonons (Chapter 4). After all, it is not unreasonable that a longitudinal sound wave should propagate in any liquid, even though we have no previous experience of superliquids.

A language has grown up to describe the low-lying excited states of a system of many atoms. These states are called **elementary excitations** and in their particle aspect the states are called **quasiparticles**. Longitudinal phonons are the elementary excitations of liquid He II. We shall give the clear-cut experimental evidence for this, but first we derive a necessary condition for superfluidity. This condition will show us why the phonon-like nature of the elementary excitations leads to the superfluid behavior of liquid He II.

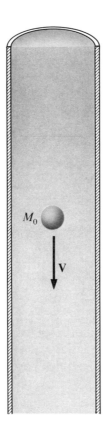

Figure 7.16 Body of mass M_0 moving with velocity $\mathbf{V}$ down a cylinder that contains liquid He II at absolute zero.

We consider in Figure 7.16 a body, perhaps a steel ball or a neutron, of mass M_0 falling with velocity $\mathbf{V}$ down a column of liquid helium at rest at absolute zero, so that initially no elementary excitations are excited. If the motion of the body generates elementary excitations, there will be a damping force on the body. In order to generate an elementary excitation of energy ε_k and momentum $\hbar \mathbf{k}$, we must satisfy the law of conservation of energy:

$$\tfrac{1}{2}M_0 V^2 = \tfrac{1}{2}M_0 V'^2 + \varepsilon_k \,, \tag{77}$$

where V' is the velocity of the body after creation of the elementary excitation. Furthermore, we must satisfy the law of conservation of momentum

$$M_0 \mathbf{V} = M_0 \mathbf{V}' + \hbar \mathbf{k}. \tag{78}$$

The two conservation laws cannot always be satisfied at the same time even if the direction of the excitation created in the process is unrestricted. To show this we rewrite (78) as

$$M_0\mathbf{V} - \hbar\mathbf{k} = M_0\mathbf{V}'$$

and take the square of both sides:

$$M_0{}^2V^2 - 2M_0\hbar\mathbf{V}\cdot\mathbf{k} + \hbar^2k^2 = M_0{}^2V'^2.$$

On multiplication by $1/2M_0$ we have

$$\tfrac{1}{2}M_0V^2 - \hbar\mathbf{V}\cdot\mathbf{k} + \frac{1}{2M_0}\hbar^2k^2 = \tfrac{1}{2}M_0V'^2. \tag{79}$$

We subtract (79) from (77) to obtain

$$\hbar\mathbf{V}\cdot\mathbf{k} - \frac{1}{2M_0}\hbar^2k^2 = \varepsilon_\mathbf{k}. \tag{80}$$

There is a lowest value of the magnitude of the velocity $\mathbf{V}$ for which this equation can be satisfied. The lowest value will occur when the direction of $\mathbf{k}$ is parallel to that of $\mathbf{V}$. This critical velocity is given by

$$V_c = \text{minimum of } \frac{\varepsilon_\mathbf{k} + \dfrac{1}{2M_0}\hbar^2k^2}{\hbar k}. \tag{81}$$

The condition is a little simpler to express if we let the mass M_0 of the body become very large, for then

$$V_c = \text{minimum of } \frac{\varepsilon_\mathbf{k}}{\hbar k}. \tag{82}$$

A body moving with a lower velocity than V_c will not be able to create excitations in the liquid, so that the motion will be resistanceless. The viscosity will appear to be zero. A body moving with higher velocity will encounter resistance because of the generation of excitations.

There is a simple geometrical construction for (82). We make a plot of the energy $\varepsilon_\mathbf{k}$ of an elementary excitation as a function of the momentum $\hbar k$ of the excitation. We construct the straight line from the origin which just touches the curve from below. The slope of this line is equal to the critical velocity. If $\varepsilon_{\mathbf{k}} = \hbar^2 k^2 / 2M$, as for the excitation of a free atom, the straight line has zero slope and the critical velocity is zero:

Free atoms: $\qquad V_c = \text{minimum of } \hbar k / 2M = 0.$ $\qquad$ (83)

The energy of a low energy phonon in liquid He II is $\varepsilon_{\mathbf{k}} = \hbar \omega_k = \hbar v_s k$ in the frequency region of sound waves where the product of wavelength and frequency is equal to the velocity of sound v_s, or where the circular frequency ω_k is equal to the product of v_s times the wavevector k. Now the critical velocity is

Phonons: $\qquad V_c = \text{minimum of } \hbar v_s k / \hbar k = v_s.$ $\qquad$ (84)

The critical velocity V_c is equal to the velocity of sound if (84) is valid for all wavevectors, which it is not in liquid helium II. The observed critical flow velocities are indeed nonzero, but considerably lower than the velocity of sound and usually lower than the solid straight line in Figure 7.17, presumably because the plot of ε_k versus $\hbar k$ may turn downward at very high $\hbar k$.

The actual spectrum of elementary excitations in liquid helium II has been determined by the observations on the inelastic scattering of slow neutrons. The experimental results are shown in Figure 7.17. The solid straight line is the Landau critical velocity for the range of wavevectors covered by the neutron experiments, and for this line the critical velocity is

$$V_c = \Delta / \hbar k_0 \approx 5 \times 10^3 \text{ cm s}^{-1}, \qquad (85)$$

where Δ and k_0 are identified on the figure.

Charged ions of helium in solution in liquid helium II under certain experimental conditions of pressure and temperature have been observed[*] to move almost like free particles and to have a limiting drift velocity near $5 \times 10^3 \text{ cm s}^{-1}$ closely equal to the calculated value of (85). Under other experimental conditions the motion of the ions is limited at a lower velocity by the creation of vortex rings. Such vortex rings are transverse modes of motion and do not appear in the longitudinal modes covered by Figure 7.17.

Our result (84) for a necessary condition for the critical velocity is more general than the calculation we have given. Our calculation demonstrates that a body will move without resistance through liquid He II at absolute zero if the velocity V of the body is less than the critical velocity V_c. However, at

---

[*] L. Meyer and F. Reif, Phys. Rev. **123**, 727 (1961); G. W. Rayfield, Phys. Rev. Letters **16**, 934 (1966).

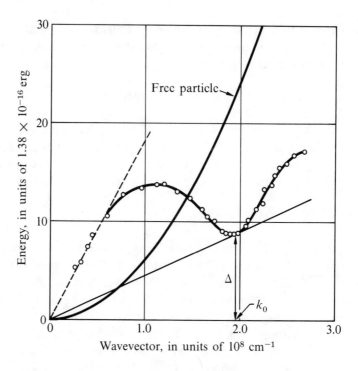

Figure 7.17 Energy ε_k versus wavevector k of elementary excitations in liquid helium at 1.12 K. The parabolic curve rising from the origin represents the theoretically calculated curve for free helium atoms at absolute zero. The open circles correspond to the energy and momentum of the measured excitations. A smooth curve has been drawn through the points. The broken curve rising linearly from the origin is the theoretical phonon branch with a velocity of sound of 237 m s^{-1}. The solid straight line gives the critical velocity, in appropriate units: The line gives the minimum of ε_k/k over the region of k covered in these experiments. After D. G. Henshaw and A. D. B. Woods, Phys. Rev. **121**, 1266 (1961).

temperatures above absolute zero, but below the Einstein temperature, there will be a normal fluid component of elementary excitations that are thermally excited. The normal fluid component is the source of resistance to the motion of the body. The superflow aspect appears first in experiments in which the liquid flows out through a fine tube in the side of a container. The normal fluid component may remain behind in the container while the superfluid component leaks out without resistance. The derivation we have given of the critical velocity also holds for this situation, with $\mathbf{V}$ as the velocity of the superfluid relative to the walls of the tube; M_0 is the mass of the fluid. Excitations would be created above V_c by the interaction between the flow of the liquid and any mechanical irregularity in the walls.

Superfluid Phases of ^{3}He

Three superfluid phases of liquid ^{3}He are known* (Figure 7.15b), but—in contrast to liquid ^{4}He—with transition temperatures of only a few millikelvin. The superfluid phases are believed to be qualitatively similar to the superconducting state of electrons in metals, where pairs of particles in orbitals near the Fermi surface form a type of bound state known as a Cooper pair. Such a pair is qualitatively like a diatomic molecule, but the radius of the molecule is much larger than the average interelectron spacing in a metal or the average interparticle spacing in liquid ^{3}He.

In metallic superconductivity the two electrons that form a Cooper pair are in a nonmagnetic (singlet) spin state. In the superfluid states of liquid ^{3}He the two atoms that form a pair are in the triplet spin states of the two ^{3}He nuclei, so that three magnetic superfluids are possible, corresponding to spin orientations $M_I = 1$, 0, and -1, or mixtures of these three states. The magnetic superfluids have been explored experimentally, and both the magnetic and superfluid properties have been confirmed.

SUMMARY

1. Compared to a classical gas, a Fermi gas at low temperature has high kinetic energy, high pressure, and low heat capacity. The entropy of the Fermi gas is zero in the ground state. The energy of the highest filled orbital in the ground state of a free particle gas of fermions of spin $\frac{1}{2}$ is

$$\varepsilon_F = \frac{\hbar^2}{2M}\left(\frac{3\pi^2 N}{V}\right)^{2/3}.$$

2. The total kinetic energy in the ground state is

$$U_0 = \tfrac{3}{5}N\varepsilon_F.$$

---

* For elementary reviews, see J. C. Wheatley, Physics Today, February 1976, p. 32; A. J. Leggett, Physics Bulletin **25**, 311 (1975); and J. R. Hook, Physics Bulletin **29**, 513 (1978). For deeper reviews, see J. C. Wheatley, Rev. Mod. Phys. **47**, 415 (1975) and A. J. Leggett, Rev. Mod. Phys. **47**, 331 (1975).

3. The density of orbitals at ε_F is

$$\mathfrak{D}(\varepsilon_F) = 3N/2\varepsilon_F.$$

4. The heat capacity of an electron gas at $\tau \ll \tau_F$ is

$$C_{el} = \tfrac{1}{3}\pi^2 \mathfrak{D}(\varepsilon_F)\tau \simeq N\tau/\tau_F \ ,$$

in fundamental units.

5. For a Bose gas at $\tau < \tau_E$ the fraction of atoms in excited orbitals is

$$N_e/N = 2.612 n_Q/n \simeq (\tau/\tau_E)^{3/2}.$$

6. The Einstein condensation temperature of a gas of noninteracting bosons is

$$\tau_E = \frac{2\pi\hbar^2}{M}\left(\frac{N}{2.612V}\right)^{2/3}.$$

PROBLEMS

1. Density of orbitals in one and two dimensions. (a) Show that the density of orbitals of a free electron in one dimension is

$$\mathfrak{D}_1(\varepsilon) = (L/\pi)(2m/\hbar^2\varepsilon)^{1/2} \ , \tag{86}$$

where L is the length of the line. (b) Show that in two dimensions, for a square of area A,

$$\mathfrak{D}_2(\varepsilon) = Am/\pi\hbar^2 \ , \tag{87}$$

independent of ε.

2. Energy of relativistic Fermi gas. For electrons with an energy $\varepsilon \gg mc^2$, where m is the rest mass of the electron, the energy is given by $\varepsilon \simeq pc$, where p is the momentum. For electrons in a cube of volume $V = L^3$ the momentum is of the form $(\pi\hbar/L)$, multiplied by $(n_x^2 + n_y^2 + n_z^2)^{1/2}$, exactly as for the nonrelativistic limit. (a) Show that in this extreme relativistic limit the Fermi energy of a gas of N electrons is given by

$$\varepsilon_F = \hbar\pi c(3n/\pi)^{1/3} \ , \tag{88}$$

where $n = N/V$. (b) Show that the total energy of the ground state of the gas is

$$U_0 = \tfrac{3}{4}N\varepsilon_F. \tag{89}$$

The general problem is treated by F. Jüttner, Zeitschrift für Physik **47**, 542 (1928).

3. Pressure and entropy of degenerate Fermi gas. (a) Show that a Fermi electron gas in the ground state exerts a pressure

$$p = \frac{(3\pi^2)^{2/3}}{5} \cdot \frac{\hbar^2}{m}\left(\frac{N}{V}\right)^{5/3} \tag{90}$$

In a uniform decrease of the volume of a cube every orbital has its energy raised: The energy of an orbital is proportional to $1/L^2$ or to $1/V^{2/3}$. (b) Find an expression for the entropy of a Fermi electron gas in the region $\tau \ll \varepsilon_F$. Notice that $\sigma \to 0$ as $\tau \to 0$.

4. Chemical potential versus temperature. Explain graphically why the initial curvature of μ versus τ is upward for a fermion gas in one dimension and downward in three dimensions (Figure 7.7). *Hint:* The $\mathfrak{D}_1(\varepsilon)$ and $\mathfrak{D}_3(\varepsilon)$ curves are different, where $\mathfrak{D}_1$ is given in Problem 1. It will be found useful to set up the integral for N, the number of particles, and to consider from the graphs the behavior of the integrand between zero temperature and a finite temperature.

5. Liquid ^{3}He as a Fermi gas. The atom ^{3}He has spin $I = \tfrac{1}{2}$ and is a fermion. (a) Calculate as in Table 7.1 the Fermi sphere parameters v_F, ε_F, and T_F for ^{3}He at absolute zero, viewed as a gas of noninteracting fermions. The density of the liquid is 0.081 g cm^{-3}. (b) Calculate the heat capacity at low temperatures $T \ll T_F$ and compare with the experimental value $C_V = 2.89Nk_BT$ as observed for $T < 0.1$ K by A. C. Anderson, W. Reese, and J. C. Wheatley, Phys. Rev. **130**, 495 (1963); see also Figure 7.18. Excellent surveys of the properties of liquid ^{3}He are given by J. Wilks, *Properties of liquid and solid helium*, Oxford, 1967, and by J. C. Wheatley, "Dilute solutions of ^{3}He in ^{4}He at low temperatures," American Journal of Physics **36**, 181–210 (1968). The principles of refrigerators based on ^{3}He–^{4}He mixtures are reviewed in Chapter 12 on cryogenics; such refrigerators produce steady temperatures down to 0.01 K in continuously acting operation.

6. Mass-radius relationship for white dwarfs. Consider a white dwarf of mass M and radius R. Let the electrons be degenerate but nonrelativistic; the protons are nondegenerate. (a) Show that the order of magnitude of the gravitational self-energy is $-GM^2/R$, where G is the gravitational constant. (If the mass density is constant within the sphere of radius R, the exact potential energy is

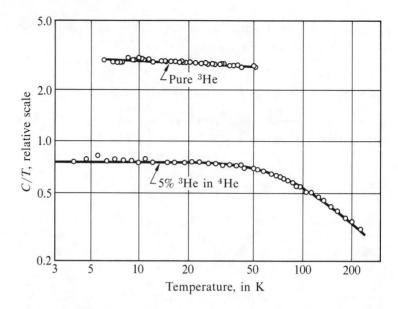

Figure 7.18   Heat capacity of liquid ^{3}He and of a 5 percent solution of
^{3}He in liquid ^{4}He. The quantity plotted on the vertical axis is C/T, and
the horizontal axis is T. Thus for a Fermi gas in the degenerate temperature
region the theoretical curves of C/T at constant volume are horizontal.
The curve for pure ^{3}He is taken at constant pressure, which accounts for
the slight slope. The curve for the solution of ^{3}He in liquid ^{4}He indicates
that the ^{3}He in solution acts as a Fermi gas; the degenerate region at low
temperature goes over to the nondegenerate region at higher temperature.
The solid line through the experimental points for the solution is drawn
for $T_F = 0.331$ K, which agrees with the calculation for free atoms if the
effective mass is taken as 2.38 times the mass of an atom of ^{3}He. Curves
after J. C. Wheatley, Amer. J. Physics **36** (1968).

$-3GM^2/5R$). (b) Show that the order of magnitude of the kinetic energy of the
electrons in the ground state is

$$\frac{\hbar^2 N^{5/3}}{mR^2} \approx \frac{\hbar^2 M^{5/3}}{mM_H{}^{5/3}R^2} \, ,$$

where m is the mass of an electron and M_H is the mass of a proton. (c) Show
that if the gravitational and kinetic energies are of the same order of magnitude
(as required by the virial theorem of mechanics), $M^{1/3}R \approx 10^{20}$ g$^{1/3}$ cm. (d) If the

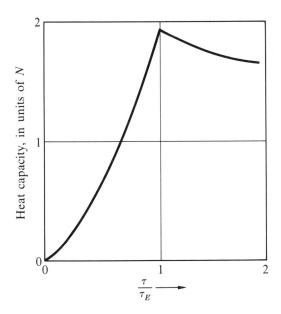

Figure 7.19 Heat capacity of an ideal Bose-Einstein gas at constant volume.

mass is equal to that of the Sun (2×10^{33} g), what is the density of the white dwarf? (e) It is believed that pulsars are stars composed of a cold degenerate gas of neutrons. Show that for a neutron star $M^{1/3}R \approx 10^{17} \, \mathrm{g}^{1/3}$ cm. What is the value of the radius for a neutron star with a mass equal to that of the Sun? Express the result in km.

7. Photon condensation. Consider a science fiction universe in which the number of photons N is constant, at a concentration of 10^{20} cm^{-3}. The number of thermally excited photons we assume is given by the result of Problem 4.1, which is $N_e = 2.404V\tau^3/\pi^2\hbar^3c^3$. Find the critical temperature in K below which $N_e < N$. The excess $N - N_e$ will be in the photon mode of lowest frequency; the excess might be described as a photon condensate in which there is a large concentration of photons in the lowest mode. In reality there is no such principle that the total number of photons be constant, hence there is no photon condensate.

8. Energy, heat capacity, and entropy of degenerate boson gas. Find expressions as a function of temperature in the region $\tau < \tau_E$ for the energy, heat capacity, and entropy of a gas of N noninteracting bosons of spin zero confined to a volume V. Put the definite integral in dimensionless form; it need not be evaluated. The calculated heat capacity above and below τ_E is shown in Figure 7.19. The experimental curve was shown in Figure 7.12. The difference between the two curves is marked: It is ascribed to the effect of interactions between the atoms.

9. Boson gas in one dimension.   Calculate the integral for $N_e(\tau)$ for a one-dimensional gas of noninteracting bosons, and show that the integral does not converge. This result suggests that a boson ground state condensate does not form in one dimension. Take $\lambda = 1$ for the calculation. (The problem should really be treated by means of a sum over orbitals on a finite line.)

10. Relativistic white dwarf stars.   Consider a Fermi gas of N electrons each of rest mass m in a sphere of radius R. Conditions in certain white dwarfs are such that the great majority of electrons have extreme relativistic kinetic energies $\varepsilon \simeq pc$, where p is the momentum. The de Broglie relation remains $\lambda = 2\pi\hbar/p$. Problem 2 gives the ground state kinetic energy of the N electrons on the assumption that $\varepsilon = pc$ for all electrons. Treat the sphere as a cube of equal volume. (a) Use the standard virial theorem argument to predict the value of N. Assume that the whole star is ionized hydrogen, but neglect the kinetic energy of the protons compared to that of the electrons. (b) Estimate the value of N. A careful treatment by Chandrasekhar leads not to a single value of N, but to a limit above which a stable white dwarf cannot exist: see D. D. Clayton, *Principles of stellar evolution and nucleosynthesis*, McGraw-Hill, 1968, p. 161; M. Harwit, *Astrophysical concepts*, Wiley, 1973.

11. Fluctuations in a Fermi gas.   Show for a single orbital of a fermion system that

$$\langle(\Delta N)^2\rangle = \langle N\rangle(1 - \langle N\rangle) \,, \tag{91}$$

if $\langle N\rangle$ is the average number of fermions in that orbital. Notice that the fluctuation vanishes for orbitals with energies deep enough below the Fermi energy so that $\langle N\rangle = 1$. By definition, $\Delta N \equiv N - \langle N\rangle$.

12. Fluctuations in a Bose gas.   If $\langle N\rangle$ as in (11) is the average occupancy of a single orbital of a boson system, then from (5.83) show that

$$\langle(\Delta N)^2\rangle = \langle N\rangle(1 + \langle N\rangle). \tag{92}$$

Thus if the occupancy is large, with $\langle N\rangle \gg 1$, the fractional fluctuations are of the order of unity: $\langle(\Delta N)^2\rangle/\langle N\rangle^2 \approx 1$, so that the actual fluctuations can be enormous. It has been said that "bosons travel in flocks." The first edition of this text has an elementary discussion of the fluctuations of photons.

13. Chemical potential versus concentration.   (a) Sketch carefully the chemical potential versus the number of particles for a boson gas in volume V at temperature τ. Include both classical and quantum regimes. (b) Do the same for a system of fermions.

14. *Two orbital boson system.* Consider a system of N bosons of spin zero, with orbitals at the single particle energies 0 and ε. The chemical potential is μ, and the temperature is τ. Find τ such that the thermal average population of the lowest orbital is twice the population of the orbital at ε. Assume $N \gg 1$ and make what approximations are reasonable.

Chapter 8

Heat and Work

Note: In (and only in) the discussion of energy conversion devices that operate in cycles, we shall define all energy, entropy, and heat transfers as positive, whether the flow is into or out of the device. This convention avoids needless difficulties with algebraic signs. The term "reversible" includes processes for which the combined entropy of the interacting systems remains constant.

ENERGY AND ENTROPY TRANSFER:
DEFINITION OF HEAT AND WORK

Heat and work are two different forms of energy transfer. Heat is the transfer of energy to a system by thermal contact with a reservoir. Work is the transfer of energy to a system by a change in the external parameters that describe the system. The parameters may include volume, magnetic field, electric field, or gravitational potential. The reason we distinguish heat from work will be clear when we discuss energy conversion processes.

The most important physical process in a modern energy-intensive civilization is the conversion of heat into work. The Industrial Revolution was made possible by the steam engine, which converts heat to work. The internal combustion engine, which seems to dominate man as much as it serves him, is a device to convert heat to work. The problem of understanding the limitations of the steam engine gave rise to much of the development of thermodynamics. Energy conversion remains one of the central applications of thermal physics because most electrical energy is generated from heat.

The fundamental difference between heat and work is the difference in the entropy transfer. Consider the energy transfer dU from a reservoir to a system with which the reservoir is in thermal contact at temperature τ; an entropy transfer $d\sigma = dU/\tau$ accompanies the energy transfer, according to the argument of Chapter 2. This energy transfer is what we defined above as heat, and we see it is accompanied by entropy transfer. Work, being energy transfer by a change in external parameters—such as the position of a piston—does not transfer any entropy to the system. There is no place for entropy to come from when only work is performed or transferred.

However, we must be careful: the total energy of two systems brought into contact is conserved, but their total entropy is not necessarily conserved and may increase. The entropy transfer between two systems in thermal contact is well defined only if the entropy of one system increases by as much as the entropy of the other decreases. Let us restrict ourselves for the present to reversible processes such that the combined entropy of the interacting systems remains constant. Later we will generalize the discussion to irreversible processes which are processes in which the total entropy of the two systems increases, as in the heat flow example in Chapter 2.

We can give a quantitative expression to the distinction between heat and work. Let dU be the energy change of a system during a reversible process; $d\sigma$ is the entropy change, and τ is the temperature. We define

$$dQ \equiv \tau d\sigma \tag{1}$$

as the heat received by the system in the process. By the principle of conservation of energy,

$$dU = dW + dQ \,, \tag{2}$$

which says that the energy change is caused partly by work done on the system and partly by heat added to the system from the reservoir. Then

$$dW = dU - dQ = dU - \tau d\sigma \tag{3}$$

is the work performed on the system in the reversible process. Our reasons for designating heat and work by dQ and dW rather than dQ and dW are explained below. For $d\sigma = 0$, we have pure work; for $dU = \tau d\sigma$, pure heat.

HEAT ENGINES: CONVERSION OF HEAT INTO WORK

Carnot inequality. Heat and work have different roles in energy conversion processes because of the difference in entropy transfer. Consider two consequences of the difference:

(a) All types of work are freely convertible into mechanical work and into each other, because the entropy transfer is zero. An ideal electrical motor, without mechanical friction or electrical resistance, is a device to convert electrical work into mechanical work. An ideal electrical generator converts mechanical work into electrical work. Because all forms of work are freely convertible, they are thermodynamically equivalent to each other and, in particular, equivalent to mechanical work. The term work denotes all types of work.

(b) Work can be completely converted into heat, but the inverse is not true: heat cannot be completely converted into work. Entropy enters the system with the heat, but does not leave the system with the work. A device that generates work from heat must necessarily strip the entropy from the heat that has been converted to work. The entropy removed from the converted input heat cannot be permitted to pile up inside the device indefinitely; this entropy must ultimately be removed from the device. The only way to do this is to provide more

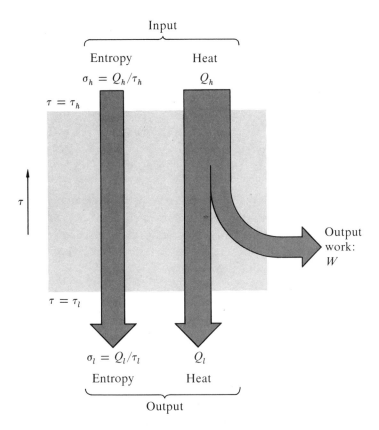

Figure 8.1   Entropy and energy flow in any continuously operating reversible device generating work from heat. The entropy outflow must equal the entropy inflow.

input heat than the amount converted to work, and to eject the excess input heat as waste heat, at a temperature lower than that of the input heat (Figure 8.1). Because $dQ/d\sigma = \tau$, the reversible heat transfer accompanying one unit of entropy is given by the temperature at which the heat is transferred. It follows that only part of the input heat need be ejected at the lower temperature to carry away all the entropy of the input heat. Only the difference between input and output heat can be converted to work. To prevent the accumulation of entropy there must be some output heat; therefore it is impossible to convert all the input heat to work!

A prohibition against unlimited entropy accumulation in a device does not mean entropy cannot accumulate temporarily, provided that it is ultimately removed. Many practical energy-conversion devices operate in cycles, and the entropy contained in the device varies periodically with time. Such a cyclic device is called a **heat engine**. The internal combustion engine is an example: The entropy contained in each cylinder is at a minimum near the beginning of the intake stroke and a maximum near the beginning of the exhaust stroke. There is a value of the entropy content to which the device returns cyclically; the entropy does not pile up indefinitely.

What fraction of the input heat Q_h taken in during one cycle at the fixed higher temperature τ_h can be converted into work? The input entropy associated with the input heat is $\sigma_h = Q_h/\tau_h$. To avoid confusing signs, we define in this discussion all energy, heat, and entropy flows as positive whether the flow is into or out of the system, rather than following the usual convention according to which a flow is positive into the system and negative out of the system. If Q_l is the waste heat leaving the system per cycle at the fixed lower temperature τ_l, the output entropy per cycle is $\sigma_l = Q_l/\tau_l$. In a reversible process this output entropy is equal to the input entropy:

$$\sigma_l = \sigma_h \qquad \text{or} \qquad Q_l/\tau_l = Q_h/\tau_h \,, \tag{4}$$

so that

$$Q_l = (\tau_l/\tau_h)Q_h. \tag{5}$$

The work generated during one cycle of a reversible process is the difference between the heat added and the waste heat extracted:

$$W = Q_h - Q_l = [1 - (\tau_l/\tau_h)]Q_h = \frac{\tau_h - \tau_l}{\tau_h} Q_h. \tag{6}$$

The ratio of the work generated to the heat added in the reversible process is called the **Carnot efficiency**:

$$\eta_C \equiv \left(\frac{W}{Q_h}\right)_{\text{rev}} = \frac{\tau_h - \tau_l}{\tau_h} = \frac{T_h - T_l}{T_h}. \tag{7}$$

This quantity is named in honor of Sadi Carnot, who derived it in 1824. It was a remarkable feat: the concept of entropy had not yet been invented, and Carnot's derivation preceded by some 15 years the recognition that heat is a form of energy.

The Carnot efficiency is the highest possible value of the **energy conversion efficiency** $\eta = W/Q_h$, the output work per unit of input heat, in any cyclic heat

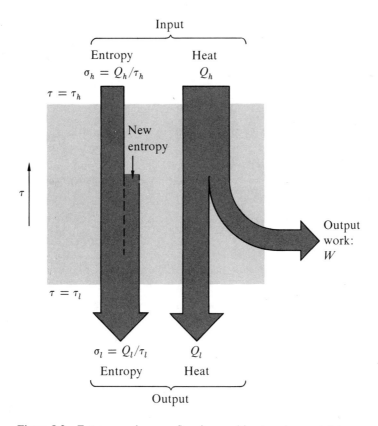

Figure 8.2 Entropy and energy flow in a real heat engine containing irreversibilities that generate new entropy inside the device. The entropy outflow at the lower temperature is larger than the entropy inflow at the higher temperature.

engine that operates between the temperatures τ_h and τ_l. Actual heat engines have lower efficiencies because the processes taking place within the device are not perfectly reversible. Entropy will be generated inside the device by irreversible processes. The energy-entropy flow diagram is modified as in Figure 8.2. We now have three inequalities

$$\sigma_l \geq \sigma_h; \tag{8}$$

$$Q_l \geq Q_h(\tau_l/\tau_h); \tag{9}$$

$$W = Q_h - Q_l \leq \frac{\tau_h - \tau_l}{\tau_h} Q_h = \eta_C Q_h. \tag{10}$$

The actual energy conversion efficiency η obeys the **Carnot inequality**

$$\eta = W/Q_h \leq 1 - (\tau_l/\tau_h) \equiv \eta_C. \tag{11}$$

We can have $\eta = \eta_C$ only in the limit of reversible operation of a device that takes in heat at τ_h and ejects heat at τ_l.

The Carnot inequality is the basic limitation on any heat engine that operates in a cyclic process. The result tells us that it is impossible to convert all input heat into work. For a given temperature ratio τ_h/τ_l the highest conversion efficiency is obtained under reversible operation. The limiting efficiency increases with increasing τ_h/τ_l, but we attain 100 percent efficiency only when $\tau_h/\tau_l \to \infty$.

The low-temperature waste heat of any heat engine must ultimately be ejected into the environment, so that τ_l cannot be below the environmental temperature, usually about 300 K. High efficiency requires an input temperature T_h high compared to 300 K. The usable temperatures in practice are unfortunately limited by various materials constraints. In power plant steam turbines, which are expected to operate continuously for years, the upper temperature is currently limited to about 600 K by problems with the strength and corrosion of steel. With $T_l = 300\,\mathrm{K}$ and $T_h = 600\,\mathrm{K}$, the Carnot efficiency is $\eta_C = \frac{1}{2}$, or 50 percent. Losses caused by unavoidable irreversibilities reduce this efficiency typically to about 40 percent. To obtain higher efficiencies is a problem in high temperature metallurgy.

Sources of irreversibility.   Figure 8.3 illustrates several common sources of irreversibility:

(a) Part of the input heat Q_h may flow directly to the low temperature, by-passing the actual energy conversion process, as in the heat flow into the cylinder walls during the combustion cycle of the internal combustion engine.

(b) Part of the temperature difference $\tau_h - \tau_l$ may not be available as temperature difference in the actual energy conversion process, because of the temperature drop across thermal resistances in the path of the heat flow.

(c) Part of the work generated may be converted back to heat by mechanical friction.

(d) Gas may expand irreversibly without doing work, as in the irreversible expansion of an ideal gas into a vacuum.

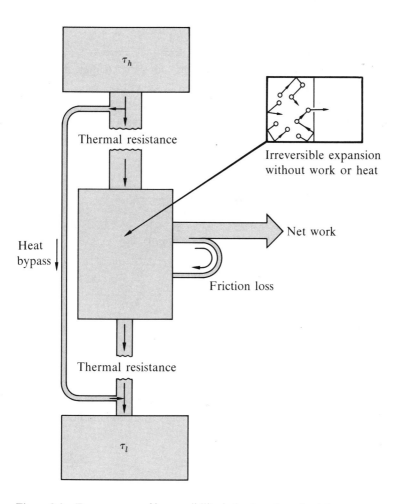

Figure 8.3   Four sources of irreversibility in heat engines: heat flow bypassing the energy conversion process, thermal resistance in the path of the heat flow, frictional losses, and entropy generation during irreversible expansions.

Refrigerators

Refrigerators are heat engines in reverse. Refrigerators consume work to move heat from a low temperature τ_l to a higher temperature τ_h. Consider the energy-entropy flow diagram of a reversible heat engine in Figure 8.1. Because no entropy is generated inside the device, its operation can be reversed, with an exact reversal of the energy and entropy flows. Equations (4) through (6) remain valid for the reversed flows.

The energy ratio of interest in a refrigerator is not the energy conversion efficiency (7), but the ratio $\gamma \equiv Q_l/W$ of the heat extracted at the low temperature to the work consumed. This ratio is called the **coefficient of refrigerator performance**; its limiting value in reversible operation is called the **Carnot coefficient** of refrigerator performance, denoted by γ_C. Do not confuse $\gamma = Q_l/W$ with $\eta \equiv W/Q_h$ for the energy conversion efficiency of a heat engine; although $\eta \leq 1$ always, γ can be >1 or <1. From Eq. (5) and $W = Q_h - Q_l$, the work consumed is

$$W = Q_h - Q_l = \frac{\tau_h - \tau_l}{\tau_l} Q_l. \tag{12}$$

The Carnot coefficient of refrigerator performance is

$$\boxed{\gamma_C = \left(\frac{Q_l}{W}\right)_{\text{rev}} = \frac{\tau_l}{\tau_h - \tau_l} = \frac{T_l}{T_h - T_l}.} \tag{13}$$

This ratio can be larger or smaller than unity.

Actual refrigerators, like actual heat engines, always contain irreversibilities that generate entropy inside the device. In a refrigerator this excess entropy is ejected at the higher temperature, as in the energy-entropy flow diagram of Figure 8.4. With the convention that all energy and entropy flows are positive, we now have

$$\sigma_h \geq \sigma_l \,, \tag{14}$$

in place of (8). Further,

$$Q_h \geq (\tau_h/\tau_l)Q_l \,, \tag{15}$$

and

$$W = Q_h - Q_l \geq [(\tau_h/\tau_l) - 1]Q_l = \frac{\tau_h - \tau_l}{\tau_l} Q_l = Q_l/\gamma_C \,, \tag{16}$$

so that

$$\boxed{\gamma = Q_l/W \leq \gamma_C.} \tag{17}$$

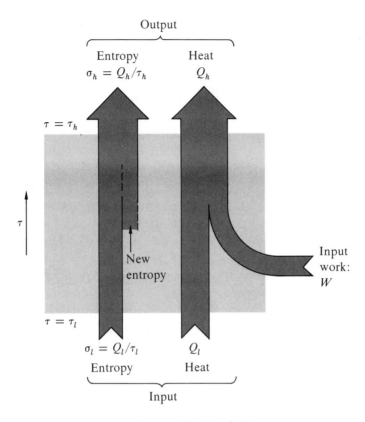

Figure 8.4 Entropy and energy flow in a refrigerator.

The Carnot coefficient γ_C is an upper limit to the actual coefficient of refrigerator performance γ, just as the Carnot efficiency η_C is an upper limit to the actual energy conversion efficiency η of a heat engine.

Both heat engines and refrigerators are subject to restrictions imposed by the law of increase of entropy, but the device design problems are totally different. In particular, the design of refrigerators to operate at the temperature of liquid helium or below is a challenging problem in thermal physics (Chapter 12).

Air Conditioners and Heat Pumps

Air conditioners are refrigerators that cool the inside of a building or an automobile; the heat is ejected to the outside environment. If we interchange the inside and outside connections, an air conditioner can be used to heat a building during the winter. Such a device is called a **heat pump**. If $\tau_h - \tau_l \ll \tau_h$ a heat pump can heat the building with a lower consumption of energy than by direct heating (Problem 1).

The limitations on the use of heat pumps are largely economical. They are much more costly to install and to maintain than are simple heaters or furnaces. Heat pumps make economic sense primarily in climatic conditions in which air conditioning is required anyway.

Carnot Cycle

The derivation of the Carnot energy conversion efficiency and of the Carnot coefficient of refrigerator performance made no statement about how to realize a process by which work is generated from heat, or about how refrigeration is achieved. The simplest and best known such process is the **Carnot cycle**. In the Carnot cycle a gas—or another working substance—is expanded and compressed in four stages, two isothermal and two isentropic, as in Figure 8.5. At point 1 the gas has the temperature τ_h and the entropy σ_L. The gas is expanded at constant τ until the entropy has increased to the value σ_H, at point 2. In the second stage the gas is further expanded, now at constant σ, until the temperature has dropped to the value τ_l, at point 3. The gas is compressed isothermally to point 4 and then compressed isentropically to the original state 1. We write σ_L and σ_H for the low and high values of the entropy *contained in the working substance*, to distinguish these values from σ_l and σ_h, which are the entropy *flows* per cycle at the low and high temperatures τ_l and τ_h. For the Carnot cycle, $\sigma_l = \sigma_h = \sigma_H - \sigma_L$.

The work done by the system in one cycle is the area of the rectangle in Figure 8.5:

$$W = (\tau_h - \tau_l)(\sigma_H - \sigma_L) , \tag{18}$$

which follows from

$$\oint dU = 0 = \oint \tau d\sigma - \oint p dV ,$$

where $\oint p dV$ is the work done by the system in one cycle. The heat taken up at $\tau = \tau_h$ during the first phase is

$$Q_h = \tau_h(\sigma_H - \sigma_L). \tag{19}$$

We combine (18) and (19) to obtain the Carnot efficiency η_C. Any process described by Figure 8.5 is called a Carnot cycle, regardless of the working substance.

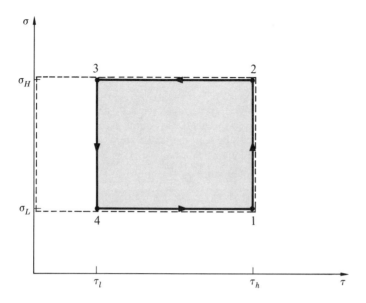

Figure 8.5   A Carnot cycle, for the conversion of heat into work, illustrated as a plot of entropy versus temperature, for an arbitrary working substance. The cycle consists of two expansion phases (1 → 2 and 2 → 3) and two compression phases (3 → 4 and 4 → 1). One of the expansion and one of the compression phases are isothermal (1 → 2 and 3 → 4), and one phase of each kind is isentropic (2 → 3 and 4 → 1). The net work done is the area of the loop. The heat consumed at τ_h is the area surrounded by the broken line.

The Carnot cycle is a point of reference to indicate what could in principle be done, rather than what in fact is done. All energy conversion cycles need a high temperature input and a low temperature output of heat, but often the heat inputs and outputs are not well-defined reservoirs at constant temperatures. Even where such reservoirs exist, as in steam turbines, there is invariably a temperature difference present between the working substance and the reservoirs. The heating and cooling processes are never truly reversible.

Example: Carnot cycle for an ideal gas.   We carry an ideal monatomic gas through a Carnot cycle. Initially the gas occupies a volume V_1 and is in thermal equilibrium with a reservoir $\mathcal{R}_h$ at the high temperature τ_h. The gas is expanded isothermally to the volume

V_2, as in Figure 8.6a. In the process the gas absorbs the heat Q_h from $\mathfrak{R}_h$ and delivers it as work W_{12} to an external mechanical system connected to the piston. For an ideal gas the heat absorbed from the reservoir is

$$Q_h = W_{12} = \int p\,dV = N\tau_h \int dV/V = N\tau_h \log(V_2/V_1). \qquad (20)$$

This work is indicated by the area labeled "12." Next, the gas is disconnected from $\mathfrak{R}_h$ and further expanded, now isentropically, until the temperature has dropped to the low temperature τ_l. In the process the additional work

$$W_{23} = U(\tau_h) - U(\tau_l) = \tfrac{3}{2}N(\tau_h - \tau_l) \qquad (21)$$

is delivered by the gas. The volume V_3 at the end of the isentropic expansion is related to V_2 by

$$\tau_l V_3{}^{2/3} = \tau_h V_2{}^{2/3}, \quad \text{or} \quad V_3/V_2 = (\tau_h/\tau_l)^{3/2}, \qquad (22)$$

from (6.63). After point 3 the gas is brought into contact with a temperature reservoir $\mathfrak{R}_l$ at the temperature τ_l, and then compressed isothermally (Figure 8.6b) to the volume V_4 chosen to satisfy

$$V_4/V_1 = (\tau_h/\tau_l)^{3/2} = V_3/V_2, \qquad (23)$$

so that $V_3/V_4 = V_2/V_1$. To accomplish this compression, the work

$$W_{34} = N\tau_l \log(V_3/V_4) = N\tau_l \log(V_2/V_1) \qquad (24)$$

must be done on the gas. This work is ejected to $\mathfrak{R}_l$ as heat:

$$Q_l = W_{34}. \qquad (25)$$

Finally, the gas is disconnected from $\mathfrak{R}_l$ and recompressed isentropically until its temperature has risen to the initial temperature τ_h. Because of the choice (23) of V_4, the gas volume at this point has returned to its initial value V_1, and the cycle is completed. In this last stage the work

$$W_{41} = \tfrac{3}{2}N(\tau_h - \tau_l) \qquad (26)$$

is performed on the gas; this cancels the work W_{23} done by the gas during the isentropic expansion $2 \to 3$, by (21).

The net work delivered by the gas during the cycle is given by the difference in shaded areas in Figures 8.6a and 8.6b, which is the enclosed area in Figure 8.6c. The isentropic curves in the p–V diagram are steeper than the isothermal curves, so that the area of the

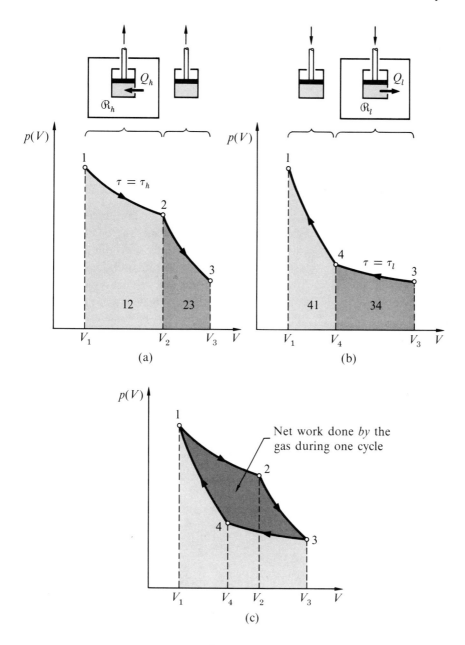

Figure 8.6 The Carnot cycle for an ideal gas, as a *p-V* plot. An ideal gas is expanded and recompressed in four stages. Two of them are isothermal, at the temperature τ_h and $\tau_l(\tau_h > \tau_l)$. Two of them are isentropic, from τ_h to τ_l, and back. The shaded areas show (a) the work done during the two expansion stages, (b) the work done during the two compression stages, and (c) the net work done during the cycle.

loop is finite and is equal to the area of the rectangle in Figure 8.5. We have

$$W = W_{12} + W_{23} - W_{34} - W_{41} = W_{12} - W_{34}$$

$$= N(\tau_h - \tau_l)\log(V_2/V_1). \tag{27}$$

The heat absorbed from $\mathcal{R}_h$ was given in (20), so that $W/Q_h = (\tau_h - \tau_l)/\tau_h$, which is just the Carnot relation (7).

Energy Conversion and the Second Law of Thermodynamics

The Carnot limits on the conversion of heat into work and on the performance of refrigerators are direct consequences of the law of increase of entropy. The second law of thermodynamics usually is formulated without mention of entropy. We stated the classical Kelvin-Planck formulation in Chapter 2: "It is impossible for any cyclic process to occur whose sole effect is the extraction of heat from a reservoir and the performance of an equivalent amount of work."

All reversible energy conversion devices that operate between the same temperatures have the same energy conversion efficiency $\eta = W/Q_h$. Were this not so, we could combine two reversible devices with different efficiencies, $\eta_1 < \eta_2$, in such a way (Figure 8.7) that device 1 with the lower efficiency is operated in reverse as a refrigerator that moves not only the entire waste heat Q_{12} from the more efficient device 2 back to the higher temperature τ_h, but an additional amount $Q(\text{in})$ of heat as well. The overall result would be the conversion of the heat $Q(\text{in})$ to work $W(\text{out})$, without any net waste heat. This would require the annihilation of entropy and would violate the law of increase of entropy.

Now that we have established that all reversible devices that operate between the same temperatures have the same energy conversion efficiency, it is sufficient to calculate this efficiency for any particular device to find the common value. The Carnot cycle device leads to $\eta_C = (\tau_h - \tau_l)/\tau_h$ for the common value.

Path Dependence of Heat and Work

We have carefully used the words heat and work to characterize energy transfer processes, and not to characterize properties of the system itself. It is not meaningful to speak of the heat content or of the work content of a system. We look at the Carnot cycle once more: Around a closed loop in the p–V plane, a net amount of work is generated by the system, and a net amount of heat is consumed. But the system—on being taken once around the loop—is returned to precisely the initial condition; no property of the system has changed. This means that there cannot exist two functions $Q(\sigma,V)$ and $W(\sigma,V)$

$$Q_h = W_1 + Q_{l2} + Q(\text{in})$$
$$Q_h = W_1 + Q_{l2} + W(\text{out})$$
$$\overline{0 = Q(\text{in}) - W(\text{out})}$$

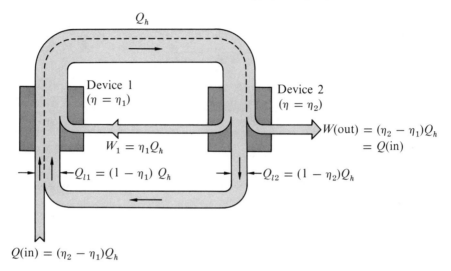

Q_h

Device 1
$(\eta = \eta_1)$

Device 2
$(\eta = \eta_2)$

$W(\text{out}) = (\eta_2 - \eta_1)Q_h$
$= Q(\text{in})$

$W_1 = \eta_1 Q_h$

$-Q_{l1} = (1 - \eta_1)\,Q_h$

$-Q_{l2} = (1 - \eta_2)Q_h$

$Q(\text{in}) = (\eta_2 - \eta_1)Q_h$

Figure 8.7 If two different reversible energy conversion devices operating between the same temperatures τ_h and τ_l could have different energy conversion efficiencies ($\eta_2 > \eta_1$), it would be possible to combine them into a single device with 100 pct efficiency by using the less efficient device 1 as a refrigerator that moves not only the entire waste heat Q_{12} of the more efficient device 2 back to the higher temperature, but an additional amount $Q(\text{in})$ of heat as well. This additional heat would then be completely converted to work.

such that the heat Q_{ab} and the work W_{ab} required to carry the system from a state (σ_a, V_a) to a state (σ_b, V_b) are given by the differences in Q and W:

$$Q_{ab} \overset{?}{=} Q(\sigma_b, V_b) - Q(\sigma_a, V_a); \qquad W_{ab} \overset{?}{=} W(\sigma_b, V_b) - W(\sigma_a, V_a).$$

If such functions existed, the net transfers of heat and of work around a closed loop necessarily would be zero, and we have shown that the transfers are not zero.

The transfers of heat and work between state (a) and state (b) depend on the path taken between the two states. This path-dependence is expressed when we say that heat and work are not state functions. Unlike temperature, entropy, and free energy, heat and work are not intrinsic attributes of the system. The increments dQ and dW that we introduced in (1) and (2) cannot be differentials

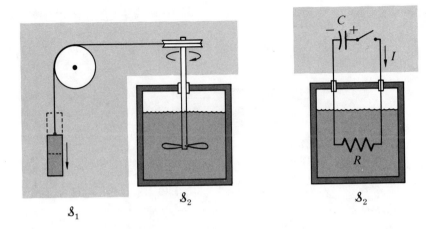

Figure 8.8   Two irreversible processes in which mechanical or electrical potential energy is used to heat a system.

of mathematical functions $Q(\sigma, V)$ and $W(\sigma, V)$. For this reason we designated the increments by dQ and dW, rather than by dQ and dW. Without the path dependence of heat and work there would not exist cyclical processes that permit the generation of work from heat.

Irreversible Work

We consider the energy transfer processes of Figure 8.8. In each process $\mathcal{S}_1$ is a purely mechanical or electrical system that delivers pure work with zero entropy change. The energy transferred to $\mathcal{S}_2$ is converted to heat, either by mechanical friction or by electrical resistance. The final state of $\mathcal{S}_2$ is the same as if the energy had been added as heat in the first place. The entropy of $\mathcal{S}_2$ is increased by $d\sigma_2 = dU_2/\tau$. This entropy is newly created entropy. Processes in which new entropy is created are irreversible because there is no way to reverse the process in order to destroy the newly created entropy. If newly created entropy arises by the conversion of work to heat, we say that irreversible work has been performed.

   If we look only at the net change in a system, there is no way to tell whether the process that led to this change was reversible or irreversible. For a change dU in energy and $d\sigma$ in entropy, we can define a reversible heat dQ_{rev} and a reversible work dW_{rev} as the amount of heat and work that would accomplish this change in a reversible process. If part of the work done on the system is irreversible, the actual work required to accomplish a given change is larger than the reversible work,

$$\partial W_{\text{irrev}} > \partial W_{\text{rev}}. \tag{28}$$

By conservation of energy

$$dU = \partial W_{\text{irrev}} + \partial Q_{\text{irrev}} = \partial W_{\text{rev}} + \partial Q_{\text{rev}}\ ,$$

so that

$$\partial Q_{\text{irrev}} < \partial Q_{\text{rev}}. \tag{29}$$

The actual heat transferred in the irreversible process must be less than the reversible heat.

Example: Sudden expansion of an ideal gas. As an example of an irreversible process we consider once more the sudden expansion of an ideal gas into a vacuum. Neither heat nor work is transferred, so that $dU = 0$ and $d\tau = 0$. The final state is identical with the state that results from a reversible isothermal expansion with the gas in thermal equilibrium with a reservoir. The work W_{rev} done on the gas in the reversible expansion from volume V_1 to V_2 is, from (6.57),

$$W_{\text{rev}} = -N\tau \log(V_2/V_1). \tag{30}$$

The work done on the gas is negative; the gas does positive work on the piston in an amount equal to the heat transfer into the system:

$$Q_{\text{rev}} = -W_{\text{rev}} > 0; \qquad W_{\text{rev}} < 0. \tag{31}$$

The entropy change is equal to Q_{rev}/τ, or

$$\sigma_2 - \sigma_1 = -W_{\text{rev}}/\tau = N\log(V_2/V_1). \tag{32}$$

In the irreversible process of expansion into the vacuum this entropy is newly created entropy because neither heat nor work flows into the system from the outside: $W_{\text{irrev}} = Q_{\text{irrev}} = 0$. From (31) we obtain

$$W_{\text{irrev}} > W_{\text{rev}}, \qquad Q_{\text{irrev}} < Q_{\text{rev}}\ , \tag{33}$$

in agreement with (28) and (29).

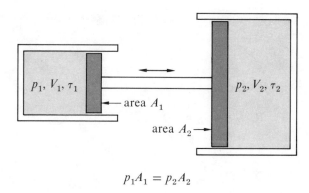

$$p_1 A_1 = p_2 A_2$$

Figure 8.9 Systems between which only work but no heat is transferred need not be at the same temperature for the process to be reversible.

In our discussion of irreversible work we assumed that the new entropy is created inside the system during the delivery of work to the system by other systems. This is not the only source of irreversibility in energy transfer. Pure heat transfer, not involving any work, is irreversible if it takes place between two systems having different temperatures. We worked out an example in Chapter 2. In this process heat is transferred from a system at τ_1 to a system at the lower temperature τ_2. We have

$$dU_1 = dQ_1 = \tau_1 \, d\sigma_1; \qquad dU_2 = dQ_2 = \tau_2 \, d\sigma_2; \qquad dQ_1 + dQ_2 = 0. \quad (34)$$

The newly created entropy is

$$d\sigma_{\text{irr}} = (d\sigma_1 + d\sigma_2) = dQ_1/\tau_1 + dQ_2/\tau_2$$

$$= (1/\tau_1 - 1/\tau_2)dQ_1 = \frac{\tau_2 - \tau_1}{\tau_1 \tau_2} dQ_1. \quad (35)$$

The heat flow is from high to low temperature: dQ_1 is negative; $\tau_2 - \tau_1$ is negative, so that $d\sigma_{\text{irr}} > 0$.

The energy transfer between two systems with different temperatures need not be irreversible if only work but no heat is transferred (Figure 8.9).

All actual energy transfer processes are invariably somewhat irreversible, but reversible processes remain the backbone of the theory of thermal physics. They constitute a natural limit, which is the equilibrium limit of vanishing entropy generation. We shall assume hereafter that the words heat and work, without a further qualifier, refer to reversible processes.

HEAT AND WORK AT CONSTANT TEMPERATURE OR CONSTANT PRESSURE

Isothermal work. We show that the total work performed on a system in a reversible isothermal process is equal to the increase in the Helmholtz free energy $F = U - \tau\sigma$ of the system. For a reversible process $dQ = \tau d\sigma = d(\tau\sigma)$, because $d\tau = 0$, so that

$$dW = dU - dQ = dU - d(\tau\sigma) = dF. \tag{36}$$

Thus in such processes the Helmholtz free energy is the natural energetic function, more appropriate than the energy U. When we treat an isothermal process in terms of the Helmholtz free energy, we automatically include the additional work that is required to make up for the heat transfer from the system to the reservoir. Often the heat transfer is the major part of the work: for the ideal gas the energy U does not change in an isothermal process, and the work done is equal to the heat transfer.

Isobaric heat and work. Many energy transfer processes—isothermal or not—take place at constant pressure, particularly those processes that take place in systems open to the atmosphere. A process at constant pressure is said to be an **isobaric process**. A simple example is the boiling of a liquid as in Figure 8.10,

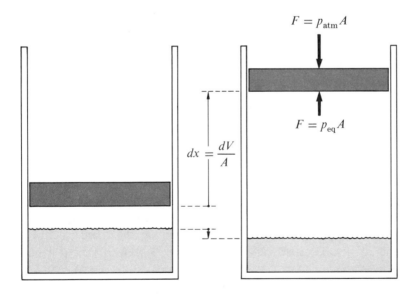

Figure 8.10 When a liquid boils under atmospheric pressure, the vapor displacing the atmosphere does work against the atmospheric pressure.

where the pressure on the piston is the external atmospheric pressure. If the system changes its volume by dV, the work $-pdV = -d(pV)$ is part of the total work done on the system. If positive, this work is provided by the environment and is in this sense "free." If negative, the work is delivered to the environment and is not extractable from the system for other purposes. For this reason it is often appropriate to subtract $-d(pV)$ from the total work. We thus obtain the **effective work** performed on the system, defined as

$$\sdW' \equiv \sdW + d(pV) = dU + d(pV) - \sdQ = dH - \sdQ \; , \qquad (37)$$

where we have defined a new function

$$H = U + pV \; , \qquad (38)$$

called the **enthalpy** which plays the role in processes at constant pressure that the energy U plays in processes at constant volume. The term pV in (38) is the work required to displace the surrounding atmosphere in order to vacate the space to be occupied by the system. Implicit in these definitions is the idea that there are other kinds of work besides that due to volume changes.

Two classes of the constant pressure processes are particularly important:

(a) Processes in which no effective work is done. The heat transfer is $\sdQ = dH$, from (37). The evaporation of a liquid (Chapter 10) from an open vessel is such a process, because no effective work is done. The heat of vaporization is the enthalpy difference between the vapor phase and the liquid phase.

(b) Processes at constant temperature and constant pressure. Then $\sdQ = \tau d\sigma = d(\tau\sigma)$, and the effective work performed on the system is, from (36) and (37),

$$dW' = dF + d(pV) = dG \; , \qquad (39)$$

where we have defined another new function

$$G = F + pV = U + pV - \tau\sigma \; , \qquad (40)$$

the **Gibbs free energy**. The effective work performed in a reversible process at constant temperature and pressure is equal to the change in the Gibbs free energy of the system. This is particularly useful in chemical reactions where the volume changes as the reaction proceeds at a constant pressure. The Gibbs free energy is used extensively in Chapter 9, and the enthalpy is used in Chapter 10.

Example: Electrolysis and fuel cells.   Electrolysis is a process that is both isothermal and isobaric. Consider an electrolyte of dilute sulfuric acid in which are immersed platinum electrodes that do not react with the acid (Figure 8.11). The sulfuric acid dissociates into H^+ and SO_4^{--} ions:

$$H_2SO_4 \rightleftharpoons 2H^+ + SO_4^{--}. \tag{41}$$

When a current is passed through the cell the hydrogen ions move to the negative electrode where they take up electrons and form molecular hydrogen gas:

$$2H^+ + 2e^- \rightarrow H_2. \tag{42}$$

The sulfate ions move to the positive electrodes where they decompose water with the release of molecular oxygen gas and electrons:

$$SO_4^{--} + H_2O \rightarrow H_2SO_4 + \tfrac{1}{2}O_2 + 2e^-. \tag{43}$$

The sum of the above three steps is the net reaction equation in the cell:

$$H_2O \rightarrow H_2 + \tfrac{1}{2}O_2. \tag{44}$$

When carried out slowly in a vessel open to the atmosphere, the process is at constant pressure and constant temperature. A negligible part of the electrical input power goes into

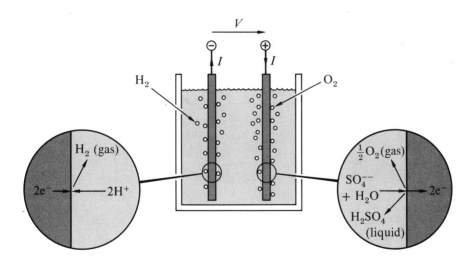

Figure 8.11   An electrolysis cell. An electrical current passes through an electrolyte, such as dilute sulfuric acid. The overall result is the decomposition of water into gaseous hydrogen and oxygen. The process is an example of work being done at constant temperature and constant pressure.

resistance heating of the electrolyte. The effective work required to decompose 1 mole of water is related to the molar Gibbs free energies of the reactants:

$$W' = \Delta G = G(H_2O) - G(H_2) - \tfrac{1}{2}G(O_2). \tag{45}$$

Chemical tables list the Gibbs free energy difference ΔG as $-237\,kJ$ per mole at room temperature.

In electrolysis this work is performed by a current I that flows under an external voltage V_0. If t is the time required to decompose one mole of water, $Q = I \times t$ is the total charge (not the heat!) flowing through the cell, and we have

$$W' = QV_0. \tag{46}$$

According to (43), there are two electrons involved in decomposing one water molecule, hence

$$Q = -2N_A e = -1.93 \times 10^5\,\text{coulomb}. \tag{47}$$

We equate (46) to (45) to obtain the condition for electrolysis to take place. This requires a minimum voltage

$$V_0 = -\Delta G/2N_A e , \tag{48}$$

or 1.229 volts. A voltage larger than V_0 must be applied to obtain a finite current flow, because V_0 alone merely reduces to zero the potential barrier between the systems on the two sides of the reaction equation (44). When $V > V_0$, the excess power $(V - V_0) \times I$ will be dissipated as heat in the electrolyte.

If $V < V_0$, the reaction (44) will proceed from right to left provided gaseous hydrogen is available at the positive electrode and gaseous oxygen at the negative electrode. In the simple setup of Figure 8.11 the gases are permitted to escape, and for $V < V_0$ nothing will happen at all. It is possible, however, to construct the electrodes as porous sponges, with hydrogen and oxygen forced through under pressure (Figure 8.12). Such a device produces a voltage V_0 between the electrodes and, if the electrodes are connected, external current will flow. This arrangement is called a hydrogen-oxygen **fuel cell**. Fuel cells were used as power sources on board the Gemini and Apollo* spacecraft and incidentally produced drinking water for the astronauts.

The principal technological limitation of fuel cells is their low current per unit electrode area. In the Apollo cell the current density was only a few hundred mA/cm^2; hence large electrode areas are required to produce reasonable currents. The current-voltage characteristic of an electrochemical cell in its two operating ranges as fuel cell and as electrolytic cell are shown in Figure 8.13.

---

* The Apollo fuel cells used Ni and NiO rather than Pt as electrodes, and KOH rather than H_2SO_4 as electrolyte. For a detailed description, and more information on fuel cells, the reader is referred to J. O. M. Bockris and S. Srinivasan, *Fuel cells: Their electrochemistry*, McGraw-Hill, New York, 1969.

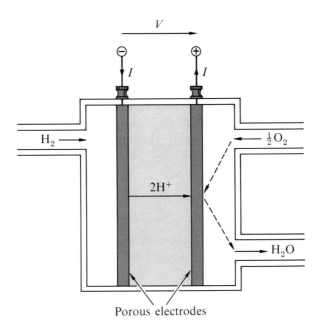

Figure 8.12 A fuel cell is an electrolysis cell operated in reverse, with hydrogen and oxygen supplied as fuels. The fuels are forced under pressure through porous electrodes separated by an electrolyte. The hydrogen and oxygen react to form water; the excess Gibbs free energy is delivered outside as electrical energy. Water forms at the positive electrode and is removed there.

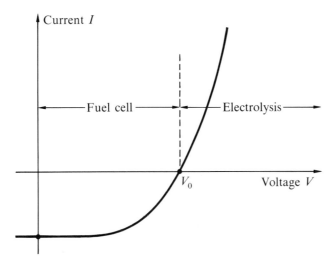

Figure 8.13 The current-voltage characteristic of an electrolytic cell or fuel cell, indicating the two operating ranges.

Chemical Work

Work performed by the transfer of particles to a system is called chemical work, because it is associated with the chemical potential.

When particles are transferred, the number of particles in the system is one of the independent variables on which the energy U depends. If $U = U(\sigma, V, N)$, then for a reversible process

$$dU = \tau d\sigma - p dV + \mu dN \, , \tag{49}$$

by the thermodynamic identity of Chapter 5. Here we have replaced the partial derivatives by their familiar equivalents (Table 5.1). By our definition of heat, the $\tau d\sigma$ term represents the transfer of heat and the $-p dV$ and μdN terms represent the performance of work, all understood to be reversible:

$$đW = -p dV + \mu dN. \tag{50}$$

The $-p dV$ term is mechanical work; the μdN term is the **chemical work**:

$$đW_c = \mu dN. \tag{51}$$

If there is no volume change, $dV = 0$. All the work is chemical.

In particle transfer there are usually two systems involved, both in contact with a heat reservoir, and the total chemical work is the sum of the contributions from both systems. In the arrangement of Figure 8.14 a pump transfers particles from system $\mathcal{S}_1$ to system $\mathcal{S}_2$. The chemical potentials are μ_1 and μ_2. If $dN = dN_2 = -dN_1$ is the number of particles transferred, the total chemical work performed is

$$đW_c = đW_{c1} + đW_{c2} = \mu_1 dN_1 + \mu_2 dN_2 = (\mu_2 - \mu_1)dN. \tag{52}$$

The work that must be supplied to the pump is $đW_c$ if there is no volume work ($dV_1 = dV_2 = 0$), and if all processes are reversible.

The result (52) gives an additional meaning of the chemical potential. We summarize the properties of the chemical potential:

(a) The chemical potential of a system is the work required to transfer one particle into the system, from a reservoir at zero chemical potential.

(b) The difference in chemical potential between two systems is equal to the net work required to move a particle from one system to the other.

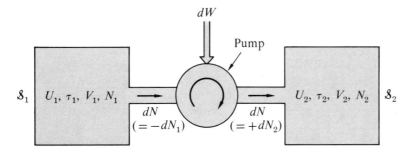

Figure 8.14   Chemical work is the work performed when particles are moved reversibly from one system to another, with the two systems having different chemical potentials. If the two volumes do not change, the work is pure chemical work; the amount per particle is the difference in chemical potentials.

(c)   If the two systems are in diffusive equilibrium they have the same chemical potential; no work is required to move a particle from one system to the other.

(d)   The difference in internal chemical potential (Chapter 5) between two systems is equal but opposite to the potential barrier that maintains the systems in diffusive equilibrium.

Example: Chemical work for an ideal gas.   We consider the work per particle required to move reversibly the atoms of a monatomic ideal gas from $\mathcal{S}_1$ with concentration n_1, to $\mathcal{S}_2$ with concentration $n_2 > n_1$, both systems being at the same temperature (Figure 8.15). If $dV = 0$, the work contains only a chemical work term, which can be calculated from the difference in chemical potential, no matter how the process is actually performed. The chemical potential difference between two ideal gas systems with different concentrations is

$$\mu_2 - \mu_1 = \tau[\log(n_2/n_Q) - \log(n_1/n_Q)] = \tau \log(n_2/n_1). \tag{53}$$

This result is equal to the mechanical work per particle required to compress the gas isothermally from the concentration n_1 to the concentration n_2. The work required to compress N particles of an ideal gas from an initial volume V_1 to a final volume V_2 is

$$W = -\int p\,dV = -N\tau \int dV/V = N\tau \log(V_1/V_2) = N\tau \log(n_2/n_1). \tag{54}$$

Hence the mechanical work per particle is $\tau \log(n_2/n_1)$, identical to the result (53). The

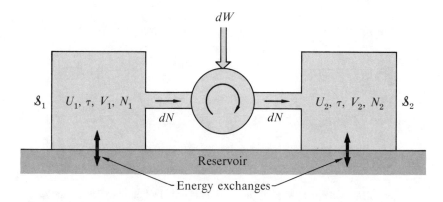

Figure 8.15   Isothermal chemical work. The amount of chemical work per particle does not change if the process is performed isothermally with both systems in thermal equilibrium with a common large reservoir.

identity of the chemical work with the isothermal compression work illustrates the equilence or convertibility of different kinds of work.

Magnetic Work and Superconductors

An important form of work is magnetic work. The most important application of magnetic work is to superconductors, and this application is treated here.

Below some critical temperature T_c that is usually less than 20 K, many electrical conductors undergo a transition from their normal state with a finite electrical conductivity to a superconducting state with an apparently infinite conductivity.

Superconductors expel magnetic fields from their interior. If the superconductor is first cooled below the critical temperature and then inserted into a magnetic field, we might expect that the infinite conductivity would shield the interior from the penetration by a magnetic field. However, the expulsion occurs even if the superconductor is cooled below T_c while in a magnetic field (Figure 8.16). This active expulsion, called the **Meissner effect**, shows that superconductivity is more than an infinite conductivity. The Meissner effect is caused by shielding currents that are spontaneously generated near the surface, in a layer about 10^{-5} cm thick. The magnetic field expulsion is not always complete. Superconductors are said to be of type II if the expulsion is incomplete, but still nonzero, in a range of fields above some low field. We shall restrict ourselves

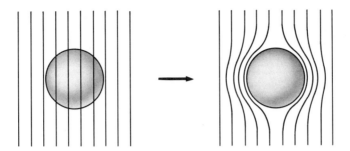

Figure 8.16 Meissner effect in a superconducting sphere cooled in a constant applied magnetic field; on passing below the transition temperature the lines of induction B are ejected from the sphere.

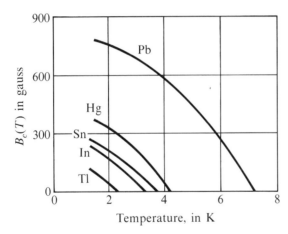

Figure 8.17 Threshold curves of the critical field versus temperature for several super-conductors. A specimen is superconducting below the curve and normal above the curve.

here to the discussion of type I superconductors, for which the field expulsion is complete up to a definite field and zero thereafter.

A sufficiently strong magnetic field will destroy superconductivity. The critical field required to do this depends on the temperature and on the super-conductor. For type I superconductors the fields are usually a few hundred gauss (Figure 8.17). In some niobium and vanadium compounds of type II, critical fields of several hundred kilogauss have been observed.

The Meissner magnetic effect shows that the normal and the superconducting states are different thermodynamic phases of the same metal, just as ice and liquid water are different phases of H_2O, except that in the superconducting

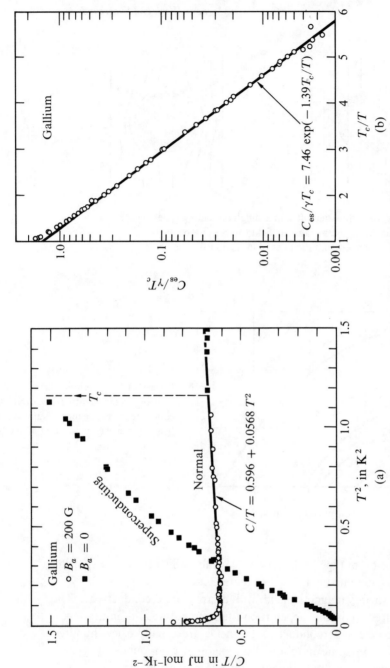

Figure 8.18 (a) The heat capacity of gallium in the normal and superconducting states. The normal state is restored by a 200 G field. In (b) the electronic part C_{es} of the heat capacity in the superconducting state is plotted on a log scale versus T_c/T: The exponential dependence on $1/T$ is evident. Here $\gamma = 0.60$ mJ mol^{-1} K^{-2}. After N. E. Phillips, Phys. Rev. **134**, 385 (1964).

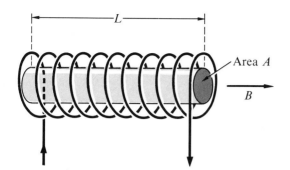

Figure 8.19 A superconductor of length L and area A in a superconducting solenoid that produces a magnetic field B.

transition it is the electronic system rather than the crystal structure of the metal that undergoes a phase transition.

The superconducting state is a distinct thermodynamic phase, as confirmed by differences in the heat capacity of the normal and the superconducting states. The heat capacity (Figure 8.18) exhibits a pronounced discontinuity at the onset of superconductivity at $\tau = \tau_c$; when superconductivity is destroyed by a magnetic field, the discontinuity disappears. The stable phase will be the phase with the lower free energy. Below $\tau = \tau_c$ in zero magnetic field the free energy of the superconducting phase is lower than that of the normal phase. The free energy of the superconducting phase increases in the magnetic field, as we show below. The free energy of the normal phase is approximately independent of the field. Eventually, as the field is increased, the free energy of the super-conducting phase will exceed that of the normal phase. The normal phase is then the stable phase, and superconductivity is destroyed.

The increase of the free energy of a superconductor in a magnetic field is calculated as the work required to reduce the magnetic field to zero in the interior of the superconductor; the zero value is required to account for the Meissner effect. Consider a superconductor in the form of a long rod of uniform cross-section inside a long solenoid that produces a uniform field B, as in Figure 8.19. The work required to reduce the field to zero inside the superconductor is equal to the work required to create within the superconductor a counteracting field $-B$ that exactly cancels the solenoid field. We know from electromagnetic theory that the work per unit volume required to create a field B is given by

(SI) $$W_{\text{mag}}/V = B^2/2\mu_0; \tag{55a}$$

or

(CGS) $$W_{\text{mag}}/V = B^2/8\pi. \tag{55b}$$

Figure 8.20   The free energy density F_N of a nonmagnetic normal metal is approximately independent of the intensity of the applied magnetic field B_a. At a temperature $\tau < \tau_c$ the metal is a superconductor in zero magnetic field, so that $F_S(\tau,0)$ is lower than $F_N(\tau,0)$. An applied magnetic field increases F_S by $B_a{}^2/2\mu_0$, in SI units (and by $B_a{}^2/8\pi$ in CGS units), so that $F_S(\tau,B_a) = F_S(\tau,0) + B_a{}^2/2\mu_0$. If B_a is larger than the critical field B_{ac} the free energy density is lower in the normal state than in the superconducting state, and now the normal state is the stable state. The origin of the vertical scale in the drawing is at $F_S(\tau,0)$. The figure equally applies to U_S and U_N at $\tau = 0$.

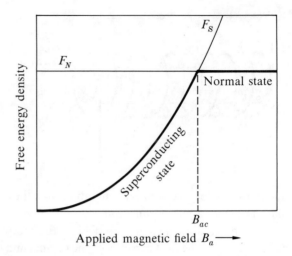

This is the amount by which the free energy density in the bulk superconductor is raised by application of an external magnetic field, in an experiment at constant temperature.

There is no comparable free energy increase for the normal conductor, because there is no screening of the applied field. Thus

$$(SI) \qquad\qquad [F_N(\tau) - F_S(\tau)]/V = B_c{}^2(\tau)/2\mu_0; \qquad\qquad (56a)$$

or

$$(CGS) \qquad\qquad [F_N(\tau) - F_S(\tau)]/V = B_c{}^2(\tau)/8\pi. \qquad\qquad (56b)$$

In a plot of the free energy density of both phases versus the magnetic field (Figure 8.20), the free energy of the superconducting phase will ultimately rise above that of the normal phase, so that in high fields the specimen will be in the normal phase, and the superconducting phase is no longer the stable phase. This is the explanation of the destruction of superconductivity by a critical magnetic field B_c.

With increasing temperature the free energy difference between normal and superconducting phase decreases as $\tau \to \tau_c$, and the critical magnetic field decreases. Everything else being equal, a high stabilization energy in a type I superconductor will lead to both a high critical temperature and a high critical field. The highest critical fields are found amongst the superconductors with the highest critical temperatures, and vice versa.

SUMMARY

1. Heat is the transfer of energy by thermal contact with a reservoir. In a reversible process $dQ = \tau d\sigma$.

2. Work is the transfer of energy by a change in the external parameters that describe the system. The entropy transfer in a reversible process is zero when only work is performed and no heat is transferred.

3. The Carnot energy conversion efficiency, $\eta_C = (\tau_h - \tau_l)/\tau_h$, is the upper limit to the ratio W/Q_h of the work generated to the heat added.

4. The Carnot coefficient of refrigerator performance, $\gamma_C = \tau_l/(\tau_h - \tau_l)$, is the upper limit to the ratio Q_l/W of the heat extracted to the work consumed.

5. The total work performed on a system at constant temperature in a reversible process is equal to the change in the Helmholtz free energy $F \equiv U - \tau\sigma$ of the system.

6. The effective work performed on a system at constant temperature and pressure in a reversible process is equal to the change in the Gibbs free energy $G \equiv U - \tau\sigma + pV$.

7. The chemical work performed on a system in the reversible transfer of dN particles to the system is μdN.

8. The change in the free energy density of a superconductor (of type I) caused by an external magnetic field B is $B^2/2\mu_0$ in SI and $B^2/8\pi$ in CGS.

PROBLEMS

1. Heat pump. (a) Show that for a reversible heat pump the energy required per unit of heat delivered inside the building is given by the Carnot efficiency (6):

$$\frac{W}{Q_h} = \eta_C = \frac{\tau_h - \tau_l}{\tau_h}.$$

What happens if the heat pump is not reversible? (b) Assume that the electricity consumed by a reversible heat pump must itself be generated by a Carnot engine operating between the temperatures τ_{hh} and τ_l. What is the ratio Q_{hh}/Q_h, of the heat consumed at τ_{hh}, to the heat delivered at τ_h? Give numerical values for $T_{hh} = 600\,\text{K}$; $T_h = 300\,\text{K}$; $T_l = 270\,\text{K}$. (c) Draw an energy-entropy flow diagram for the combination heat engine-heat pump, similar to Figures 8.1, 8.2 and 8.4, but involving no external work at all, only energy and entropy flows at three temperatures.

2. *Absorption refrigerator.* In absorption refrigerators the energy driving the process is supplied not as work, but as heat from a gas flame at a temperature $\tau_{hh} > \tau_h$. Mobile home and cabin refrigerators may be of this type, with propane fuel. (a) Give an energy-entropy flow diagram similar to Figures 8.2 and 8.4 for such a refrigerator, involving no work at all, but with energy and entropy flows at the three temperatures $\tau_{hh} > \tau_h > \tau_l$. (b) Calculate the ratio Q_l/Q_{hh}, for the heat extracted at $\tau = \tau_l$, where Q_{hh} is the heat input at $\tau = \tau_{hh}$. Assume reversible operation.

3. *Photon Carnot engine.* Consider a Carnot engine that uses as the working substance a photon gas. (a) Given τ_h and τ_l as well as V_1 and V_2, determine V_3 and V_4. (b) What is the heat Q_h taken up and the work done by the gas during the first isothermal expansion? Are they equal to each other, as for the ideal gas? (c) Do the two isentropic stages cancel each other, as for the ideal gas? (d) Calculate the total work done by the gas during one cycle. Compare it with the heat taken up at τ_h and show that the energy conversion efficiency is the Carnot efficiency.

4. *Heat engine—refrigerator cascade.* The efficiency of a heat engine is to be improved by lowering the temperature of its low-temperature reservoir to a value τ_r, below the environmental temperature τ_l, by means of a refrigerator. The refrigerator consumes part of the work produced by the heat engine. Assume that both the heat engine and the refrigerator operate reversibly. Calculate the ratio of the net (available) work to the heat Q_h supplied to the heat engine at temperature τ_h. Is it possible to obtain a higher net energy conversion efficiency in this way?

5. *Thermal pollution.* A river with a water temperature $T_l = 20°C$ is to be used as the low temperature reservoir of a large power plant, with a steam temperature of $T_h = 500°C$. If ecological considerations limit the amount of heat that can be dumped into the river to 1500 MW, what is the largest electrical output that the plant can deliver? If improvements in hot-steam technology would permit raising T_h by 100°C, what effect would this have on the plant capacity?

6. *Room air conditioner.* A room air conditioner operates as a Carnot cycle refrigerator between an outside temperature T_h and a room at a lower temperature T_l. The room gains heat from the outdoors at a rate $A(T_h - T_l)$; this heat is removed by the air conditioner. The power supplied to the cooling unit is P. (a) Show that the steady state temperature of the room is

$$T_l = (T_h + P/2A) - [(T_h + P/2A)^2 - T_h^2]^{1/2}.$$

(b) If the outdoors is at 37°C and the room is maintained at 17°C by a cooling power of 2 kW, find the heat loss coefficient A of the room in $W\,K^{-1}$. A good discussion of room air conditioners is given by H. S. Leff and W. D. Teeters, Amer. J. Physics **46**, 19 (1978). In a realistic unit the cooling coils may be at 282 K and the outdoor heat exchanger at 378 K.

7. Light bulb in a refrigerator. A 100 W light bulb is left burning inside a Carnot refrigerator that draws 100 W. Can the refrigerator cool below room temperature?

8. Geothermal energy. A very large mass M of porous hot rock is to be utilized to generate electricity by injecting water and utilizing the resulting hot steam to drive a turbine. As a result of heat extraction, the temperature of the rock drops, according to $dQ_h = -MC\,dT_h$, where C is the specific heat of the rock, assumed to be temperature independent. If the plant operates at the Carnot limit, calculate the total amount W of electrical energy extractable from the rock, if the temperature of the rock was initially $T_h = T_i$, and if the plant is to be shut down when the temperature has dropped to $T_h = T_f$. Assume that the lower reservoir temperature T_l stays constant.

At the end of the calculation, give a numerical value, in kWh, for $M = 10^{14}$ kg (about 30 km^3), $C = 1$ J g^{-1} K^{-1}, $T_i = 600°$C, $T_f = 110°$C, $T_l = 20°$C. Watch the units and explain all steps! For comparison: The total electricity produced in the world in 1976 was between 1 and 2 times 10^{14} kWh.

9. Cooling of nonmetallic solid to $T = 0$. We saw in Chapter 4 that the heat capacity of nonmetallic solids at sufficiently low temperatures is proportional to T^3, as $C' = aT^3$. Assume it were possible to cool a piece of such a solid to $T = 0$ by means of a reversible refrigerator that uses the solid specimen as its (varying!) low-temperature reservoir, and for which the high-temperature reservoir has a fixed temperature T_h equal to the initial temperature T_i of the solid. Find an expression for the electrical energy required.

10. Irreversible expansion of a Fermi gas. Consider a gas of N noninteracting, spin $\frac{1}{2}$ fermions of mass M, initially in a volume V_i at temperature $\tau_i = 0$. Let the gas expand irreversibly into a vacuum, without doing work, to a final volume V_f. What is the temperature of the gas after expansion if V_f is sufficiently large for the classical limit to apply? Estimate the factor by which the gas should be expanded for its temperature to settle to a constant final value. Give numerical values for the final temperature in kelvin for two cases: (a) a particle mass equal to the electron mass, and $N/V = 10^{22}$ cm^{-3}, as in metals; (b) a particle mass equal to a nucleon, and $N/V = 10^{30}$, as in white dwarf stars.

Chapter 9

Gibbs Free Energy
and Chemical Reactions

GIBBS FREE ENERGY

The Helmholtz free energy F introduced in Chapter 3 describes a system at constant volume and temperature. But many experiments, and in particular many chemical reactions, are performed at constant pressure, often one atmosphere. It is useful to introduce another function to treat the equilibrium configuration at constant pressure and temperature. As in Chapter 8, we define the **Gibbs free energy** G as

$$G \equiv U - \tau\sigma + pV. \tag{1}$$

Chemists often call this the free energy, and physicists often call it the thermodynamic potential.

The most important property of the Gibbs free energy is that it is a minimum for a system $\mathscr{S}$ in equilibrium at constant pressure when in thermal contact with a reservoir $\mathscr{R}$. The differential of G is

$$dG = dU - \tau d\sigma - \sigma d\tau + pdV + Vdp.$$

Consider a system (Figure 9.1) in thermal contact with a heat reservoir $\mathscr{R}_1$ at temperature τ and in mechanical contact with a pressure reservoir $\mathscr{R}_2$ that maintains the pressure p, but cannot exchange heat. Now $d\tau = 0$ and $dp = 0$, so that the differential dG of the system in the equilibrium configuration becomes

$$dG_{\mathscr{S}} = dU_{\mathscr{S}} - \tau d\sigma_{\mathscr{S}} + pdV_{\mathscr{S}}. \tag{2}$$

The thermodynamic identity (5.39) is

$$\tau d\sigma_{\mathscr{S}} = dU_{\mathscr{S}} - \mu dN_{\mathscr{S}} + pdV_{\mathscr{S}}, \tag{3}$$

so that (2) becomes $dG_{\mathscr{S}} = \mu dN_{\mathscr{S}}$. But $dN_{\mathscr{S}} = 0$, whence

$$dG_{\mathscr{S}} = 0, \tag{4}$$

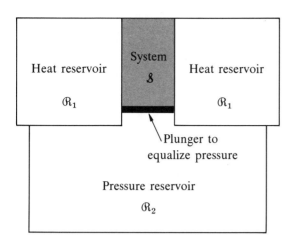

Figure 9.1 A system in thermal equilibrium with a heat reservoir and in mechanical equilibrium with a barystat or pressure reservoir which maintains a constant pressure on the system. The barystat is thermally insulated.

which is the condition for G_δ to be an extremum with respect to system variations at constant pressure, temperature, and particle number. These are, therefore, the natural variables for $G(N,\tau,p)$.

That the extremum of G_δ must be a minimum, rather than a maximum, follows directly from the minus sign associated with the entropy in (1): Any irreversible change taking place entirely within δ will increase σ and thus decrease G_δ.

With (2),

$$dG = \mu dN - \sigma d\tau + V dp. \tag{5}$$

The differential (5) may be written as

$$dG = \left(\frac{\partial G}{\partial N}\right)_{\tau,p} dN + \left(\frac{\partial G}{\partial \tau}\right)_{N,p} d\tau + \left(\frac{\partial G}{\partial p}\right)_{N,\tau} dp. \tag{6}$$

Comparison of (5) and (6) gives the relations

$$(\partial G/\partial N)_{\tau,p} = \mu; \tag{7}$$

$$(\partial G/\partial \tau)_{N,p} = -\sigma; \tag{8}$$

$$(\partial G/\partial p)_{N,\tau} = V. \tag{9}$$

Three Maxwell relations may be obtained from these by cross-differentiation; see Problem 1.

In the Gibbs free energy $G \equiv U - \tau\sigma + pV$ the variables τ and p are **intensive quantities**: they do not change value when two identical systems are put together. But U, σ, V, and G are linear in the number of particles N: their value doubles when two identical systems are put together, apart from interface effects. We say that U, σ, V, N and G are **extensive quantities**. Assume that only one particle species is present. If G is directly proportional to N, we must be able to write

$$G = N\varphi(p,\tau) , \tag{10}$$

where φ is independent of N because it is a function only of the intensive quantities p and τ. If two identical volumes of gas at equal pressure and temperature, each with $\frac{1}{2}N$ molecules, are put together, the Gibbs free energy

$$G = \tfrac{1}{2}N\varphi(p,\tau) + \tfrac{1}{2}N\varphi(p,\tau) = N\varphi(p,\tau)$$

does not change in the process. It follows from this argument that

$$(\partial G/\partial N)_{p,\tau} = \varphi(p,\tau). \tag{11}$$

We saw in (7) that

$$(\partial G/\partial N)_{p,\tau} = \mu , \tag{12}$$

so that φ must be identical with μ, and (10) becomes

$$\boxed{G(N,p,\tau) = N\mu(p,\tau).} \tag{13}$$

Thus the chemical potential for a single-component system is equal to the Gibbs free energy per particle, G/N. For G for an ideal gas, see (21) below.

If more than one chemical species is present, (13) is replaced by a sum over all species:

$$G = \sum_j N_j\mu_j. \tag{14}$$

The thermodynamic identity becomes

$$\tau d\sigma = dU + pdV - \sum \mu_j dN_j;\tag{15}$$

and (5) becomes

$$dG = \sum \mu_j dN_j - \sigma d\tau + Vdp.\tag{16}$$

We shall develop the theory of chemical equilibria by exploiting the property that $G = \sum N_j \mu_j$ is a minimum with respect to changes in the distribution of reacting molecules at constant τ, p. No new atoms come into the system in a reaction; the atoms that are present redistribute themselves from one molecular species to another molecular species.

Example: Comparison of G with F.   Let us see what is different about the two relations

$$(\partial F/\partial N)_{\tau,V} = \mu(N,\tau,V)\tag{17}$$

and

$$(\partial G/\partial N)_{\tau,p} = \mu(\tau,p).\tag{18}$$

We found in (6.18) that for an ideal gas

$$\mu(N,\tau,V) = \tau \log(N/Vn_Q) \ ,\tag{19}$$

so that $\mu(N,\tau,V)$ is not independent of N and therefore we cannot write $F = N\mu(\tau,V)$ as the integral of (17).

That is, F is not directly proportional to N if the system is kept at constant volume as the number of particles is increased. Instead, from (6.24),

$$F(\tau,V,N) = N\tau[\log(N/Vn_Q) - 1].\tag{20}$$

But the Gibbs free energy for the ideal gas is

$$\begin{aligned}G(\tau,p,N) &= F + pV = N\tau[\log(p/\tau n_Q) - 1] + N\tau\\ &= N\tau \log(p/\tau n_Q) \ ,\end{aligned}\tag{21}$$

by use of the ideal gas law in the form $N/V = p/\tau$. We readily identify in (21) the chemical potential as

$$\mu(\tau,p) = \tau \log(p/\tau n_Q) \ ,\tag{22}$$

by reference to the result $G = N\mu(\tau, p)$. We see that N appears unavoidably in $\mu(\tau, V)$ in (19), but not in $\mu(\tau, p)$ in (22). The chemical potential is the Gibbs free energy per particle, but it is not the Helmholtz free energy per particle. Of course, we are free to write μ as either (19) or (22), as is convenient.

EQUILIBRIUM IN REACTIONS

We may write the equation of a chemical reaction as

$$v_1 A_1 + v_2 A_2 + \cdots + v_l A_l = 0 , \tag{23}$$

or

$$\sum_j v_j A_j = 0 , \tag{24}$$

where the A_j denote the chemical species, and the v_j are the coefficients of the species in the reaction equation. Here v is the Greek letter nu. For the reaction $H_2 + Cl_2 = 2HCl$ we have

$$A_1 = H_2; \quad A_2 = Cl_2; \quad A_3 = HCl; \quad v_1 = 1; \quad v_2 = 1; \quad v_3 = -2. \tag{25}$$

The discussion of chemical equilibria is usually presented for reactions under conditions of constant pressure and temperature. In equilibrium the Gibbs free energy is a minimum with respect to changes in the proportions of the reactants. The differential of G is

$$dG = \sum_j \mu_j dN_j - \sigma d\tau + V dp. \tag{26}$$

Here μ_j is the chemical potential of species j, as defined by $\mu_j \equiv (\partial G / \partial N_j)_{\tau, p}$. At constant pressure $dp = 0$ and at constant temperature $d\tau = 0$; then (26) reduces to

$$dG = \sum_j \mu_j dN_j. \tag{27}$$

The change in the Gibbs free energy in a reaction depends on the chemical potentials of the reactants. In equilibrium G is an extremum and dG must be zero.

The change dN_j in the number of molecules of species j is proportional to the coefficient v_j in the chemical equation $\sum v_j A_j = 0$. We may write dN_j in the form

$$dN_j = v_j \, d\hat{N} \, , \tag{28}$$

where $d\hat{N}$ indicates how many times the reaction (24) takes place. The change dG in (27) becomes

$$dG = \left(\sum_j v_j \mu_j \right) d\hat{N}. \tag{29}$$

In equilibrium $dG = 0$, so that

$$\boxed{\sum_j v_j \mu_j = 0.} \tag{30}$$

This is the condition for equilibrium in a transformation of matter at constant pressure and temperature.*

Equilibrium for Ideal Gases

We obtain a simple and useful form of the general equilibrium condition $\sum v_j \mu_j = 0$ when we assume that each of the constituents acts as an ideal gas. We utilize (6.48) to write the chemical potential of species j as

$$\mu_j = \tau(\log n_j - \log c_j) \, , \tag{31}$$

where n_j is the concentration of species j and

$$c_j \equiv n_{Qj} Z_j(\text{int}) \, , \tag{32}$$

which depends on the temperature but not on the concentration. Here $Z_j(\text{int})$ is the internal partition function, (6.44). Then (30) can be rearranged as

$$\sum_j v_j \log n_j = \sum_j v_j \log c_j \, , \tag{33a}$$

---

* But the result is more general: once equilibrium is reached, the reaction does not proceed further, and there is no further change in the thermal average values of the concentrations. The volume at equilibrium will be known, so that the condition (30) applies as well when V and τ are specified as when p and τ are specified.

or as

$$\sum_j \log n_j^{\nu_j} = \sum_j \log c_j^{\nu_j}. \tag{33b}$$

The left-hand side can be rewritten as

$$\sum_j \log n_j^{\nu_j} = \log \prod_j n_j^{\nu_j} , \tag{33c}$$

and the right-hand side can be expressed as

$$\log \prod_j c_j^{\nu_j} \equiv \log K(\tau). \tag{33d}$$

Here $K(\tau)$, called the **equilibrium constant**, is a function only of the temperature. With (32) we have

$$K(\tau) \equiv \prod_j n_{Qj}^{\nu_j} \exp[-\nu_j F_j(\text{int})/\tau] , \tag{34}$$

because the internal free energy is $F_j(\text{int}) = -\tau \log Z_j(\text{int})$. From (33c,d) and (34) we have

$$\prod_j n_j^{\nu_j} = K(\tau) , \tag{35}$$

known as the **law of mass action**. The result says that the indicated product of the concentrations of the reactants is a function of the temperature alone. A change in the concentration of any one reactant will force a change in the equilibrium concentration of one or more of the other reactants.

To calculate the equilibrium constant $K(\tau)$ in (34), it is essential to choose in a consistent way the zero of the internal energy of each reactant. We need consistency here because the value of each partition function $Z_j(\text{int})$ depends on our choice of the zero of the energy eigenstates. The different zeros for the different reactants must be related to give properly the energy or free energy difference in the reaction. It is not difficult to arrange this, but it does not happen without a conscious effort on our part. For a dissociation reaction such as $H_2 \rightleftharpoons 2H$, the simplest procedure is to choose the zero of the internal energy of each composite particle (here the H_2 molecule) to coincide with the energy of the dissociated particles (here 2H) at rest. Accordingly, we place the energy of the ground state of the composite particle at $-E_B$, where E_B is the energy required in the reaction to dissociate the composite particle into its constituents and is taken to be positive.

Example: Equilibrium of atomic and molecular hydrogen.   The statement of the law of mass action for the reaction $H_2 = 2H$ or $H_2 - 2H = 0$ for the dissociation of molecular hydrogen into atomic hydrogen is

$$[H_2][H]^{-2} = \frac{[H_2]}{[H]^2} = K(\tau). \tag{36}$$

Here $[H_2]$ denotes the concentration of molecular hydrogen, and $[H]$ the concentration of atomic hydrogen. It follows that

$$\frac{[H]}{[H_2]} = \frac{1}{[H_2]^{1/2} K^{1/2}}; \tag{37}$$

that is, the relative concentration of atomic hydrogen at a given temperature is inversely proportional to the square root of the concentration of molecular hydrogen. The equilibrium constant K is given by

$$\log K = \log n_Q(H_2) - 2 \log n_Q(H) - F(H_2)/\tau , \tag{38}$$

in terms of the internal free energy of H_2, per molecule. Spin factors are absorbed in $F(H_2)$. Here the zero of energy is taken for an H atom at rest. The more tightly bound is H_2, the more negative is $F(H_2)$, and the higher is K, leading to a higher proportion of H_2 in the mixture. The energy to dissociate H_2 is 4.476 eV per molecule, at absolute zero.

It may be said that the dissociation of molecular hydrogen into atomic hydrogen is an example of entropy dissociation: The gain in entropy associated with the decomposition of H_2 into two independent particles compensates the loss in binding energy. It is believed that most of the hydrogen in intergalactic space is present as H and not H_2: The reaction equilibrium is thrown in the direction of H by the low values of the concentration of H_2. Hydrogen is very dilute in intergalactic space.

Example: pH and the ionization of water.   In liquid water the ionization process

$$H_2O \leftrightarrow H^+ + OH^- \tag{39}$$

proceeds to a slight extent. At room temperature the reaction equilibrium is described approximately by the concentration product

$$[H^+][OH^-] = 10^{-14} \, mol^2 l^{-2} , \tag{40}$$

where the ionic concentrations are given in moles per liter. In pure water $[H^+] = [OH^-] = 10^{-7} \, mol \, l^{-1}$. An acid is said to act as a proton donor. The concentration of H^+ ions is increased by adding an acid to the water and the concentration of OH^- ions will decrease as required to maintain the product $[H^+][OH^-]$ constant. Similarly, the concentration of OH^- ions can be increased by adding a base to the water, and the H^+ concentration will decrease accordingly. The physical state of water is more complicated than the equation of the ionization process suggests—the H^+ ions are not bare protons, but are associated with groups* of H_2O molecules. This does not significantly affect the validity of the reaction equation.

It is often convenient to express the acidity or alkalinity of a solution in terms of the pH, defined as

$$pH \equiv -\log_{10}[H^+]. \tag{41}$$

The pH of a solution is the negative of the logarithm base ten of the hydrogen ion concentration in moles per liter of solution. The pH of pure water is 7 because $[H^+] = 10^{-7} \, mol \, l^{-1}$. The strongest acidic solutions have pH near 0 or even negative; an apple may have pH ~ 3. Human blood plasma has a pH of 7.3 to 7.5; it is slightly basic.

Example: Kinetic model of mass action.   Suppose that atoms A and B combine to form a molecule AB. We suppose that AB is formed in a biatomic collision of A and B. Let n_A, n_B, n_{AB} denote the concentrations of A, B, and AB respectively. The rate of change of n_{AB} is

$$dn_{AB}/dt = Cn_A n_B - Dn_{AB} , \tag{42}$$

where the rate constant C describes the formation of AB in a collision of A with B, and the rate constant D describes the reverse process, the thermal decay of AB into its component atoms A and B. In thermal equilibrium the concentrations of all constitutents are constant, so that $dn_{AB}/dt = 0$ and

$$Cn_A n_B = Dn_{AB}; \qquad n_A n_B/n_{AB} = D/C , \tag{43}$$

a function of temperature only. This result is consistent with the law of mass action that we derived earlier by standard thermodynamics.

Suppose AB is not formed principally by the bimolecular collision of A and B, but is formed by some catalytic process such as

$$A + E \leftrightarrow AE; \qquad AE + B \leftrightarrow AB + E. \tag{44}$$

---

* The dominant species present is most likely $H^+ \cdot 4H_2O$, a complex of 4 water molecules surrounding one proton. A review is given by M. Eigen and L. De Maeyer, Proc. Roy. Soc. (London) **A247**, 505 (1958).

Here E is the catalyst which is returned to its original state at the end of the second step. So long as the intermediate product AE is so short lived that no significant quantity of A is tied up as AE, the ratio $n_A n_B / n_{AB}$ in equilibrium must be the same as if AB were formed in the direct process $A + B \leftrightarrow AB$ treated above. No matter by what route the reaction actually proceeds, the equilibrium must be the same. The rates, however, may differ.

The equality in equilibrium of the direct and inverse reaction rates is called the **principle of detailed balance**.

Comment: Reaction rates. The law of mass action expresses the condition satisfied by the concentrations once a reaction has gone to equilibrium. It tells us nothing about how fast the reaction proceeds. A reaction $A + B = C$ may evolve energy ΔH as it proceeds, but before the reaction can occur A and B may have to negotiate a potential barrier, as in Figure 9.2. The barrier height is called the activation energy. Only molecules on the high energy end of their energy distribution will be able to react; others will not be able to get over the potential hill. A catalyst speeds up a reaction by offering an alternate reaction path with a lower energy of activation, but it does not change the equilibrium concentrations.

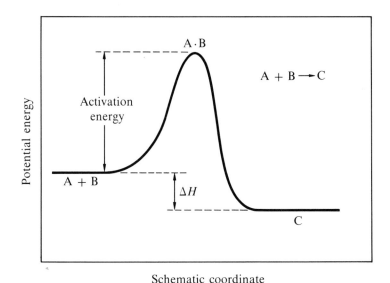

Schematic coordinate

Figure 9.2 The quantity ΔH measures the energy evolved in the reaction and determines the equilibrium·concentration ratio $[A][B]/[C]$. The activation energy is the height of the potential barrier to be negotiated before the reaction can proceed, and it determines the rate at which the reaction takes place.

SUMMARY

1. The Gibbs free energy

$$G \equiv U - \tau\sigma + pV$$

is a minimum in thermal equilibrium at constant temperature and pressure.

2. $(\partial G/\partial \tau)_{N,p} = -\sigma;$ $(\partial G/\partial p)_{N,\tau} = V;$ $(\partial G/\partial N)_{\tau,p} = \mu.$

3. $G(\tau,p,N) = N\mu(\tau,p)$

4. The law of mass action for a chemical reaction is that

$$\prod n_j^{\,\nu_j} = K(\tau) \ ,$$

a function of the temperature alone.

PROBLEMS

1. Thermal expansion near absolute zero.   (a) Prove the three Maxwell relations

$$(\partial V/\partial \tau)_p = -(\partial\sigma/\partial p)_\tau \ , \tag{45a}$$

$$(\partial V/\partial N)_p = +(\partial\mu/\partial p)_N \ , \tag{45b}$$

$$(\partial\mu/\partial \tau)_N = -(\partial\sigma/\partial N)_\tau. \tag{45c}$$

Strictly speaking, (45a) should be written

$$(\partial V/\partial \tau)_{p,N} = -(\partial\sigma/\partial p)_{\tau,N} \ ,$$

and two subscripts should appear similarly in (45b) and (45c). It is common to omit those subscripts that occur on both sides of these equalities. (b) Show with the help of (45a) and the third law of thermodynamics that the volume coefficient of thermal expansion

$$\alpha = \frac{1}{V}\left(\frac{\partial V}{\partial \tau}\right)_p \tag{46}$$

approaches zero as $\tau \to 0$.

2. Thermal ionization of hydrogen. Consider the formation of atomic hydrogen in the reaction e + H$^+$ ⇌ H, where e is an electron, as the adsorption of an electron on a proton H$^+$. (a) Show that the equilibrium concentrations of the reactants satisfy the relation

$$[e][H^+]/[H] \cong n_Q \exp(-I/\tau) , \qquad (47)$$

where I is the energy required to ionize atomic hydrogen, and $n_Q \equiv (m\tau/2\pi\hbar^2)^{3/2}$ refers to the electron. Neglect the spins of the particles; this assumption does not affect the final result. The result is known as the Saha equation. If all the electrons and protons arise from the ionization of hydrogen atoms, then the concentration of protons is equal to that of the electrons, and the electron concentration is given by

$$[e] = [H]^{1/2}n_Q{}^{1/2}\exp(-I/2\tau). \qquad (48)$$

A similar problem arises in semiconductor physics in connection with the thermal ionization of impurity atoms that are donors of electrons.

Notice that:

(1) The exponent involves $\frac{1}{2}I$ and not I, which shows that this is not a simple "Boltzmann factor" problem. Here I is the ionization energy.

(2) The electron concentration is proportional to the square root of the hydrogen atom concentration.

(3) If we add excess electrons to the system, then the concentration of protons will decrease.

(b) Let [H(exc)] denote the equilibrium concentration of H atoms in the first excited electronic state, which is $\frac{3}{4}I$ above the ground state. Compare [H(exc)] with [e] for conditions at the surface of the Sun, with [H] $\simeq 10^{23}$ cm^{-3} and $T \simeq 5000$ K.

3. Ionization of donor impurities in semiconductors. A pentavalent impurity (called a donor) introduced in place of a tetravalent silicon atom in crystalline silicon acts like a hydrogen atom in free space, but with e^2/ϵ playing the role of e^2 and an effective mass m^* playing the role of the electron mass m in the description of the ionization energy and radius of the ground state of the impurity atom, and also for the free electron. For silicon the dielectric constant $\epsilon = 11.7$ and, approximately, $m^* = 0.3$ m. If there are 10^{17} donors per cm^3, estimate the concentration of conduction electrons at 100 K.

4. Biopolymer growth. Consider the chemical equilibrium of a solution of linear polymers made up of identical units. The basic reaction step is monomer + Nmer = (N + 1)mer. Let K_N denote the equilibrium constant for this reaction. (a) Show from the law of mass action that the concentrations $[\cdots]$ satisfy

$$[N + 1] = [1]^{N+1}/K_1 K_2 K_3 \cdots K_N \tag{49}$$

(b) Show from the theory of reactions that for ideal gas conditions (an ideal solution):

$$K_N = \frac{n_Q(N) n_Q(1)}{n_Q(N + 1)} \exp[(F_{N+1} - F_N - F_1)/\tau]. \tag{50}$$

Here

$$n_Q(N) = (2\pi\hbar^2/M_N \tau)^{-3/2} , \tag{51}$$

where M_N is the mass of the Nmer molecule, and F_N is the free energy of one Nmer molecule. (c) Assume $N \gg 1$, so that $n_Q(N) \simeq n_Q(N + 1)$. Find the concentration ratio $[N + 1]/[N]$ at room temperature if there is zero free energy change in the basic reaction step: that is, if $\Delta F = F_{N+1} - F_N - F_1 = 0$. Assume $[1] = 10^{20}\,\mathrm{cm}^{-3}$, as for amino acid molecules in a bacterial cell. The molecular weight of the monomer is 200. (d) Show that for the reaction to go in the direction of long molecules we need $\Delta F < -0.4\,\mathrm{eV}$, approximately. This condition is not satisfied in Nature, but an ingenious pathway is followed that simulates the condition. An elementary discussion is given by C. Kittel, Am. J. Phys. **40**, 60 (1972).

5. Particle-antiparticle equilibrium.   (a) Find a quantitative expression for the thermal equilibrium concentration $n = n^+ = n^-$ in the particle-antiparticle reaction $A^+ + A^- = 0$. The reactants may be electrons and positrons; protons and antiprotons; or electrons and holes in a semiconductor. Let the mass of either particle be M; neglect the spins of the particles. The minimum energy release when A^+ combines with A^- is Δ. Take the zero of the energy scale as the energy with no particles present. (b) Estimate n in cm^{-3} for an electron (or a hole) in a semiconductor $T = 300\,\mathrm{K}$ with a Δ such that $\Delta/\tau = 20$. The hole is viewed as the antiparticle to the electron. Assume that the electron concentration is equal to the hole concentration; assume also that the particles are in the classical regime. (c) Correct the result of (a) to let each particle have a spin of $\frac{1}{2}$. Particles that have antiparticles are usually fermions with spins of $\frac{1}{2}$.

Chapter 10

Phase Transformations

Note: In the first section s denotes σ/N, the entropy per atom. In the section on ferromagnetism, μ is the magnetic moment of an atom.

VAPOR PRESSURE EQUATION

The curve of pressure versus volume for a quantity of matter at constant temperature is determined by the free energy of the substance. The curve is called an **isotherm**. We consider the isotherms of a real gas in which the atoms or molecules interact with one another and under appropriate conditions can associate together in a liquid or solid phase. A **phase** is a portion of a system that is uniform in composition.

Two phases may coexist, with a definite boundary between them. An isotherm of a real gas may show a region in the p–V plane in which liquid and gas coexist in equilibrium with each other. As in Figure 10.1, part of the volume contains atoms in the gas phase. There are isotherms at low temperatures for which solid and liquid coexist and isotherms for which solid and gas coexist. Everything we say for the liquid-gas equilibrium holds also for the solid-gas equilibrium and the solid-liquid equilibrium.

Liquid and vapor* may coexist on a section of an isotherm only if the temperature of the isotherm lies below a **critical temperature** τ_c. Above the critical temperature only a single phase—the fluid phase—exists, no matter how great the pressure. There is no more reason to call this phase a gas than a liquid, so we avoid the issue and call it a fluid. Values of the critical temperature for several gases are given in Table 10.1.

Liquid and gas will never coexist along the entire extent of an isotherm from zero pressure to infinite pressure; they coexist at most only along a section of the isotherm. For a fixed temperature and fixed number of atoms, there will be a volume above which all atoms present are in the gas phase. A small drop of water placed in an evacuated sealed bell jar at room temperature will evaporate entirely, leaving the bell jar filled with H_2O gas at some pressure. A drop of water exposed to air not already saturated with moisture may evaporate entirely. There is a concentration of water, however, above which the atoms from the vapor will bind themselves into a liquid drop. The volume relations are suggested by Figure 10.1.

The thermodynamic conditions for the coexistence of two phases are the conditions for the equilibrium of two systems that are in thermal, diffusive,

---

* Vapor is a term used for a gas when the gas is in equilibrium with its liquid or solid form.

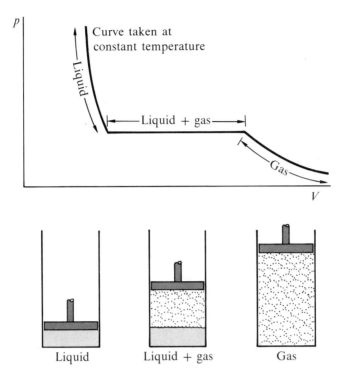

Figure 10.1 Pressure-volume isotherm of a real gas at a temperature such that liquid and gas phases may coexist, that is, $\tau < \tau_c$. In the two-phase region of liquid + gas the pressure is constant, but the volume may change. At a given temperature there is only a single value of the pressure for which a liquid and its vapor are in equilibrium. If at this pressure we move the piston down, some of the gas is condensed to liquid, but the pressure remains unchanged as long as any gas remains.

Table 10.1 Critical temperatures of gases

	T_c, in K		T_c, in K
He	5.2	H_2	33.2
Ne	44.4	N_2	126.0
Ar	151	O_2	154.3
Kr	210	H_2O	647.1
Xe	289.7	CO_2	304.2

and mechanical contact. These conditions are that $\tau_1 = \tau_2; \mu_1 = \mu_2; p_1 = p_2$ or, for liquid and gas,

$$\tau_l = \tau_g; \qquad \mu_l = \mu_g; \qquad p_l = p_g , \tag{1}$$

where the subscripts l and g denote the liquid and gas phases. Note that the chemical potentials of the same chemical species in the two phases must be equal if the phases coexist. The chemical potentials are evaluated at the common pressure and common temperature of the liquid and gas, so that

$$\boxed{\mu_l(p,\tau) = \mu_g(p,\tau).} \tag{2}$$

At a general point in the p–τ plane the two phases do not coexist: If $\mu_l < \mu_g$ the liquid phase alone is stable, and if $\mu_g < \mu_l$ the gas phase alone is stable. Metastable phases may occur, by supercooling or superheating. A metastable phase may have a transient existence, sometimes brief, sometimes long, at a temperature for which another and more stable phase of the same substance has a lower chemical potential.

Derivation of the Coexistence Curve, p Versus τ

Let p_0 be the pressure for which two phases, liquid and gas, coexist at the temperature τ_0. Suppose that the two phases also coexist at the nearby point $p_0 + dp; \tau_0 + d\tau$. The curve in the p, τ plane along which the two phases coexist divides the p, τ plane into a phase diagram, as given in Figure 10.2 for H_2O. It is a condition of coexistence that

$$\mu_g(p_0,\tau_0) = \mu_l(p_0,\tau_0) , \tag{3}$$

and also that

$$\mu_g(p_0 + dp, \tau_0 + d\tau) = \mu_l(p_0 + dp, \tau_0 + d\tau). \tag{4}$$

Equations (3) and (4) give a relationship between dp and $d\tau$.

We make a series expansion of each side of (4) to obtain

$$\mu_g(p_0,\tau_0) + \left(\frac{\partial \mu_g}{\partial p}\right)_\tau dp + \left(\frac{\partial \mu_g}{\partial \tau}\right)_p d\tau + \cdots$$

$$= \mu_l(p_0,\tau_0) + \left(\frac{\partial \mu_l}{\partial p}\right)_\tau dp + \left(\frac{\partial \mu_l}{\partial \tau}\right)_p d\tau + \cdots . \tag{5}$$

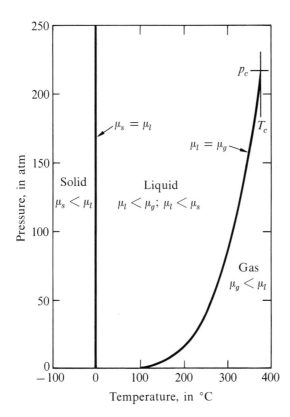

Figure 10.2 Phase diagram of H_2O. The relationships of the chemical potentials μ_s, μ_l, and μ_g in the solid, liquid, and gas phases are shown. The phase boundary here between ice and water is not exactly vertical; the slope is actually negative, although very large. After *International Critical Tables*, Vol. 3, and P. W. Bridgman, Proc. Am. Acad. Sci. **47**, 441 (1912); for the several forms of ice, see Zemansky, p. 375.

In the limit as dp and $d\tau$ approach zero,

$$\left(\frac{\partial\mu_g}{\partial p}\right)_\tau dp + \left(\frac{\partial\mu_g}{\partial\tau}\right)_p d\tau = \left(\frac{\partial\mu_l}{\partial p}\right)_\tau dp + \left(\frac{\partial\mu_l}{\partial\tau}\right)_p d\tau, \tag{6}$$

by (3) and (5). This result may be rearranged to give

$$\frac{dp}{d\tau} = \frac{\left(\frac{\partial\mu_l}{\partial\tau}\right)_p - \left(\frac{\partial\mu_g}{\partial\tau}\right)_p}{\left(\frac{\partial\mu_g}{\partial p}\right)_\tau - \left(\frac{\partial\mu_l}{\partial p}\right)_\tau}, \tag{7}$$

which is the differential equation of the coexistence curve or vapor pressure curve.

The derivatives of the chemical potential which occur in (7) may be expressed in terms of quantities accessible to measurement. In the treatment of the Gibbs free energy in Chapter 9 we found the relations

$$G = N\mu(p,\tau); \qquad \left(\frac{\partial G}{\partial p}\right)_{N,\tau} = V; \qquad \left(\frac{\partial G}{\partial \tau}\right)_{N,p} = -\sigma. \qquad (8)$$

With the definitions

$$v \equiv V/N, \qquad s \equiv \sigma/N \qquad (9)$$

for the volume and entropy per molecule in each phase, we have

$$\frac{1}{N}\left(\frac{\partial G}{\partial p}\right)_{N,\tau} = \frac{V}{N} = v = \left(\frac{\partial \mu}{\partial p}\right)_{\tau}; \qquad \frac{1}{N}\left(\frac{\partial G}{\partial \tau}\right)_{N,p} = -\frac{\sigma}{N} = -s = \left(\frac{\partial \mu}{\partial \tau}\right)_{p}. \qquad (10)$$

Then (7) for $dp/d\tau$ becomes

$$\frac{dp}{d\tau} = \frac{s_g - s_l}{v_g - v_l}. \qquad (11)$$

Here $s_g - s_l$ is the increase of entropy of the system when we transfer one molecule from the liquid to the gas, and $v_g - v_l$ is the increase of volume of the system when we transfer one molecule from the liquid to the gas.

It is essential to understand that the derivative $dp/d\tau$ in (11) is not simply taken from the equation of state of the gas. The derivative refers to the very special interdependent change of p and τ in which the gas and liquid continue to coexist. The number of molecules in each phase will vary as the volume is varied, subject only to $N_l + N_g = N$, a constant. Here N_l and N_g are the numbers of molecules in the liquid and gas phases, respectively.

The quantity $s_g - s_l$ is related directly to the quantity of heat that must be added to the system to transfer one molecule reversibly from the liquid to the gas, while keeping the temperature of the system constant. (If heat is not added to the system from outside in the process, the temperature will decrease when the molecule is transferred to the gas.) The quantity of heat added in the transfer is

$$đQ = \tau(s_g - s_l), \qquad (12)$$

by virtue of the connection between heat and the change of entropy in a reversible process. The quantity

$$L \equiv \tau(s_g - s_l) \qquad (13)$$

defines the **latent heat of vaporization**, and is easily measured by elementary calorimetry.

We let

$$\Delta v = v_g - v_l \tag{14}$$

denote the change of volume when one molecule is transferred from the liquid to the gas. We combine (11), (13), and (14) to obtain

$$\boxed{\frac{dp}{d\tau} = \frac{L}{\tau \, \Delta v}.} \tag{15}$$

This is known as the **Clausius-Clapeyron equation** or the **vapor pressure equation**. The derivation of this equation was a remarkable early accomplishment of thermodynamics. Both sides of (15) are easily determined experimentally, and the equation has been verified to high precision.

We obtain a particularly useful form of (15) if we make two approximations:

(a) We assume that $v_g \gg v_l$: the volume occupied by an atom in the gas phase is very much larger than in the liquid (or solid) phase, so that we may replace Δv by v_g:

$$\Delta v \cong v_g = V_g/N_g. \tag{16}$$

At atmospheric pressure $v_g/v_l \approx 10^3$, and the approximation is very good.

(b) We assume that the ideal gas law $pV_g = N_g\tau$ applies to the gas phase, so that (16) may be written as

$$\Delta v \cong \tau/p. \tag{17}$$

With these approximations the vapor pressure equation becomes

$$\frac{dp}{d\tau} = \frac{L}{\tau^2} p; \qquad \frac{d}{d\tau} \log p = \frac{L}{\tau^2}, \tag{18}$$

where L is the latent heat per molecule. Given L as a function of temperature, this equation may be integrated to find the coexistence curve.

If, in addition, the latent heat L is independent of temperature over the temperature range of interest, we may take $L = L_0$ outside the integral. Thus when we integrate (18) we obtain

$$\int \frac{dp}{p} = L_0 \int \frac{d\tau}{\tau^2}, \tag{19}$$

whence

$$\log p = -L_0/\tau + \text{constant}; \qquad p(\tau) = p_0 \exp(-L_0/\tau) , \qquad (20)$$

where p_0 is a constant. We defined L_0 as the latent heat of vaporization of one molecule. If L_0 refers instead to one mole, then

$$p(T) = p_0 \exp(-L_0/RT) , \qquad (21)$$

where R is the gas constant, $R \equiv N_0 k_B$, where N_0 is the **Avogadro constant**. For water the latent heat at the liquid-gas transition is 2485 J g^{-1} at 0°C and 2260 J g^{-1} at 100°C, a substantial variation with temperature.

The vapor pressure of water and of ice is plotted in Figure 10.3 as $\log p$ versus $1/T$. The curve is linear over substantial regions, consistent with the

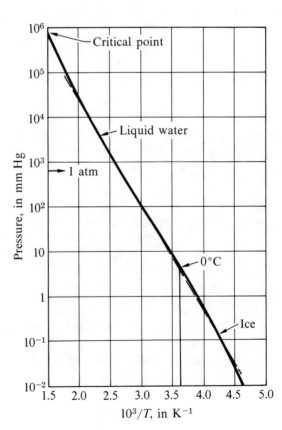

Figure 10.3   Vapor pressure of water and of ice plotted versus $1/T$. The vertical scale is logarithmic. The dashed line is a straight line.

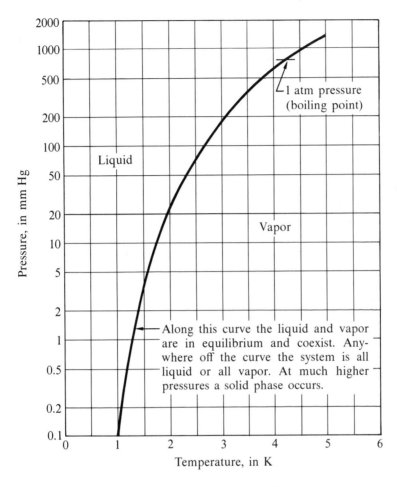

Figure 10.4 Vapor pressure versus temperature for ^{4}He. After H. van Dijk et al., Journal of Research of the National Bureau of Standards **63A**, 12 (1959).

approximate result (20). The vapor pressure of ^{4}He, plotted in Figure 10.4, is widely used in the measurement of temperatures between 1 and 5 K.

The phase diagram of ^{4}He at low temperatures was shown in Figure 7.14. Notice that the liquid-solid coexistence curve is closely horizontal below 1.4 K. We infer from this and (11) that the entropy of the liquid is very nearly equal to the entropy of the solid in this region. It is remarkable that the entropies should be so similar, because a normal liquid is much more disordered than a solid, so that the entropy of a normal liquid is considerably higher than that of a normal solid. But ^{4}He is a quantum liquid. For another quantum liquid, ^{3}He, the slope of the liquid-solid curve is negative at low temperatures (Figure 7.15), and in this region the entropy of the liquid is less than the entropy of the solid. The solid has more accessible states than the liquid! Liquid ^{3}He has a relatively low entropy for a liquid because it approximates a Fermi gas, which generally has a low entropy when $\tau \ll \tau_F$ because a large proportion of the atoms have their momenta ordered into the Fermi sphere of Chapter 7.

Triple point.　　The triple point t of a substance is that point p_t, τ_t in the p–τ plane at which all three phases, vapor, liquid, and solid, are in equilibrium. Here $\mu_g = \mu_l = \mu_s$. Consider an equilibrium mixture of liquid and solid phases enclosed in a volume somewhat larger than that occupied by the mixture alone. The remaining volume will contain only the vapor, in equilibrium with both condensed phases, and at a pressure equal to the common equilibrium vapor pressure of both phases. This pressure is the triple point pressure.

The triple point temperature is not identical with the melting temperature of the substance at atmospheric pressure. Melting temperatures depend somewhat on pressure; the triple point temperature is the melting temperature under the common equilibrium vapor pressure of the two condensed phases.

For water the triple point temperature is 0.01 K above the atmospheric pressure melting temperature: $T_t = 0.01°C = 273.16$ K. The Kelvin scale is defined such that the triple point of water is exactly 273.16 K; see Appendix B.

Latent heat and enthalpy.　　The latent heat of a phase transformation, as from the liquid phase to the gas phase, is equal to τ times the entropy difference of the two phases at constant pressure. The latent heat is also equal to the difference of $H \equiv U + pV$ between the two phases, where H is called the **enthalpy**. The differential is $dH = dU + pdV + Vdp$. When we cross the coexistence curve, the thermodynamic identity applies:

$$\tau d\sigma = dU + pdV - (\mu_g - \mu_l)dN , \tag{22}$$

On the coexistence curve $\mu_g = \mu_l$. Thus at constant pressure

$$L = \tau \Delta \sigma = \Delta U + p \Delta V = \Delta H = H_g - H_l. \tag{23}$$

Values of H are tabulated; they are found by integration of the heat capacity at constant pressure:

$$C_p = \tau \left(\frac{\partial \sigma}{\partial \tau} \right)_p = \left(\frac{\partial U}{\partial \tau} \right)_p + p \left(\frac{\partial V}{\partial \tau} \right)_p = \left(\frac{\partial H}{\partial \tau} \right)_p , \tag{24}$$

or

$$H = \int C_p\, d\tau. \tag{25}$$

Example: Model system for gas-solid equilibrium. We construct a simple model to describe a solid in equilibrium with a gas, as in Figure 10.5. We can easily derive the vapor pressure curve for this model. Roughly the same model would apply to a liquid.

Imagine the solid to consist of N atoms, each bound as a harmonic oscillator of frequency ω to a fixed center of force. The binding energy of each atom in the ground state is ε_0; that is, the energy of an atom in its ground state is $-\varepsilon_0$ referred to a free atom at rest. The energy states of a single oscillator are $n\hbar\omega - \varepsilon_0$, where n is a positive integer or zero (Figure 10.6). For the sake of simplicity we suppose that each atom can oscillate only in one dimension. The result for oscillators in three dimensions is left as a problem.

The partition function of a single oscillator in the solid is

$$Z_s = \sum_n \exp[-(n\hbar\omega - \varepsilon_0)/\tau] = \exp(\varepsilon_0/\tau) \sum_n \exp(-n\hbar\omega/\tau) = \frac{\exp(\varepsilon_0/\tau)}{1 - \exp(-\hbar\omega/\tau)}. \tag{26}$$

The free energy F_s is

$$F_s = U_s - \tau\sigma_s = -\tau \log Z_s. \tag{27}$$

The Gibbs free energy in the solid is, per atom,

$$G_s = U_s - \tau\sigma_s + pv_s = F_s + pv_s = \mu_s. \tag{28}$$

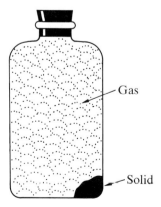

Figure 10.5 Atoms in a solid in equilibrium with atoms in the gas phase. The equilibrium pressure is a function of temperature. The energy of the atoms in the solid phase is lower than in the gas phase, but the entropy of the atoms tends to be higher in the gas phase. The equilibrium configuration is determined by the counterplay of the two effects. At low temperature most of the atoms are in the solid; at high temperature all or most of the atoms may be in the gas.

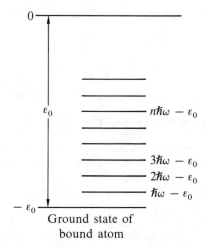

Figure 10.6 States of an atom bound as a harmonic oscillator of frequency ω. The ground state is assumed to be ε_0 below that of a free atom at rest in the gas phase.

The pressure in the solid is equal to that of the gas with which it is in contact, but the volume v_s per atom in the solid phase is much smaller than the volume v_g per atom in the gas phase: $v_s \ll v_g$.

If we neglect the term pv_s we have for the chemical potential of the solid $\mu_s \cong F_s$, whence the absolute activity is

$$\lambda_s \equiv \exp(\mu_s/\tau) \simeq \exp(F_s/\tau) = \exp(-\log Z_s)$$

$$= \frac{1}{Z_s} = \exp(-\varepsilon_0/\tau)[1 - \exp(-\hbar\omega/\tau)]. \tag{29}$$

We make the ideal gas approximation to describe the gas phase, and we take the spin of the atom to be zero. Then, from Chapter 6,

$$\lambda_g = \frac{n}{n_Q} = \frac{p}{\tau n_Q} = \frac{p}{\tau}\left(\frac{2\pi\hbar^2}{M\tau}\right)^{3/2}. \tag{30}$$

The gas is in equilibrium with the solid when $\lambda_g = \lambda_s$, or

$$p = \tau n_Q \exp(-\varepsilon_0/\tau)[1 - \exp(-\hbar\omega/\tau)]. \tag{31}$$

If we insert n_Q from (3.63):

$$p = \left(\frac{M}{2\pi\hbar^2}\right)^{3/2} \tau^{5/2} \exp(-\varepsilon_0/\tau)[1 - \exp(-\hbar\omega/\tau)]. \tag{32}$$

VAN DER WAALS EQUATION OF STATE

The simplest model of a liquid-gas phase transition is that of van der Waals, who modified the ideal gas equation $pV = N\tau$ to take into account approximately the interactions between atoms or molecules. By the argument that we give below, he was led to a modified equation of state of the form

$$(p + N^2a/V^2)(V - Nb) = N\tau ,   (33)$$

known as the **van der Waals equation of state**. This is written for N atoms in volume V. The a, b are interaction constants to be defined; the constant a is a measure of the long range attractive part of the interaction between two molecules, and the constant b is a measure of their short range repulsion (Figure 10.7). We shall derive (33) with the help of the general relation $p = -(\partial F/\partial V)_{\tau,N}$. We shall then treat the thermodynamic properties of the model in order to exhibit the liquid-gas transition.

For an ideal gas we have, from (6.24),

$$F(\text{ideal gas}) = -N\tau[\log(n_Q/n) + 1].   (34)$$

The hard core repulsion at short distances can be treated approximately as if the gas had available not the volume V, but the free volume $V - Nb$, when b is the volume per molecule. We therefore replace the concentration $n = N/V$ in (34) by $N/(V - Nb)$. Thus, instead of (34), we have

$$F = -N\tau\{\log[n_Q(V - Nb)/N] + 1\}.   (35)$$

To this we now add a correction for the intermolecular attractive forces.

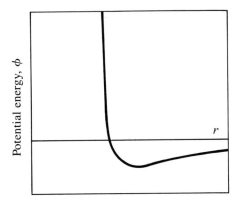

Figure 10.7 The interaction energy between two molecules consists of a short range repulsion plus a long range attraction. The short range repulsion can be described approximately by saying that each molecule has a hard, impenetrable core.

Mean Field Method

There exists a simple approximate method, called the mean field method, for taking into account the effect of weak long range interactions among the particles of a system. The most widely known applications of the method are to gases and to ferromagnets. Let $\varphi(r)$ denote the potential energy of interaction of two atoms separated by a distance r. When the concentration of atoms in the gas is n, the average value of the total interaction of all other atoms on the atom at $r = 0$ is

$$\int_b^\infty dV\, \varphi(r)n = n \int_b^\infty dV\, \varphi(r) = -2na \;, \qquad (36)$$

where $-2a$ denotes the value of the integral $\int dV\, \varphi(r)$. The factor of two is a useful convention. We exclude the hard core sphere of volume b from the volume of integration. In writing (36) we assume that the concentration n is constant throughout the volume accessible to the molecules of the gas. That is, we use the mean value of n. This assumption is the essence of the mean field approximation. By assuming uniform concentration we ignore the increase of concentration in regions of strong attractive potential energy. In modern language we say that the mean field method neglects correlations between interacting molecules.

From (36) it follows that the interactions change the energy and the free energy of a gas of N molecules in volume V by

$$\Delta F \simeq \Delta U = -\tfrac{1}{2}(2Nna) = -N^2 a/V. \qquad (37)$$

The factor $\tfrac{1}{2}$ is common to self-energy problems; it arranges that an interaction "bond" between two molecules is counted only once in the total energy. The exact number of bonds is $\tfrac{1}{2}N(N - 1)$, which we approximate as $\tfrac{1}{2}N^2$.

We add (37) to (35) to obtain the van der Waals approximation for the Helmholtz free energy of a gas:

$$F(\text{vdW}) = -N\tau\{\log[n_Q(V - Nb)/N] + 1\} - N^2 a/V. \qquad (38)$$

The pressure is

$$p = -(\partial F/\partial V)_{\tau,N} = \frac{N\tau}{V - Nb} - \frac{N^2 a}{V^2} \qquad (39)$$

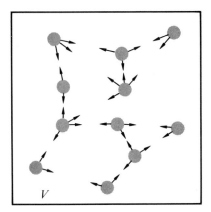

Figure 10.8 Directions of intermolecular forces that act on molecules near the boundary of a volume V. The van der Waals argument suggests that these forces contribute an internal pressure $N^2 a/V^2$ which is to be added to the external pressure p, so that $p + N^2 a/V^2$ should be used as the pressure in the gas law.

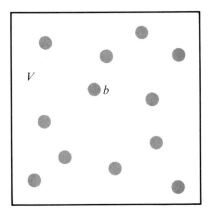

Figure 10.9 The container of volume V has N molecules, each of volume b. The volume not occupied by molecules is $V - Nb$. Intuition suggests that this free volume should be used in the gas law in place of the container volume V.

or

$$(p + N^2 a/V^2)(V - Nb) = N\tau , \qquad (40)$$

the **van der Waals equation of state**. The terms in a and b are interpreted in Figures 10.8 and 10.9.

Critical Points for the van der Waals Gas

We define the quantities

$$p_c = a/27b^2; \qquad V_c = 3Nb; \qquad \tau_c = 8a/27b. \qquad (41)$$

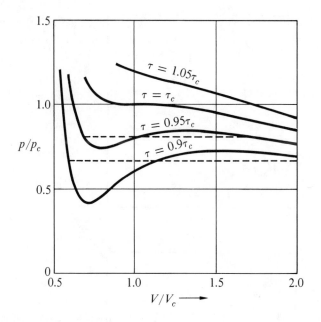

Figure 10.10   The van der Waals equation of state near the critical temperature. Courtesy of R. Cahn.

In terms of these quantities the van der Waals equation becomes

$$\left(\frac{p}{p_c} + \frac{3}{(V/V_c)^2}\right)\left(\frac{V}{V_c} - \frac{1}{3}\right) = \frac{8\tau}{3\tau_c}. \tag{42}$$

This equation is plotted in Figure 10.10 for several temperatures near the temperature τ_c. The equation may be written in terms of the dimensionless variables

$$\hat{p} \equiv p/p_c; \qquad \hat{V} \equiv V/V_c; \qquad \hat{\tau} \equiv \tau/\tau_c, \tag{43}$$

as

$$\left(\hat{p} + \frac{3}{\hat{V}^2}\right)\left(\hat{V} - \frac{1}{3}\right) = \frac{8}{3}\hat{\tau}; \qquad \hat{p} = \frac{\frac{8}{3}\hat{\tau}}{\hat{V} - \frac{1}{3}} - \frac{3}{\hat{V}^2}. \tag{44}$$

This result is known as the **law of corresponding states**. In terms of $\hat{p}$, $\hat{V}$, $\hat{\tau}$, all gases look alike—if they obey the van der Waals equation. Values of a

and b are usually obtained by fitting to the observed p_c and τ_c. States of two substances at the same $\hat{p}$, $\hat{V}$, $\hat{\tau}$ are called corresponding states of the substances. Real gases do not obey the equation to high accuracy.

At one point, the **critical point**, the curve of $\hat{p}$ versus $\hat{V}$ at constant $\hat{\tau}$ has a horizontal point of inflection. Here the local maximum and minimum of the p–V curve coincide, and there is no separation between the vapor and liquid phases. At a horizontal point of inflection

$$\left(\frac{\partial \hat{p}}{\partial \hat{V}}\right)_{\hat{\tau}} = 0; \qquad \left(\frac{\partial^2 \hat{p}}{\partial \hat{V}^2}\right)_{\hat{\tau}} = 0. \tag{45}$$

These conditions are satisfied by (44) if $\hat{p} = 1$; $\hat{V} = 1$; $\hat{\tau} = 1$. We call p_c, V_c, and τ_c the critical pressure, critical volume, and critical temperature, respectively. Above τ_c no phase separation exists.

Gibbs Free Energy of the van der Waals Gas

The Gibbs free energy of the van der Waals gas exhibits the characteristics of the liquid-gas phase transition at constant pressure. With $G = F + pV$, we have from (38) and (39) the result

$$G(\tau,V,N) = \frac{N\tau V}{V - Nb} - \frac{2N^2 a}{V} - N\tau\{\log[n_Q(V - Nb)/N] + 1\}. \tag{46}$$

This equation gives G as a function of V, τ, N; the natural variables for G are p, τ, N. Unfortunately we cannot conveniently put G into an analytic form as a function of pressure instead of volume. We want $G(\tau,p,N)$ because we can then obtain $\mu(\tau,p)$ as $G(\tau,p,N)/N$ by (9.13). It is μ that determines the phase coexistence relation $\mu_l = \mu_g$. The results of numerical calculations of G versus p are plotted in Figure 10.11 for temperatures below and at the critical temperature. At any temperature the lowest branch represents the stable phase; the other branches represent unstable phases. The pressure at which the branches cross determines the transition between gas and liquid; this pressure is called the **equilibrium vapor pressure**. Results for G versus τ are plotted in Figure 10.12.

Figure 10.13 shows, on a p–V diagram, the region $V < V_1$ in which only the liquid phase exists and the region $V > V_2$ in which only the gas phase exists. The phases coexist between V_1 and V_2. The value of V_1 or V_2 is determined by the condition that $\mu_l(\tau,p) = \mu_g(\tau,p)$ along the horizontal line between V_1 and V_2. This will occur if the shaded area below the line is equal to the shaded area

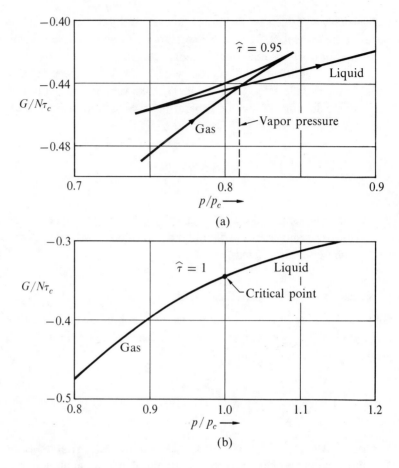

Figure 10.11 (a) Gibbs free energy versus pressure for van der Waals equation of state: $\tau = 0.95\tau_c$. Courtesy of R. Cahn. (b) Gibbs free energy versus pressure for van der Waals equation of state: $\tau = \tau_c$.

above the line. To see this, consider

$$dG = -\sigma d\tau + Vdp + \mu dN. \tag{47}$$

We have $dG = Vdp$ at constant τ and constant total number of particles. The difference of G between V_1 and V_2 is

$$G_g - G_l = \int Vdp , \tag{48}$$

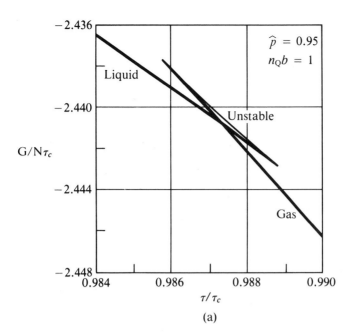

Figure 10.12a Gibbs free energy versus temperature for van der Waals equation of state at $p = 0.95\,p_c$. Courtesy of A. Manoliu.

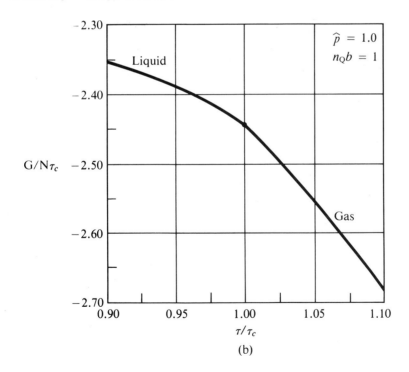

Figure 10.12b Gibbs free energy versus temperature for van der Waals equation of state at the critical pressure p_c.

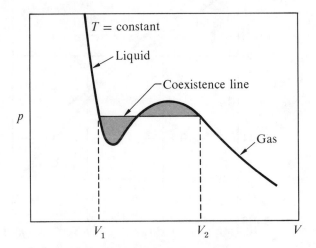

Figure 10.13   Isotherm of van der Waals gas at a temperature below the critical temperature. For volumes less than V_1 only the liquid phase exists; for volumes above V_2 only the gas phase exists. Between V_1 and V_2 the system in stable equilibrium lies along the coexistence line and is an inhomogeneous mixture of two phases. The liquid and gas phases coexist. The proportion of the liquid and gas phases must be such that the sum of their volumes equals the volume V that is available.

but the integral is just the sum of the shaded areas, one negative and one positive. When the magnitudes of the areas are equal, $G_g(\tau,p) = G_l(\tau,p)$ and $\mu_g(\tau,p) = \mu_l(\tau,p)$ along the horizontal coexistence line drawn in the figure. In equilibrium we require $\mu_g = \mu_l$.

Nucleation.   Let $\Delta\mu = \mu_g - \mu_l$ be the chemical potential difference between the vapor surrounding a small liquid droplet and the liquid in bulk (an infinitely large drop). If $\Delta\mu$ is positive, the bulk liquid will have a lower free energy than the gas and thus the liquid will be more stable than the gas. However, the surface free energy of a liquid drop is positive and tends to increase the free energy of the liquid. At small drop radii the surface can be dominant and the drop can be unstable with respect to the gas. We calculate the change in Gibbs free energy when a drop of radius R forms. If n_l is the concentration of molecules in the liquid,

$$\Delta G = G_l - G_g = -(4\pi/3)R^3 n_l \Delta\mu + 4\pi R^2 \gamma \qquad (49)$$

where γ is the surface free energy per unit area, or surface tension. The liquid drop will grow when $G_l < G_g$. An unstable maximum of ΔG is attained when

$$d\,\Delta G/dR = 0 = -4\pi R^2 n_l \Delta\mu + 8\pi R\gamma \; , \qquad (50)$$

or

$$R_c = 2\gamma/n_l \Delta\mu. \qquad (51)$$

This is the **critical radius for nucleation** of a drop. At smaller R the drop will tend to evaporate spontaneously because that will lower the free energy. At larger R the drop will tend to grow spontaneously because that, too, will lower the free energy.

The free energy barrier (Figure 10.14) that must be overcome by a thermal fluctuation in order for a nucleus to grow beyond R_c is found by substitution of (51) in (49):

$$(\Delta G)_c = (16\pi/3)[\gamma^3/n_l{}^2(\Delta\mu)^2]. \qquad (52)$$

If we assume that the vapor behaves like an ideal gas, we can use Chapter 5 to express $\Delta\mu$ as

$$\Delta\mu = \tau\log(p/p_{eq}) \; ,$$

where p is the vapor pressure in the gas phase and p_{eq} the equilibrium vapor pressure of the bulk liquid ($R \rightarrow \infty$). We use $\gamma = 72\,\mathrm{erg\,cm^{-2}}$ to estimate R_c for water at 300 K and $p = 1.1p_{eq}$ to be $1 \times 10^{-6}\,\mathrm{cm}$.

Ferromagnetism

A ferromagnet has a spontaneous magnetic moment, which means a magnetic moment even in zero applied magnetic field. We develop the mean field approximation to the temperature dependence of the magnetization, defined as the magnetic moment per unit volume. The central assumption is that each magnetic atom experiences an effective field B_E proportional to the magnetization:

$$B_E = \lambda M \; , \qquad (53)$$

where λ is a constant. We take the external applied field as zero.

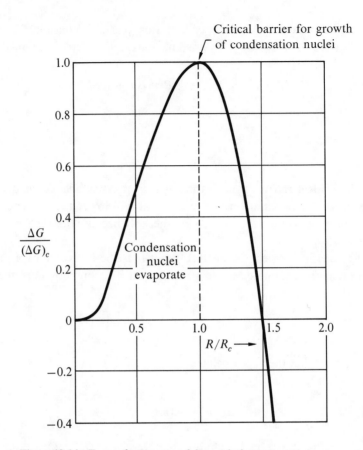

Figure 10.14 Excess free energy of drop relative to gas, as function of drop radius R, both in reduced units. The gas is supersaturated because the liquid has the lower free energy for this curve as drawn, but the surface energy of small drops creates an energy barrier that inhibits the growth of nuclei of the liquid phase. Thermal fluctuations eventually may carry nuclei over the barrier.

Consider a system with a concentration n of magnetic atoms, each of spin $\frac{1}{2}$ and of magnetic moment μ. In Chapter 3 we found an exact result for the magnetization in a field B:

$$M = n\mu \tanh(\mu B/\tau). \qquad (54)$$

In the mean field approximation (53) this becomes, for a ferromagnet,

$$M = n\mu \tanh(\mu\lambda M/\tau) , \qquad (55)$$

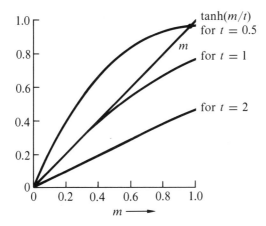

Figure 10.15 Graphical solution of Eq. (56) for the reduced magnetization m as a function of temperature. The reduced magnetization is defined as $m = M/n\mu$. The left-hand side of Eq. (56) is plotted as a straight line m with unit slope. The right-hand side is $\tanh(m/t)$ and is plotted versus m for three different values of the reduced temperature $t = \tau/n\mu^2\lambda = \tau/\tau_c$. The three curves correspond to the temperatures $2\tau_c$, τ_c, and $0.5\tau_c$. The curve for $t = 2$ intersects the straight line m only at $m = 0$, as appropriate for the paramagnetic region (there is no external applied magnetic field). The curve for $t = 1$ (or $\tau = \tau_c$) is tangent to the straight line m at the origin; this temperature marks the onset of ferromagnetism. The curve for $t = 0.5$ is in the ferromagnetic region and intersects the straight line m at about $m = 0.94 \, n\mu$. As $t \to 0$ the intercept moves up to $m = 1$, so that all magnetic moments are lined up at absolute zero.

a transcendental equation for M. We shall see that solutions of this equation with nonzero M exist in the temperature range between 0 and τ_c. To solve (55) we write it in terms of the reduced magnetization $m \equiv M/n\mu$ and the reduced temperature $t \equiv \tau/n\mu^2\lambda$, whence

$$m = \tanh(m/t). \tag{56}$$

We plot the right and left sides of this equation separately as functions of m, as in Figure 10.15. The intercept of the two curves gives the value of m at the temperature of interest. The critical temperature is $t = 1$, or $\tau_c = n\mu^2\lambda$. The curves of M versus τ obtained in this way reproduce roughly the features of the

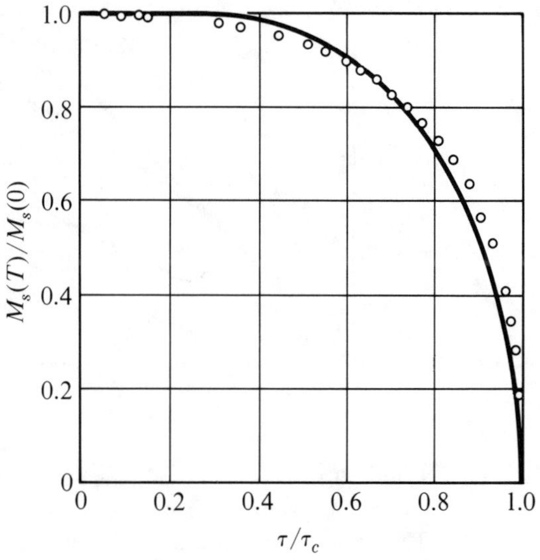

Figure 10.16 Saturation magnetization of nickel as a function of temperature, together with the theoretical curve for spin $\frac{1}{2}$ on the mean field theory.

experimental results, as shown in Figure 10.16 for nickel. As τ increases the magnetization decreases smoothly to zero at $\tau = \tau_c$, called the **Curie temperature**.

LANDAU THEORY OF PHASE TRANSITIONS

Landau gave a systematic formulation of the mean field theory of phase transitions applicable to a large variety of systems exhibiting such transitions. We consider systems at constant volume and temperature, so that their Helmholtz free energy $F = U - \tau\sigma$ is a minimum in equilibrium. The big question is, a minimum with respect to what variables? It is not helpful to consider all possible variables. We suppose here that the system can be described by a single **order parameter** ξ, the Greek xi, which might be the magnetization in a ferromagnetic system, the dielectric polarization in a ferroelectric system, the fraction of superconducting electrons in a superconductor, or the fraction of neighbor A–B bonds to total bonds in an alloy AB. In thermal equilibrium the order parameter will have a certain value $\xi = \xi_0(\tau)$. In the Landau theory we imagine that ξ can be independently specified, and we consider the Landau free energy function

$$F_L(\xi,\tau) \equiv U(\xi,\tau) - \tau\sigma(\xi,\tau) \ , \qquad (57)$$

where the energy and entropy are taken when the order parameter has the specified value ξ, not necessarily ξ_0. The equilibrium value $\xi_0(\tau)$ is the value of

ξ that makes F_L a minimum, at a given τ, and the actual Helmholtz free energy $F(\tau)$ of the system at τ is equal to that minimum:

$$F(\tau) = F_L(\xi_0,\tau) \le F_L(\xi,\tau) \quad \text{if} \quad \xi \ne \xi_0. \tag{58}$$

Plotted as a function of ξ for constant τ, the Landau free energy may have more than one minimum. The lowest of these determines the equilibrium state. In a first order phase transition another minimum becomes the lowest minimum as τ is increased.

We restrict ourselves to systems for which the Landau function is an even function of ξ in the absence of applied fields. Most ferromagnetic and ferro-electric systems are examples of this. We also assume that $F_L(\xi,\tau)$ is a sufficiently well-behaved function of ξ that it can be expanded in a power series in ξ—something that should not be taken for granted. For an even function of ξ, as assumed,

$$F_L(\xi,\tau) = g_0(\tau) + \tfrac{1}{2}g_2(\tau)\xi^2 + \tfrac{1}{4}g_4(\tau)\xi^4 + \tfrac{1}{6}g_6(\tau)\xi^6 + \cdots. \tag{59}$$

The entire temperature dependence of $F_L(\xi,\tau)$ is contained in the expansion coefficients g_0, g_2, g_4, g_6. These coefficients are matters for experiment or theory.

The simplest example of a phase transition occurs when $g_2(\tau)$ changes sign at a temperature τ_0, with g_4 positive and the higher terms negligible. For simplicity we take $g_2(\tau)$ linear in τ:

$$g_2(\tau) = (\tau - \tau_0)\alpha , \tag{60}$$

over the temperature range of interest, and we take g_4 as constant in that range. With these idealizations,

$$F_L(\xi,\tau) = g_0(\tau) + \tfrac{1}{2}\alpha(\tau - \tau_0)\xi^2 + \tfrac{1}{4}g_4\xi^4. \tag{61}$$

The form (60) cannot be accurate over a very wide temperature range, and it certainly fails at low temperatures because such a linear dependence on temperature is not consistent with the third law.

The equilibrium value of ξ is found at the minimum of $F_L(\xi;\tau)$ with respect to ξ:

$$(\partial F_L/\partial \xi)_\tau = (\tau - \tau_0)\alpha\xi + g_4\xi^3 = 0 , \tag{62}$$

which has the roots

$$\xi = 0 \quad \text{and} \quad \xi^2 = (\tau_0 - \tau)(\alpha/g_4). \tag{63}$$

With α and g_4 positive, the root $\xi = 0$ corresponds to the minimum of the free energy function (61) at temperatures above τ_0; here the Helmholtz free energy is

$$F(\tau) = g_0(\tau). \tag{64}$$

The other root, $\xi^2 = (\alpha/g_4)(\tau_0 - \tau)$ corresponds to the minimum of the free energy function at temperatures below τ_0; here the Helmholtz free energy is

$$F(\tau) = g_0(\tau) - (\alpha^2/4g_4)(\tau - \tau_0)^2. \tag{65}$$

The variation of $F(\tau)$ with temperature is shown in Figure 10.17. The variation of $F_L(\xi;\tau)$ as a function of ξ^2 for three representative temperatures is shown in Figure 10.18, and the temperature dependence of the equilibrium value of ξ is shown in Figure 10.19.

Our model describes a phase transition in which the value of the order parameter goes continuously to zero as the temperature is increased to τ_0. The entropy

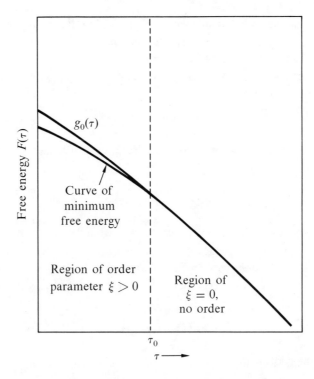

Figure 10.17 Temperature dependence of the free energy for an idealized phase transition of the second order.

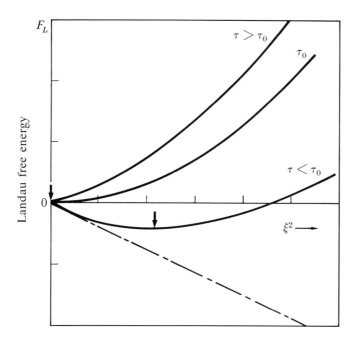

Figure 10.18 Landau free energy function versus ξ^2 at representative temperatures. As the temperature drops below τ_0 the equilibrium value of ξ gradually increases, as defined by the position of the minimum of the free energy.

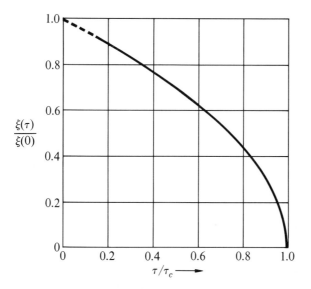

Figure 10.19 Spontaneous polarization versus temperature, for a second-order phase transition. The curve is not realistic at low temperatures because of the use of Eq. (60): the third law of thermodynamics requires that $d\xi/d\tau \to 0$ as $\tau \to 0$.

$-\partial F/\partial \tau$ is continuous at $\tau = \tau_0$, so that there is no latent heat at the transition temperature τ_0. Such a transition is by definition a second order transition. Transitions with a nonzero latent heat are called first order transitions; we discuss them presently. The real world contains a remarkable diversity of second order transitions; the best examples are ferromagnets and superconductors.

Example: Ferromagnets. In the mean field approximation, ferromagnets satisfy the Landau theory. To show this, consider an atom of magnetic moment μ in a magnetic field B, which we shall set equal to the mean field λM as in (53). The interaction energy density is

$$U(M) = -\tfrac{1}{2}M \cdot B = -\tfrac{1}{2}\lambda M^2 \ , \tag{66}$$

where the factor $\tfrac{1}{2}$ is common to self-energy problems. The entropy density is given approximately by Problem 2.2 as

$$\sigma(M) = \text{constant} - M^2/2n\mu^2 \ , \tag{67}$$

in the regime in which $M \ll n\mu$. Thus the free energy function per unit volume is

$$F_L(M) = \text{constant} - \tfrac{1}{2}M^2\!\left(\lambda - \frac{\tau}{n\mu^2}\right) + \text{terms of higher order.} \tag{68}$$

At the transition temperature the coefficient of M^2 vanishes, so that

$$\tau_0 = n\mu^2\lambda \ , \tag{69}$$

in agreement with the discussion following (56).

First Order Transitions

A latent heat characterizes a first order phase transition. The liquid-gas transition at constant pressure is a first order transition. In the physics of solids first order transitions are common in ferroelectric crystals and in phase transformations in metals and alloys. The Landau function describes a first order transition when the expansion coefficient g_4 is negative and g_6 is positive. We consider

$$F_L(\xi;\tau) = g_0(\tau) + \tfrac{1}{2}\alpha(\tau - \tau_0)\xi^2 - \tfrac{1}{4}|g_4(\tau)|\xi^4 + \tfrac{1}{6}g_6\xi^6 + \cdots. \tag{70}$$

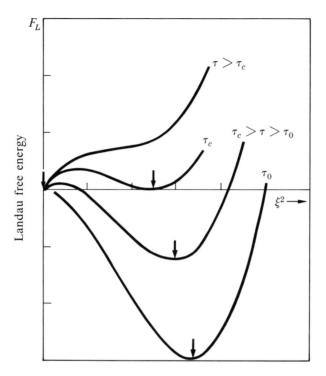

Figure 10.20   Landau free energy function versus ξ^2 in a first order transition, at representative temperatures. At τ_c the Landau function has equal minima at $\xi = 0$ and at a finite ξ as shown. For τ below τ_c the absolute minimum is at larger values of ξ; as τ passes through τ_c there is a discontinuous change in the position of the absolute minimum. The arrows mark the minima.

The extrema of this function are given by the roots of $\partial F_L / \partial \xi = 0$ as in Figure 10.20:

$$\alpha(\tau - \tau_0)\xi - |g_4(\tau)|\xi^3 + g_6\xi^5 = 0. \qquad (71)$$

Either $\xi = 0$ or

$$\alpha(\tau - \tau_0) - |g_4(\tau)|\xi^2 + g_6\xi^4 = 0. \qquad (72)$$

At the transition temperature τ_c the free energies will be equal for the phases with $\xi = 0$ and with the root $\xi \neq 0$. The value of τ_c will not be equal to τ_0,

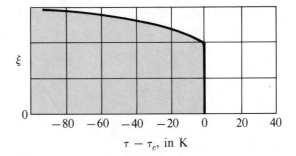

Figure 10.21   Dependence of ξ on $\tau - \tau_c$ for a typical first order phase transition.

and the order parameter ξ (Figure 10.21) does not go continuously to zero at τ_c. These results differ from those in the second order phase transition treated earlier, where ξ went to zero continuously at $\tau_0 = \tau_c$. A first order transformation may show hysteresis, as in supercooling or supersaturation, but no hysteresis exists in a second order transition.

SUMMARY

1. The coexistence curve in the p–τ plane between two phases must satisfy the Clausius-Clapeyron equation:

$$\frac{dp}{d\tau} = \frac{L}{\tau \, \Delta v} \ ,$$

where L is the latent heat and Δv is the volume difference per atom between the two phases.

2. The latent heat $L = H_1 - H_2$, where $H \equiv U + pV$ is the enthalpy.

3. The van der Waals equation of state is

$$(p + N^2 a/V^2)(V - Nb) = N\tau.$$

4. In the Landau free energy function

$$F_L(\xi,\tau) \equiv U(\xi,\tau) - \tau\sigma(\xi,\tau) \ ,$$

the energy and entropy are taken when the order parameter has the specified value ξ, not necessarily the thermal equilibrium value ξ_0. The function F_L is a minimum with respect to ξ when the system is in thermal equilibrium.

5. A first order phase transition is characterized by a latent heat and by hysteresis.

PROBLEMS

1. Entropy, energy, and enthalpy of van der Waals gas. (a) Show that the entropy of the van der Waals gas is

$$\sigma = N\{\log[n_Q(V - Nb)/N] + \tfrac{5}{2}\}. \qquad (73)$$

(b) Show that the energy is

$$U = \tfrac{3}{2}N\tau - N^2a/V. \qquad (74)$$

(c) Show that the enthalpy $H \equiv U + pV$ is

$$H(\tau,V) = \tfrac{5}{2}N\tau + N^2b\tau/V - 2N^2a/V; \qquad (75)$$

$$H(\tau,p) = \tfrac{5}{2}N\tau + Nbp - 2Nap/\tau. \qquad (76)$$

All results are given to first order in the van der Waals correction terms a, b.

2. Calculation of dT/dp for water. Calculate from the vapor pressure equation the value of dT/dp near $p = 1$ atm for the liquid-vapor equilibrium of water. The heat of vaporization at $100°C$ is $2260\,\mathrm{J\,g^{-1}}$. Express the result in kelvin/atm.

3. Heat of vaporization of ice. The pressure of water vapor over ice is 3.88 mm Hg at $-2°C$ and 4.58 mm Hg at $0°C$. Estimate in $\mathrm{J\,mol^{-1}}$ the heat of vaporization of ice at $-1°C$.

4. Gas-solid equilibrium. Consider a version of the example (26)–(32) in which we let the oscillators in the solid move in three dimensions. (a) Show that in the high temperature regime ($\tau \gg \hbar\omega$) the vapor pressure is

$$p \cong \left(\frac{M}{2\pi}\right)^{3/2} \frac{\omega^3}{\tau^{1/2}} \exp(-\varepsilon_0/\tau). \qquad (77)$$

(b) Explain why the latent heat per atom is $\varepsilon_0 - \tfrac{1}{2}\tau$.

5. Gas-solid equilibrium. Consider the gas-solid equilibrium under the extreme assumption that the entropy of the solid may be neglected over the temperature range of interest. Let $-\varepsilon_0$ be the cohesive energy of the solid, per atom.

Treat the gas as ideal and monatomic. Make the approximation that the volume accessible to the gas is the volume V of the container, independent of the much smaller volume occupied by the solid. (a) Show that the total Helmholtz free energy of the system is

$$F = F_s + F_g = -N_s\varepsilon_0 + N_g\tau[\log(N_g/Vn_Q) - 1] ,\qquad(78)$$

where the total number of atoms, $N = N_s + N_g$ is constant. (b) Find the minimum of the free energy with respect to N_g; show that in the equilibrium condition

$$N_g = n_Q V \exp(-\varepsilon_0/\tau).\qquad(79)$$

(c) Find the equilibrium vapor pressure.

6. Thermodynamics of the superconducting transition. (a) Show that

$$(\sigma_S - \sigma_N)/V = \frac{1}{2\mu_0}\frac{d(B_c{}^2)}{d\tau} = \frac{B_c}{\mu_0}\frac{dB_c}{d\tau} ,\qquad(80)$$

in SI units for B_c. Because B_c decreases with increasing temperature, the right side is negative. The superconducting phase has the lower entropy: it is the more ordered phase. As $\tau \to 0$, the entropy in both phases will go to zero, consistent with the third law. What does this imply for the shape of the curve of B_c versus τ? (b) At $\tau = \tau_c$, we have $B_c = 0$ and hence $\sigma_S = \sigma_N$. Show that this result has the following consequences: (1) The two free energy curves do not cross at τ_c but merge, as shown in Figure 10.22. (2) The two energies are the same: $U_S(\tau_c) = U_N(\tau_c)$. (3) There is no latent heat associated with the transition at $\tau = \tau_c$. What is the latent heat of the transition when carried out in a magnetic field, at $\tau < \tau_c$? (c) Show that C_S and C_N, the heat capacities per unit volume, are related by

$$\Delta C = C_S - C_N = \frac{\tau}{2\mu_0}\frac{d^2(B_c{}^2)}{d\tau^2}.\qquad(81)$$

Figure 8.18 is a plot of C/T vs T^2 and shows that C_S decreases much faster than linearly with decreasing τ, while C_N decreases as $\gamma\tau$. For $\tau \ll \tau_c$, ΔC is dominated by C_N. Show that this implies

$$\gamma = -\frac{1}{\mu_0} B_c \frac{d^2 B_c}{d\tau^2}\bigg|_{\tau=0}.\qquad(82)$$

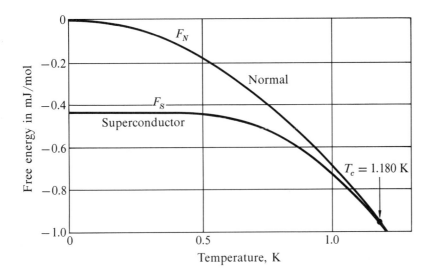

Figure 10.22 Experimental values of the free energy as a function of temperature for aluminum in the superconducting state and in the normal state. Below the transition temperature $T_e = 1.180$ K the free energy is lower in the superconducting state. The two curves merge at the transition temperature, so that the phase transition is second order (there is no latent heat of transition at T_e). The curve F_S is measured in zero magnetic field, and F_N is measured in a magnetic field sufficient to put the specimen in the normal state. Courtesy of N. E. Phillips.

7. Simplified model of the superconducting transition. The $B_c(\tau)$ curves of most superconductors have shapes close to simple parabolas. Suppose that

$$B_c(\tau) = B_{c0}[1 - (\tau/\tau_c)^2]. \tag{83}$$

Assume that C_S vanishes faster than linearly as $\tau \to 0$. Assume also that C_N is linear in τ, as for a Fermi gas (Chapter 7). Draw on the results of Problem 6 to calculate and plot the τ dependences of the two entropies, the two heat capacities, and the latent heat of the transition. Show that $C_S(\tau_c)/C_N(\tau_c) = 3$.

8. First order crystal transformation. Consider a crystal that can exist in either of two structures, denoted by α and β. We suppose that the α structure is the stable low temperature form and the β structure is the stable high temperature form of the substance. If the zero of the energy scale is taken as the state of separated atoms at infinity, then the energy density $U(0)$ at $\tau = 0$ will be negative. The phase stable at $\tau = 0$ will have the lower value of $U(0)$; thus $U_\alpha(0) < U_\beta(0)$. If the velocity of sound v_β in the β phase is lower than v_α in the α phase, corresponding to lower values of the elastic moduli for β, then the thermal excitations in the β phase will have larger amplitudes than in the α phase. The larger the thermal excitation, the larger the entropy and the lower the free energy. Soft systems tend to be stable at high temperatures, hard systems at low. (a) Show from Chapter 4 that the free energy density contributed by the phonons in a solid at a temperature much less than the Debye temperature is given by $-\pi^2\tau^4/30v^3\hbar^3$, in the Debye approximation with v taken as the velocity of all phonons. (b) Show that at the transformation temperature

$$\tau_c^{\,4} = (30\hbar^3/\pi^2)[U_\beta(0) - U_\alpha(0)]/(v_\beta^{\,-3} - v_\alpha^{\,-3}). \tag{84}$$

There will be a finite real solution if $v_\beta < v_\alpha$. This example is a simplified model of a class of actual phase transformations in solids. (c) The latent heat of transformation is defined as the thermal energy that must be supplied to carry the system through the transformation. Show that the latent heat for this model is

$$L = 4[U_\beta(0) - U_\alpha(0)]. \tag{85}$$

In (84) and (85), U refers to unit volume.

Chapter 11

Binary Mixtures

Note: All composition percentages are stated in atomic percent.

Many applications of materials science, and large parts of chemistry and biophysics, are concerned with the properties of multicomponent systems that have two or more phases in coexistence. Beautiful, unexpected, and important physical effects occur in such systems. We treat the fundamentals of the subject in this chapter, with examples drawn from simple situations.

SOLUBILITY GAPS

Mixtures are systems of two or more different chemical species. **Binary mixtures** have only two constituents. Mixtures with three and four constituents are called ternary and quaternary mixtures. If the constituents are atoms, and not molecules, the mixture is called an **alloy**.

A mixture is homogeneous when its constituents are intermixed on an atomic scale to form a single phase, as in a solution. A mixture is heterogeneous when it contains two or more distinct phases, such as oil and water. The everyday expression "oil and water do not mix" means that their mixture does not form a single homogeneous phase.

The properties of mixtures differ from the properties of pure substances. The melting and solidification properties of mixtures are of special interest. Heterogeneous mixtures may melt at lower temperatures than their constituents. Consider a gold-silicon alloy: pure Au melts at 1063°C and pure Si at 1404°C, but an alloy of 69 pct Au and 31 pct Si melts (and solidifies) at 370°C. This is not the result of the formation of any low-melting Au–Si compound: microscopic investigation of the solidified mixture shows a two phase mixture of almost pure Au side by side with almost pure Si (Figure 11.1). Mixtures with such properties are common, and they are of practical importance precisely because of their lowered melting points.

What determines whether two substances form a homogeneous or a heterogeneous mixture? What is the composition of the phases that are in equilibrium with each other in a heterogeneous mixture? The properties of mixtures can be understood from the principle that any system at a fixed temperature will evolve to the configuration of minimum free energy. Two substances will dissolve in each other and form a homogeneous mixture if that is the configuration of lowest free energy accessible to the components. The substances will

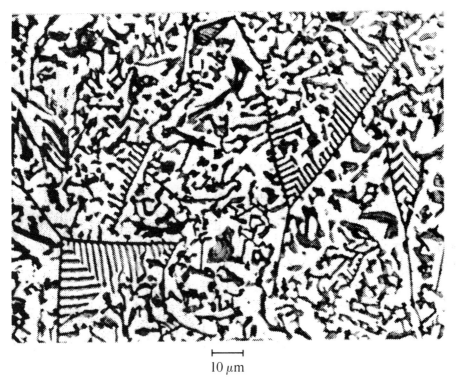

10 μm

Figure 11.1 Heterogeneous gold-silicon alloy. When a mixture of 69 pct Au and 31 pct Si is melted and then solidified, the mixture segregates into a phase of almost pure Au (light phase) coexistent with a phase of almost pure Si (dark phase). Magnified about 800 times. The composition given is that of the lowest-melting Au-Si mixture, the so-called eutectic mixture, a concept explained later in the text. Photograph courtesy of Stephan Justi.

form a heterogeneous mixture if the combined free energy of the two separate phases side by side is lower than the free energy of the homogeneous mixture: then we say that the mixture exhibits a **solubility gap**.

A heterogeneous mixture will melt at a lower temperature than the separate substances if the free energy of the homogeneous melt is lower than the combined free energies of the two separate solid phases.

Throughout this chapter we assume for simplicity that the external pressure may be neglected, and we set $pV = 0$. Then volume changes do not involve work, and the appropriate free energy is the Helmholtz free energy F rather than the Gibbs free energy G. We will usually simply speak of the free energy.

We discuss binary mixtures of constituents that do not form well-defined compounds with each other. Our principal interest is in binary alloys. Consider a mixture of N_A atoms of substance A and N_B atoms of substance B. The total number of atoms is

$$N = N_A + N_B. \tag{1}$$

We express the composition of the system in terms of the fraction x of B atoms:

$$x = N_B/N; \qquad 1 - x = N_A/N. \tag{2}$$

Suppose the system forms a homogeneous solution, with an average free energy per atom given by

$$f = F/N. \tag{3}$$

Suppose further that $f(x)$ has the functional form shown in Figure 11.2. Because this curve contains a range in which the second derivative d^2f/dx^2 is negative, we can draw a line tangent to the curve at two points, at $x = x_\alpha$ and $x = x_\beta$. Free energy curves of this shape are common, and we will see later what may cause this shape. Any homogeneous mixture in the composition range

$$x_\alpha < x < x_\beta \tag{4}$$

is unstable with respect to two separate phases of composition x_α and x_β. We shall show that the average free energy per atom of the segregated mixture is given by the point i on the straight line connecting the points α and β. Thus in the entire composition range (4) the segregated system has a lower free energy than the homogeneous system.

Proof: The free energy of a segregated mixture of the two phases α and β is

$$F = N_\alpha f(x_\alpha) + N_\beta f(x_\beta) , \tag{5}$$

where N_α and N_β are the total numbers of atoms in phases α and β, respectively. These numbers satisfy the relations

$$N_\alpha + N_\beta = N; \qquad x_\alpha N_\alpha + x_\beta N_\beta = N_B , \tag{6}$$

which may be solved for N_α and N_β:

$$N_\alpha = \frac{x_\beta - x}{x_\beta - x_\alpha} N; \qquad N_\beta = \frac{x - x_\alpha}{x_\beta - x_\alpha} N. \tag{7}$$

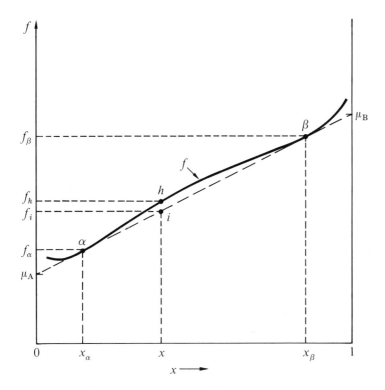

Figure 11.2   Free energy per atom as a function of composition, for a
system with a solubility gap. If the free energy per atom of a
homogeneous mixture has a shape such that a tangent can be drawn
that touches the curve at two different points α and β, the composition
range between the two points is unstable. Any mixture with a
composition in this range will decompose into two phases with the
composition x_α and x_β. The free energy of the two phase mixture is
given by the point i on the straight line, below the point h.

From (5) we obtain

$$f_i(x) = \frac{F}{N} = \frac{1}{x_\beta - x_\alpha}\left[(x_\beta - x)f(x_\alpha) + (x - x_\alpha)f(x_\beta)\right], \tag{8}$$

for the free energy of the two phase system. This result is linear in x and is a
straight line in the f–x plane. If we set $x = x_\alpha$ or x_β, we see that the line does go
through the points α and β. Thus f_i in the interval between x_α and x_β is given by
the point i on the straight line connecting α and β.

We have not yet made use of the assumption that the straight line is tangent to $f(x)$ at the points α and β, and therefore our result holds for any straight line that has two points α and β in common with $f(x)$. But for a given value of x, the lowest free energy is obtained by drawing the lowest possible straight line that has two points in common with $f(x)$, on opposite sides of x. The lowest possible straight line is the two-point tangent shown. The compositions x_α and x_β are the limits of the solubility gap of the system.

Once the system has reached its lowest free energy, the two phases must be in diffusive equilibrium with respect to both atomic species, so that their chemical potentials satisfy

$$\mu_{A\alpha} = \mu_{A\beta}; \qquad \mu_{B\alpha} = \mu_{B\beta}. \tag{9}$$

We show in Problem 1 that μ_A and μ_B are given by the intercepts of the two-point tangent with the two vertical edges of the $f(x)$ plot at $x = 0$ and $x = 1$, as in Figure 11.2.

ENERGY AND ENTROPY OF MIXING

The Helmholtz free energy $F \equiv U - \tau\sigma$ has contributions from the energy and from the entropy. We treat the effect of mixing two components A and B on both terms. Let u_A and u_B be the energy per atom of the pure substances A and B, referred to separated atoms at infinity. The average energy per atom of the constituents is

$$u = (u_A N_A + u_B N_B)/N = u_A + (u_B - u_A)x , \tag{10}$$

which defines a straight line in the u–x plane, Figure 11.3. The average energy per atom of the homogeneous mixture may be larger or smaller than for the separate constituents. In the example of Figure 11.3, the energy of the homogeneous mixture is larger than the energy of the separate constituents. The energy excess is called the **energy of mixing**.

If the $-\tau\sigma$ term in the free energy is negligible, as at $\tau = 0$, a positive mixing energy means that a homogeneous mixture is not stable. Any such mixture will then separate into two phases. But at a finite temperature the $-\tau\sigma$ term in the free energy of the homogeneous mixture always tends to lower the free energy.

The entropy of a mixture contains a contribution, called the **entropy of mixing**, that is not present in the entropies of the separate components. The mixing entropy arises when atoms of the different species are interchanged in position; this operation generates a different state of the system. Because of such inter-

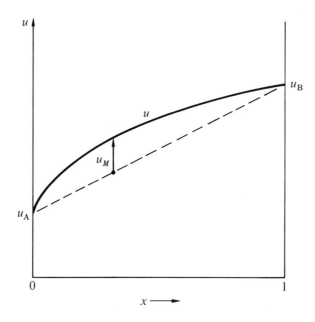

Figure 11.3 Energy per atom as a function of composition in a system with a positive mixing energy. A simple example for which a solubility gap may occur is that of a system in which the energy per atom of the homogeneous mixture is greater than that of the separate phases, so that $d^2u/dx^2 < 0$ for all compositions. The mixing energy is the difference between the $u(x)$ curve and the straight line.

changes a mixture has more accessible states than the two separate substances, and hence the mixture has the higher entropy.

In (3.80) we calculated the mixing entropy σ_M of a homogeneous alloy $A_{1-x}B_x$, to find

$$\sigma_M = -N[(1 - x)\log(1 - x) + x\log x] , \tag{11}$$

as plotted in Figure 11.4. The curve of σ_M versus x has the important property that the slope at the ends of the composition range is vertical. We have

$$\frac{1}{N}\frac{d\sigma_M}{dx} = \log(1 - x) - \log x = \log\frac{1 - x}{x} , \tag{12}$$

which goes to $+\infty$ as $x \to 0$ and to $-\infty$ as $x \to 1$.

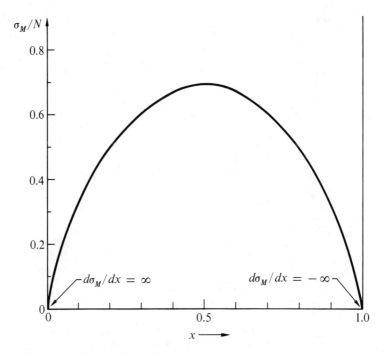

Figure 11.4   Mixing entropy. In any mixture of two constituents an interchange of two atoms of different species leads to a new state of the system. The logarithm of the number of states related in this way is the mixing entropy.

Consider now the quantity

$$f_0(x) = u(x) - (\sigma - \sigma_M)\tau/N , \qquad (13)$$

which is the free energy per atom without the mixing entropy contribution. The non-mixing part of the entropy, $\sigma - \sigma_M$, is usually nearly the same for the mixture as for the separate components, so that $(\sigma - \sigma_M)\tau$ is nearly a linear function of the composition x. If we assume this, the $f_0(x)$ curve has the same shape as the $u(x)$ curve, but offset vertically.

If we add the mixing entropy contribution $-\tau\sigma_M/N$ to $f_0(x)$, we obtain at various temperatures the $f(x)$ curves shown in Figure 11.5. In drawing the figure we have ignored the temperature dependence of $f_0(x)$ itself, because for

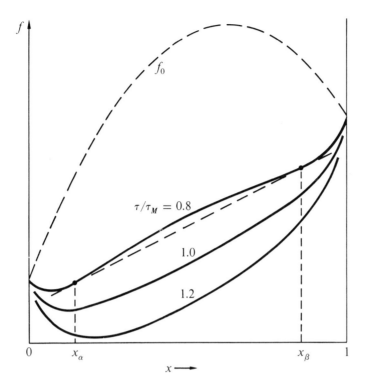

Figure 11.5   Free energy per atom versus composition, at three
temperatures. The curve f_0 is the free energy per atom without the
mixing entropy contribution. For illustration a parabolic composition
dependence is assumed, and the temperature dependence of f_0 is
neglected. The three solid curves represent the free energy including
the mixing entropy, for the temperatures $0.8 \, \tau_M$, $1.0 \, \tau_M$, and $1.2 \, \tau_M$,
where τ_M is the maximum temperature for which there is a solubility
gap. The phase separation at $0.8 \, \tau_M$ is apparent.

our argument this is irrelevant. Three important deductions follow from the
construction of the $f(x)$ curves:

(a)   At all finite temperatures $f(x)$ turns up at both ends of the composition
      range, because of the infinite slope of the mixing entropy contribution.

(b)   Below a certain temperature τ_M there is a composition range within which
      the negative second derivative of the $f_0(x)$ curve is stronger than the
      positive second derivative of the $-\tau \sigma_M$ contribution, thereby making it
      possible to draw a common tangent to $f(x)$ at two different values of x.

(c)   Above τ_M the curve has a positive second derivative at all compositions.

We conclude that the A–B system with positive mixing energy will exhibit a solubility gap below the temperature τ_M. The composition range of the gap widens with decreasing temperature, but the gap can reach the edges of the composition range only as $\tau \to 0$. At any finite temperature there is a finite solubility of A in B and of B in A, a result obtained earlier in Chapter 3. The new result is that the mutual solubility is limited only below τ_M. Positive mixing energies arise in different ways. We now discuss three examples.

Example: Binary alloy with nearest-neighbor interactions.   Consider an alloy $A_{1-x}B_x$ in which the attractive interaction between unlike atoms is weaker than the attractive interaction between like atoms. For simplicity we speak of the interactions as bonds. There are three different bonds: A–A, A–B, and B–B. Let u_{AA}, u_{AB} and u_{BB} be the potential energies of each bond. These binding energies will usually be negative with respect to separated atoms.

We assume the atoms are randomly distributed among the lattice sites. The average energy of the bonds surrounding an A atom is

$$u_A = (1 - x)u_{AA} + xu_{AB} , \tag{14}$$

where $(1 - x)$ is the proportion of A and x is the proportion of B. This result is written in the mean field approximation of Chapter 10. Similarly, for B atoms,

$$u_B = (1 - x)u_{AB} + xu_{BB}. \tag{15}$$

The total energy is obtained by summing over both atom types. If each atom has p nearest neighbors, the average energy per atom is

$$
\begin{aligned}
u &= \tfrac{1}{2}p[(1 - x)u_A + xu_B] \\
&= \tfrac{1}{2}p[(1 - x)^2 u_{AA} + 2x(1 - x)u_{AB} + x^2 u_{BB}].
\end{aligned} \tag{16}
$$

The factor $\tfrac{1}{2}$ arises because each bond is shared by the two atoms it connects. The result (16) can be written as

$$u = \tfrac{1}{2}p[(1 - x)u_{AA} + xu_{BB}] + u_M. \tag{17}$$

Here

$$u_M = px(1 - x)[u_{AB} - \tfrac{1}{2}(u_{AA} + u_{BB})] \tag{18}$$

is the mixing energy. On this model the mixing energy as a function of x is a parabola, as in Figure 11.5.

A solubility gap occurs whenever $d^2f/dx^2 < 0$, that is, when

$$\frac{d^2u}{dx^2} = \frac{d^2u_M}{dx^2} < \tau \frac{d^2\sigma}{dx^2} \cong \tau \frac{d^2\sigma_M}{dx^2}. \tag{19}$$

From (18),

$$\frac{d^2u_M}{dx^2} = -2p[u_{AB} - \tfrac{1}{2}(u_{AA} + u_{BB})]. \tag{20}$$

From (12),

$$\frac{\tau}{N}\frac{d^2\sigma_M}{dx^2} = -\frac{\tau}{x(1-x)} \le -\frac{\tau}{4}. \tag{21}$$

The equal sign holds for $x = \tfrac{1}{2}$. With these results (19) yields

$$\tau_M = \tfrac{1}{2}p[u_{AB} - \tfrac{1}{2}(u_{AA} + u_{BB})] \tag{22}$$

as the lower limit of the temperature for a solubility gap.

There are many reasons why mixed bonds may be weaker than the bonds of the separate constituents. If the constituent atoms of an alloy differ in radius, the difference introduces elastic strains that raise the energy. Water and oil "do not mix" because water molecules carry a large electric dipole moment that leads to a strong electrostatic attraction between water molecules. This attraction is absent in water-oil bonds, which are only about as strong as the weaker oil-oil bonds.

Example: Mixture of two solids with different crystal structures. Consider a homogeneous crystalline mixture of gold and silicon. The stable crystal structure of gold is the face-centered cubic structure in which every atom is surrounded by twelve equidistant nearest neighbors. The stable crystal structure of silicon is the diamond structure in which every atom is surrounded by only four equidistant nearest neighbors. If in pure Au we replace a small fraction x of the atoms by Si, we obtain a homogeneous mixture $Au_{1-x}Si_x$ with the fcc crystal structure of Au. Similarly, if in pure Si we replace a small fraction $1 - x$ of the atoms by Au, we obtain a homogeneous mixture $Au_{1-x}Si_x$, but with the diamond crystal structure of Si. There are two different free energies, one for each crystal structure (Figure 11.6). The two curves must cross somewhere in the composition range, or else pure Au and Si would not crystallize in different structures. The equilibrium curve consists of the lower of the two curves, with a kink at the crossover point. Such a system exhibits a solubility gap on either side of the crossover composition. The curves shown in the figure are schematic; in the actual Au–Si system the unstable range extends so close to the edges of the diagram that it cannot be represented on a full-scale plot extending from $x = 0$ to $x = 1$.

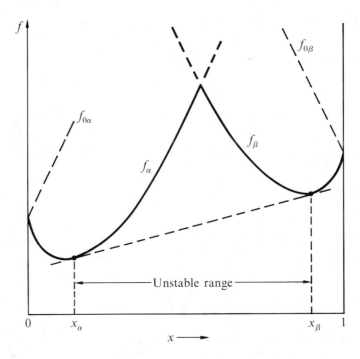

Figure 11.6   Free energy versus composition for crystalline homogeneous mixtures for which the two constituents of the mixture crystallize in different crystal structures. Two different free energy curves are involved, one for each crystal structure.

Different crystal structures for the pure constituents are an important cause of solubility gaps in crystalline solid mixtures. Our argument applies to mixtures of this kind, provided the two structures do not transform continuously into each other with changing composition. This is a tacit assumption in our discussion, an assumption not always satisfied when the two crystal structures are closely similar. The other assumption we make throughout this chapter is that no stable compound formation should occur. In the presence of compound formation the behavior of the mixture may be more complex.

Example: Liquid ^3He–^4He mixtures at low temperatures.   The most interesting liquid mixture with a solubility gap is the mixture of the two helium isotopes ^{3}He and ^{4}He, atoms of the former being fermions and of the latter bosons. There is a solubility gap in the mixture below 0.87 K, as in Figure 11.7. This property is utilized in the helium dilution refrigerator (Chapter 12). The mixing energy must be positive to have a solubility gap. The origin of the positive mixing energy is the following: ^{4}He atoms are bosons. At sufficiently low temperatures almost all ^{4}He atoms occupy the ground state orbital of the system, where they have

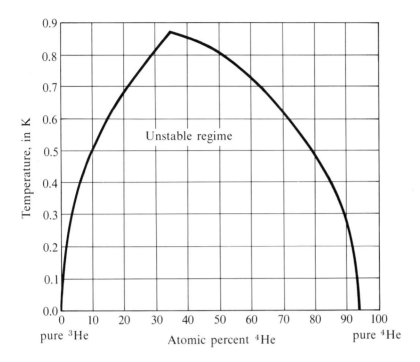

Figure 11.7 Liquid mixtures of ^{3}He and ^{4}He.

zero kinetic energy. Almost the entire kinetic energy of the mixture is contributed by the ^{3}He atoms, which are fermions. The energy per atom of a degenerate Fermi gas increases with concentration as $n^{2/3}$, as in Chapter 7. This energy has a negative second derivative $\partial^2 \varepsilon / \partial n^2 \propto \partial^2 u / \partial x^2$, which by (19) is equivalent to a positive mixing energy.

Phase Diagrams for Simple Solubility Gaps

A **phase diagram** represents the temperature dependence of solubility gaps, as in Figure 11.8. The two compositions x_α and x_β are plotted horizontally, the corresponding temperature vertically. The x_α and x_β branches merge at the maximum temperature τ_M for which a solubility gap exists. At a given temperature any mixture whose overall composition falls within the range enclosed by the curve is unstable as a homogeneous mixture. The phase diagrams of actual mixtures with solubility gaps may be more complex, according to the actual form of the free energy relation $f(x)$, but the underlying principles are the same.

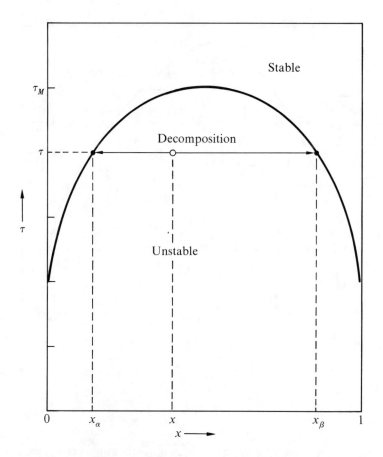

Figure 11.8   Phase diagram for a binary system with a solubility gap. A homogeneous mixture of composition x will be unstable at temperature τ if the point (x,τ) falls below the stability boundary curve. The mixture will then form two separate phases of the compositions given by the intersections of the stability boundary curve with the horizontal line for temperature τ. The stability boundary curve shown here was calculated quantitatively for the system of Figure 11.5, with a parabolic $f_0(x)$.

PHASE EQUILIBRIA BETWEEN LIQUID AND SOLID MIXTURES

When a small fraction of a homogeneous liquid mixture freezes, the composition of the solid that forms is almost always different from that of the liquid. The phenomenon is readily understood from the free energies for liquid and solid mixtures. We consider a simple model, under two assumptions: (a) Neither the solid nor the liquid has a solubility gap. (b) The melting temperature τ_A of pure constituent A is lower than the melting temperature τ_B of pure constituent B. We consider a temperature between τ_A and τ_B.

The free energies per atom, $f_S(x)$ for the solid and $f_L(x)$ for the liquid, are shown qualitatively in Figure 11.9a. The two curves intersect at some composition. Let us draw a tangent common to both curves, touching f_S at $x = x_S$ and f_L at $x = x_L$. We can define three composition ranges, each with different internal equilibria:

(a) When $x < x_L$, the system in equilibrium is a homogeneous liquid.

(b) When $x_L < x < x_S$, the system in equilibrium consists of two phases, a solid phase of composition x_S and a liquid phase of composition x_L.

(c) When $x > x_S$ the system in equilibrium is a homogeneous solid.

The compositions x_S and x_L of a solid and a liquid phase in equilibrium are temperature dependent. As the temperature decreases the free energy of the solid decreases more rapidly than that of the liquid. The tangential points in Figure 11.9a move to the left. This behavior is represented by a phase diagram similar to the earlier representation of the equilibrium composition curves for mixtures with phase separation. In Figure 11.9b the curve for x_L is called the **liquidus curve**; the curve for x_S is the **solidus curve**.

The phase diagrams have been determined experimentally for vast numbers of binary mixtures. Those for most of the possible binary alloys are known.* For most metal alloys the phase diagrams are more complicated than Figure 11.9b, which was drawn for a simple system, germanium-silicon.

When the temperature is lowered in a binary liquid mixture with the phase diagram of Figure 11.9b, solidification takes place over a finite temperature range, not just at a fixed temperature. To see this, consider a liquid with the initial composition x_{iL} shown in Figure 11.10. As the temperature is lowered, solidification begins at $\tau = \tau_i$. The composition of the solid formed is given by x_{iS}, so that the composition of the remaining liquid is changed. In the example $x_{iS} > x_{iL}$, so that the liquid moves towards lower values of x, where the solidification temperature is lower. The temperature has to be lowered if solidification is to continue. The composition of the liquid moves along the liquidus curve until the solidification is completed at $\tau = \tau_A$. The solid formed is nonuniform in composition and is not in equilibrium. The solid may homogenize afterward by atomic diffusion, particularly if the temperature remains high for a long time. But for many solids atomic diffusion is too slow, and the inhomogeneity remains "frozen in" indefinitely.

---

* The standard tabulations are by M. Hansen, *Constitution of binary alloys*, McGraw-Hill, 1958; R. P. Elliott, *Constitution of binary alloys, first supplement*, McGraw-Hill, 1965; F. A. Shunk, *Constitution of binary alloys, second supplement*, McGraw-Hill, 1969.

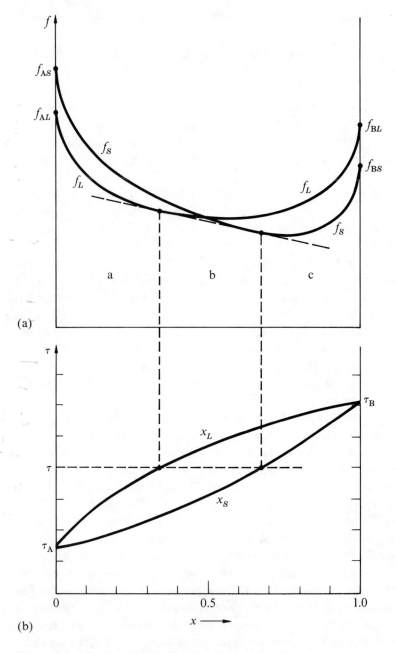

Figure 11.9 Phase equilibrium between liquid and solid mixtures. In this example neither phase exhibits a solubility gap. We assume $\tau_A < \tau < \tau_B$. The upper figure (a) shows the free energies for the two phases; the lower figure (b) shows the corresponding phase diagram. The curves x_L and x_S in the phase diagram are called the liquidus and the solidus curves. The phase diagram is the Ge-Si phase diagram, with $T_{Ge} = 940°C$ and $T_{Si} = 1412°C$.

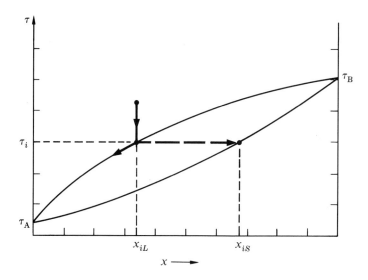

Figure 11.10 Most liquid mixtures do not solidify at a sharp temperature, but over a finite temperature range from τ_i to τ_A. The higher-melting constituent precipitates first, thereby enriching the lower-melting constituent in the liquid phase and thus lowering the solidification temperature of the liquid.

Advanced Treatment: Eutectics.   There are many binary systems in which the liquid phase remains a liquid down to temperatures significantly below the lower melting temperature of the constituents. The gold-silicon alloy is such a system: a mixture of 69 pct Au and 31 pct Si starts to solidify at 370°C. At other compositions solidification starts at a higher temperature. When we plot the temperature of the onset of solidification as a function of alloy composition, we obtain the two-branch liquidus curve in Figure 11.11. Mixtures with two liquidus branches are called **eutectics**. The minimum solidification temperature is the eutectic temperature, where the composition is the eutectic composition.

The solidified solid at the eutectic composition is a two phase solid, with nearly pure gold side by side with nearly pure silicon, as in Figure 11.1. In the solid Au–Si mixture there is a very wide solubility gap. The low melting point occurs for the eutectic composition because the free energy of the homogeneous melt is lower than the free energy of the two phase solid, for temperatures at or above the eutectic temperature.

Such behavior is common among systems that exhibit a solubility gap in the solid but not in the liquid. The behavior of eutectics can be understood from the free energy plots in Figure 11.12a. We assume $f_S(x)$ for the solid as in Figure 11.6,

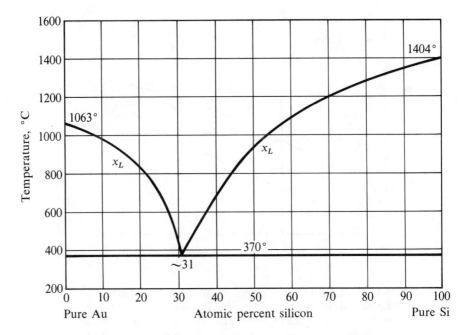

Figure 11.11 Eutectic phase diagram of gold-silicon alloys. The liquidus consists of two branches that come together at the eutectic temperature $T_e = 370°C$. The horizontal line and the experimental data points at 370°C indicate that throughout the entire composition range the mixture does not complete its solidification until the eutectic temperature is reached.

corresponding to different crystal structures α and β for the two pure constituents. Figure 11.12a is constructed for a temperature above the eutectic temperature but below the melting temperature of either constituent, so that the free energy of the liquid reaches below the common tangent to the solid phase curves. We can draw two new two-point tangents that give even lower free energies. We now distinguish five different composition ranges:

(a) and (e). For $x < x_{\alpha S}$ or $x > x_{\beta S}$, the equilibrium state of the system is a homogeneous solid. In the first range the solid will have the crystal structure α; in the second range the structure is β.

(c). For $x_{\alpha L} < x < x_{\beta L}$, the equilibrium state is a homogeneous liquid.

(b) and (d). For $x_{\alpha S} < x < x_{\alpha L}$ or $x_{\beta L} < x < x_{\beta S}$, a liquid phase is in equilibrium with a solid phase.

As the temperature is lowered, $f_{\alpha S}$ and $f_{\beta S}$ decrease more rapidly than f_L, and the range of the homogeneous liquid becomes narrower. Figure 11.12b shows the corresponding phase diagram, including the two solidus curves.

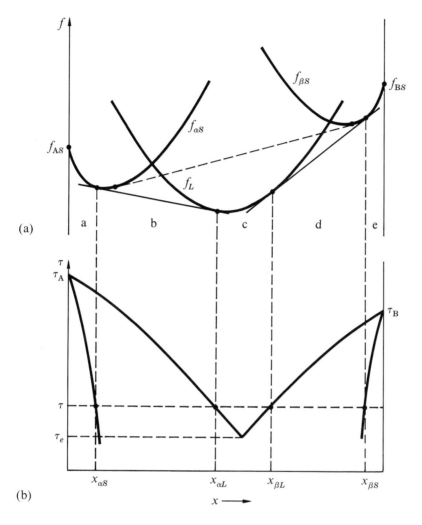

Figure 11.12   Free energies (a) and phase diagram (b) in a simple eutectic system.

At the eutectic temperature τ_e the free energy of the liquid phase is tangential to the common tangent to $f_{\alpha S}$ and $f_{\beta S}$, as in Figure 11.13. The composition at which f_L touches the tangent is the eutectic composition. At $\tau < \tau_e$, the free energy f_L lies above the tangent, although f_L may be below the free energy of a homogeneous solid.

A mixture of composition equal to the eutectic composition solidifies and melts at a single temperature, just like a pure substance. The solidification of

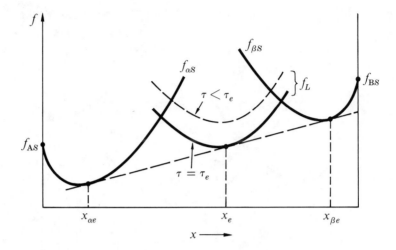

Figure 11.13   Free energies in a eutectic system at $\tau = \tau_e$ and at $\tau < \tau_e$.

compositions away from the eutectic composition starts at a higher temperature and ends at the eutectic temperature. Melting starts at the eutectic temperature and ends at a higher temperature.

The minimum property of the melting temperature of eutectics is widely utilized. The Au–Si eutectic plays a large role in semiconductor device technology: the eutectic permits low temperature welding of electrical contact wires made of gold to silicon devices. Lead-tin alloys exhibit a eutectic (Figure 11.14) at 183°C to give solder a melting temperature below that of pure tin, 232°C. According to whether a sharp melting temperature or a melting range is desired, either the exact eutectic composition (26 pct lead) or a different composition is employed. Salt sprinkled on ice melts the ice because of the low eutectic temperature $-21.2°C$ of the H_2O–NaCl eutectic at 8.17 mol pct NaCl.

The solidus curves of eutectic systems vary greatly in character. For the Pb–Sn system (Figure 11.14) the solid phases in equilibrium with the melt contain an appreciable fraction of the minority constituent, and this fraction increases with decreasing temperature. In other systems this fraction may be small or may decrease with decreasing temperature, or both. The Au–Si system is an example: The relative concentration of Au in solid Si in equilibrium with an Au–Si melt reaches a maximum value of only 2×10^{-6} around 1300°C, and it drops off rapidly at lower temperature.

In our discussion of the free energy curves of Figures 11.12 and 11.13 we assumed that the composition at which the liquid phase free energy touches the

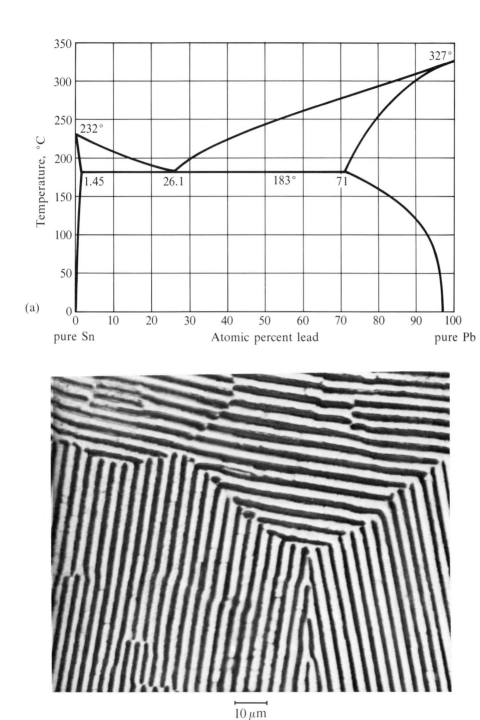

Figure 11.14 (a) Phase diagram of the Pb-Sn system, after Hansen. (b) Microphotograph of the Pb-Sn eutectic, magnified about 800 times. Courtesy of J. D. Hunt and K. A. Jackson.

329

tangent to the solid phase curves lies between the compositions $x_{\alpha S}$ and $x_{\beta S}$. In some systems this point lies outside the interval, as if either $f_{\alpha S}$ and f_L or $f_{\beta S}$ and f_L were interchanged in Figure 11.12a. Such systems are called **peritectic systems**.

SUMMARY

1. A mixture exhibits a solubility gap when the combined free energy of two separate phases side by side is lower than the free energy of the homogeneous mixture.

2. The mixing entropy arises when atoms of different species are interchanged in position. For the alloy $A_{1-x}B_x$, we have

$$\sigma_M = -N[(1-x)\log(1-x) + x\log x].$$

3. The mixing energy for nearest-neighbor interactions is

$$u_M = px(1-x)[u_{AB} - \tfrac{1}{2}(u_{AA} + u_{BB})] \; ,$$

for p nearest neighbors.

4. The liquidus is the composition curve x_L versus τ for a liquid phase in equilibrium with a solid. The solidus is the composition curve x_S versus τ for a solid phase in equilibrium with a liquid.

5. Mixtures with two branches to the liquidus curve are called eutectics. The minimum solidification temperature is called the eutectic temperature.

PROBLEMS

1. Chemical potentials in two-phase equilibrium.   Show that the chemical potentials μ_A and μ_B of the two atomic species A and B of an equilibrium two phase mixture are given by the intercepts of the two-point tangent in Figure 11.2 with the vertical edges of the diagram at $x = 0$ and $x = 1$.

2. Mixing energy in ^3He–^4He and Pb–Sn mixtures.   The phase diagram of liquid ^3He–^4He mixtures in Figure 11.8 shows that the solubility of ^3He in ^4He remains finite (about 6 pct) as $\tau \to 0$. Similarly, the Pb–Sn phase diagram of Figure 11.14 shows a finite residual solubility of Pb in solid Sn with decreasing

τ. What do such finite residual solubilities imply about the form of the function $u(x)$?

3. Segregation coefficient of impurities. Let B be an impurity in A, with $x \ll 1$. In this limit the non-mixing parts of the free energy can be expressed as linear functions of x, as $f_0(x) = f_0(0) + xf_0'(0)$, for both liquid and solid phases. Assume that the liquid mixture is in equilibrium with the solid mixture. Calculate the equilibrium concentration ratio $k = x_S/x_L$, called the **segregation coefficient**. For many systems $k \ll 1$, and then a substance may be purified by melting and partial resolidification, discarding a small fraction of the melt. This principle is widely used in the purification of materials, as in the zone refining of semi-conductors. Give a numerical value for k for $f_{0S}' - f_{0L}' = 1\,\text{eV}$ and $T = 1000\,\text{K}$.

4. Solidification range of a binary alloy. Consider the solidification of a binary alloy with the phase diagram of Figure 11.10. Show that, regardless of the initial composition, the melt will always become fully depleted in component B by the time the last remnant of the melt solidifies. That is, the solidification will not be complete until the temperature has dropped to T_A.

5. Alloying of gold into silicon. (a) Suppose a 1000 Å layer of Au is evaporated onto a Si crystal, and subsequently heated to 400°C. From the Au–Si phase diagram, Figure 11.11, estimate how deep the gold will penetrate into the silicon crystal. The densities of Au and Si are 19.3 and 2.33 g cm^{-3}. (b) Redo the estimate for 800°C.

Chapter 12

Cryogenics

Cryogenics is the physics and technology of the production of low temperatures. We discuss the physical principles of the most important cooling methods, down to the lowest temperatures.

The dominant principle of low temperature generation down to 10 mK is the cooling of a gas by letting it do work against a force during an expansion. The gas employed may be a conventional gas; the free electron gas in a semiconductor; or the virtual gas of ^{3}He atoms dissolved in liquid ^{4}He. The force against which work is done may be external or internal to the gas. Below 10 mK the dominant cooling principle is the isentropic demagnetization of a paramagnetic substance.

We discuss the cooling methods in the order in which they occur in a laboratory cooling chain that starts by liquefying helium and proceeds from there to the lowest laboratory temperatures, usually 10 mK, sometimes 1 μK. Household cooling appliances and automobile air conditioners utilize the same evaporation cooling method that is used in the laboratory for cooling liquid helium below its boiling temperature, to about 1 K.

COOLING BY EXTERNAL WORK
IN AN EXPANSION ENGINE

In the isentropic expansion of a monatomic ideal gas from pressure p_1 to a lower pressure p_2, the temperature drops according to

$$T_2 = T_1(p_2/p_1)^{2/5}, \tag{1}$$

by (6.64). Suppose $p_1 = 32$ atm; $p_2 = 1$ atm; and $T_1 = 300$ K; then the temperature will drop to $T_2 = 75$ K. We are chiefly interested in helium as the working gas, and for helium (1) is an excellent approximation if the cooling process is reversible.

The problems in implementing expansion cooling arise from the partial irreversibility of actual expansion processes. The problems are compounded by the nonexistence of good low temperature lubricants. Actual expansion cooling cycles follow Figure 12.1. The compression and expansion parts of

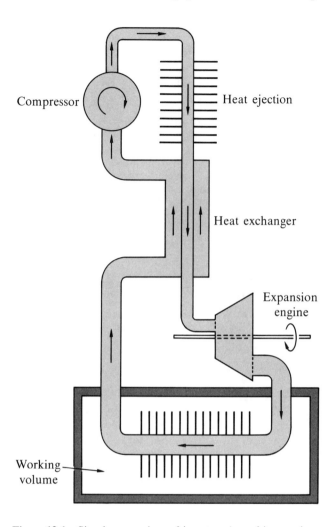

Figure 12.1 Simple expansion refrigerator. A working gas is compressed; the heat of compression is ejected into the environment. The compressed room temperature gas is precooled further in the counterflow heat exchanger. It then does work in an expansion engine, where it cools to a temperature below that of the working volume. After extracting heat from the working volume, the gas returns to the compressor via the heat exchanger.

the cycle are separated. The compression is performed at or above room temperature. The hot compressed gas is cooled to near room temperature by ejecting heat into the environment. The gas is further precooled in a **counter-flow heat exchanger** by contact with the cold return gas stream at the low temperature of the cooling load. The gas is then cooled to its lowest temperature in the expansion engine, usually a low friction turbine. The cold gas extracts heat from the cooling load and then returns to the compressor via the heat exchanger. The heat exchanger greatly reduces the cooling requirements imposed on the expansion engine. The design of the heat exchanger is as important as the design of the expansion engine.

The work extracted by the expansion engine is the enthalpy difference between the input and output gas: The total energy flowing into the expansion engine is the internal energy U_1 of the gas plus the displacement work $p_1 V_1$ done by the compressor, where both U_1 and V_1 refer to a given mass of gas. The total energy leaving the engine with the gas is the energy U_2 of the gas plus the work $p_2 V_2$ required to move the gas against the pressure p_2. The work extracted by the engine is the difference

$$W = (U_1 + p_1 V_1) - (U_2 + p_2 V_2) = H_1 - H_2. \tag{2}$$

For a monatomic ideal gas $U = \frac{3}{2} N\tau$ and $pV = N\tau$, hence $H = \frac{5}{2} N\tau$. The work performed on the engine by the gas is

$$W = \frac{5}{2} N(\tau_1 - \tau_2). \tag{3}$$

The counterflow heat exchanger is an enthalpy exchange device: it is an expansion engine which extracts no external work.

Most gas liquefiers use expansion engines to precool the gas close to its liquefaction temperature. It is impractical to carry the expansion cooling to the point of liquefaction: the formation of a liquid phase inside expansion engines causes mechanical operating difficulties. The final liquefaction stage is usually a Joule-Thomson stage, discussed below. Helium and hydrogen liquefiers usually contain two or more expansion engines at successive temperatures, with multiple heat exchangers.

The principle of cooling by isentropic expansion of an ideal gas is applicable to the electron gas in semiconductors. When electrons flow from a semiconductor with high electron concentration into a semiconductor with a lower electron concentration, the electron gas expands and does work against the potential barrier between the two substances that equalizes the two chemical potentials. The resulting electronic cooling, called the **Peltier effect**, is used down to about 195 K quite routinely; in multistage units temperatures down to 135 K have been achieved.

Gas Liquefaction by the Joule-Thomson Effect

Intermolecular attractive interactions cause the condensation of all gases. At temperatures slightly above the condensation temperature the interactions are strong enough that work against them during expansion causes significant cooling of the gas. If the cooling is sufficient, part of the gas will condense. This process is Joule-Thomson liquefaction.

The practical implementation is simple. Gas at pressure p_1 is forced through a constriction called an expansion valve into space with a lower pressure p_2, as in Figure 12.2. The work is the difference between the displacement work $-p_1 \, dV_1$ done on the gas in pushing it through the expansion valve and the displacement work $+p_2 \, dV_2$ recovered from the gas on the downstream side. Here dV_1 is negative and dV_2 is positive.

The overall process is at constant enthalpy. To see this, notice that the expansion valve acts as an expansion engine that extracts zero work. With $W = 0$ in (2), we have $H_1 = H_2$ in the Joule-Thomson effect. For an ideal gas $H = \frac{5}{2}N\tau$, so that $\tau_1 = \tau_2$ in the expansion. There is zero cooling effect for an ideal gas.

In real gases a small temperature change occurs because of the internal work done by the molecules during expansion. The sign of the temperature

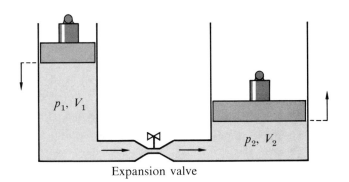

Expansion valve

Figure 12.2 The Joule-Thomson effect. A gas is pushed through an expansion value. If the gas is nonideal, there will be a temperature change during the expansion because of work done against the intermolecular forces. If the temperature is initially below a certain inversion temperature, τ_{inv}, the gas will cool on Joule-Thomson expansion.

Table 12.1   Liquefaction data for low boiling gases

Gas	T_b, K	T_c, K	T_{inv}, K	ΔH, kJ/mol	V_l, cm^3/mol	$\Delta H/V_l$ watt hr/liter
CO_2	195	304	(2050)	25.2	22.3	314
CH_4	112	191	(1290)	8.18	34.4	66
O_2	90.2	155	893	6.82	28.1	67
N_2	77.3	126	621	5.57	34.6	45
H_2	20.4	33.3	205	0.90	28.6	8.7
^{4}He	4.18	5.25	51	0.082	32.0	0.71
^{3}He	3.20	3.35	(23)	0.025	50.8	0.14

NOTE: T_b = atmospheric-pressure boiling temperature; T_c = critical temperature; T_{inv} = Joule-Thomson inversion temperature; ΔH = molar latent heat of vaporization; V_l = molar volume of the liquid. The last column, $\Delta H/V_l$ indicates the heat in watts that can be taken up for a refrigerant consumption of 1 liter per hour; T_{inv} values in parentheses are van der Waals values calculated from T_c and not measured value.

Carbon dioxide solidifies when cooled at atmospheric pressure, because its triple point occurs above atmospheric pressure. Solid CO_2 is known as dry ice. Methane, CH_4, is the principal constituent of natural gas, which is liquefied in huge quantities for shipping as LNG fuel. Liquid oxygen and nitrogen are separated in the liquefaction of air. For helium, we give data both for the common isotope ^{4}He and for ^{3}He.

change during a Joule-Thomson expansion depends on the initial temperature. All gases have an inversion temperature τ_{inv} below which such an expansion cools, above which it heats the gas. Inversion temperatures for common gases are listed in Table 12.1.

Example: Joule-Thomson effect for van der Waals gas.   We found in (10.75) that

$$H = \tfrac{5}{2}N\tau + (N^2/V)(b\tau - 2a) \tag{4}$$

for a van der Waals gas, where a and b are positive constants. The last two terms are the corrections caused by the short range repulsion and the long range attraction. The corrections have opposite signs. The total correction changes sign at the temperature

$$\tau_{inv} = 2a/b = \tfrac{27}{4}\tau_c, \tag{5}$$

where τ_c is the critical temperature, defined by (10.46).

The temperature τ_{inv} is the inversion temperature. For $\tau < \tau_{inv}$ the enthalpy at fixed temperature increases as the volume increases; here in expansion the work done against the attractive interactions between molecules is dominant. In a process at constant enthalpy this increase is compensated by a decrease of the $\tfrac{5}{2}N\tau$ term, that is, by cooling the gas. For

$\tau > \tau_{\mathrm{inv}}$ the enthalpy at a fixed temperature decreases because now the work done by the strong short range repulsive interactions is dominant: at the higher temperature the molecules penetrate farther into the repulsive region.

Linde cycle. In gas liquefiers the Joule-Thomson expansion is combined with a counterflow heat exchanger, as shown in Figure 12.3. The combination is called a Linde cycle, after Carl von Linde who used such a cycle in 1895 to liquefy air starting from room temperature. In our discussion we assume that the expanded gas returning from the heat exchanger is at the same temperature as the compressed gas entering it. We neglect any pressure difference between the output of the heat exchanger and the pressure above the liquid.

To and from compressor
or precooling stages

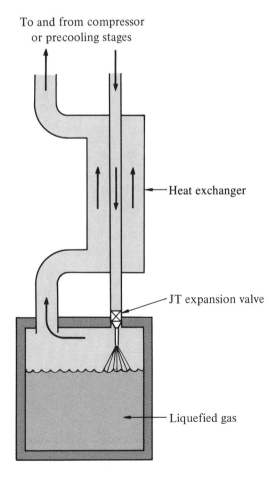

Heat exchanger

JT expansion valve

Liquefied gas

Figure 12.3 The Linde cycle. Gas is liquefied by combining Joule-Thomson expansion with a counterflow heat exchanger.

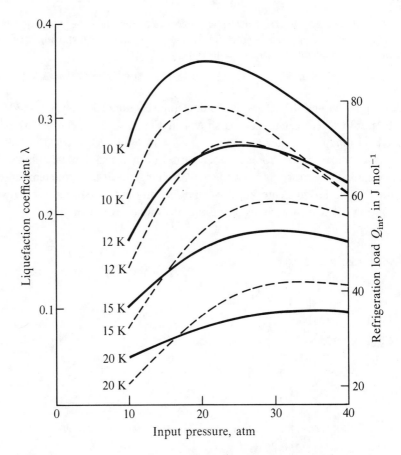

Figure 12.4 Performance of helium liquefiers operating by the Linde cycle, as a function of the input pressure, for an output pressure of 1 atm and for various values of the input temperature. The solid curves give the liquefaction coefficient. The broken curves give $Q_{int} = H_{out} - H_{in}$, the internal refrigeration load available at 4.2 K if the load is placed inside the liquefier and the still cold helium gas boiled off by the load is returned through the heat exchanger rather than boiled off into the atmosphere. See Problem 3. After A. J. Croft in *Advanced cryogenics* (C. A. Bailey, ed.), Plenum, 1971, p. 187.

The combination heat exchanger–expansion valve is a constant enthalpy arrangement. Let one mole of gas enter the combination; suppose that the fraction λ is liquefied. Constant enthalpy requires that

$$H_{in} = \lambda H_{liq} + (1 - \lambda)H_{out}. \tag{6}$$

Here $H_{in} = H(T_{in}, p_{in})$ and $H_{out} = H(T_{in}, p_{out})$ are the enthalpies per mole of gas at the input and output pressures, both at the common upper temperature of the heat exchanger. H_{liq} is the enthalpy per mole of liquid at its boiling temperature under the pressure p_{out}. From (6) we obtain the fraction

$$\lambda = \frac{H_{out} - H_{in}}{H_{out} - H_{liq}}, \tag{7}$$

called the liquefaction coefficient.

Liquefaction takes place when $H_{out} > H_{in}$; that is, when

$$H(T_{in}, p_{out}) > H(T_{in}, p_{in}). \tag{8}$$

Only the enthalpies at the input temperature of the heat exchanger matter. If the Joule-Thomson expansion at this temperature cools the gas, liquefaction will take place.

The three enthalpies in (7) are known experimentally. Figure 12.4 shows the liquefaction coefficient calculated from them for helium. The liquefaction coefficient drops rapidly with increasing T_{in}, because of the decrease of the numerator in (7) and the increase of the denominator. To obtain useful liquefaction, say $\lambda > 0.1$, input temperatures below one-third of the inversion temperature are usually required. For many gases this requires precooling of the gas by an expansion engine. The combination of an expansion engine and a Linde cycle is called a **Claude cycle**. The expansion engine is invariably preceded by another heat exchanger, as in Figure 12.1.

Evaporation Cooling: Pumped Helium, to 0.3 K

Starting from liquid helium, the simplest route to lower temperatures is by evaporation cooling of the liquid helium, by pumping away helium vapor. The latent heat of vaporization of the liquid helium is extracted along with the vapor. The heat extraction causes the further cooling: work is done against the interatomic forces that caused the helium to liquefy in the first place. In Joule-Thomson cooling the initial state is a gas, while in evaporation cooling the initial state is a liquid.

Table 12.2   Temperatures, in kelvin, at which the vapor pressures of ^{4}He and ^{3}He reach specified values

p (torr)	10^{-4}	10^{-3}	10^{-2}	10^{-1}	1	10	100
^{4}He	0.56	0.66	0.79	0.98	1.27	1.74	2.64
^{3}He	0.23	0.28	0.36	0.47	0.66	1.03	1.79

The lowest temperature accessible by evaporation cooling of liquid helium is a problem in vacuum technology (Chapter 14). As the temperature drops, the equilibrium vapor pressure drops (Table 12.2) and so does the rate at which helium gas and its heat of vaporization can be extracted from the liquid helium bath.

Evaporation cooling is the dominant cooling principle in everyday cooling devices such as household refrigerators and freezers and in air conditioners. The only difference is in the working substance.

Helium Dilution Refrigerator: Millidegrees

Once the equilibrium vapor pressure of liquid ^{3}He has dropped to 10^{-3} torr, classical refrigeration principles lose their utility. The temperature range from 0.6 K to 0.01 K is dominated by the helium dilution refrigerator, which is an evaporation refrigerator in a very clever quantum disguise.*

We saw in Chapter 7 that ^{4}He atoms are bosons, while ^{3}He atoms are fermions. This distinction is not important at temperatures appreciably higher than the superfluid transition temperature of ^{4}He, 2.17 K. However, the two isotopes behave as altogether different substances at lower temperatures. Below 0.87 K liquid ^{3}He and ^{4}He are immiscible over a wide composition range, like oil and water. This was discussed in Chapter 11 and is shown in the phase diagram of ^{3}He–^{4}He mixtures in Figure 11.7. A mixture with composition in the range labeled unstable will decompose into two separate phases whose compositions are given by the two branches of the curve enclosing that area. The concentrated ^{3}He phase floats on top of the dilute ^{3}He phase.

As $T \to 0$, the ^{3}He concentration of the phase dilute in ^{3}He drops to about 6 pct, and the phase rich in ^{3}He becomes essentially pure ^{3}He. Consider a liquid

---

* For good reviews, see D. S. Betts, Contemporary Physics **9**, 97 (1968); J. C. Wheatley, Am. J. Phys. **36**, 181 (1968); for a general review of cooling techniques below 1 K see W. J. Huiskamp and O. V. Lounasmaa, Repts. Prog. Phys. **36**, 423 (1973); O. V. Lounasmaa, *Experimental principles and methods below 1 K*, Academic Press, New York, 1974. A very elementary account is O. V. Lounasmaa, Scientific American **221**, 26 (1969).

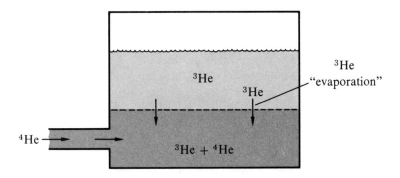

Figure 12.5 Cooling principle of the helium dilution refrigerator. Liquid ^{3}He is in equilibrium with a ^{3}He-^{4}He mixture. When ^{4}He is added to the mixture, ^{3}He evaporates from the pure ^{3}He fluid and absorbs heat in the process.

^{3}He–^{4}He mixture with more than 6 pct ^{3}He at a temperature in the millidegree range, near the bottom of Figure 11.7. At these temperatures almost all the ^{4}He atoms have condensed into the ground state orbital. Their entropy is negligible compared to that of the remaining ^{3}He atoms, which then behave as if they were present alone, as a gas occupying the volume of the mixture. If the ^{3}He concentration exceeds 6 pct, the excess condenses into concentrated liquid ^{3}He and latent heat is liberated. If concentrated liquid ^{3}He is evaporated into the ^{4}He rich phase, the latent heat is consumed. The principle of evaporation cooling can again be applied: this is the basis of the helium dilution refrigerator.

To see how the solution of ^{3}He can be employed to obtain refrigeration, consider the equilibrium between the concentrated ^{3}He liquid phase and the dilute ^{3}He gas-like phase (Figure 12.5). Suppose that the ^{3}He:^{4}He ratio of the dilute phase is decreased, as by dilution with pure ^{4}He. In order to restore the equilibrium concentration, ^{3}He atoms will evaporate from the concentrated ^{3}He liquid. Cooling will result.

To obtain a cyclic process the ^{3}He–^{4}He mixture must be separated again. The most common method is by distillation, using the different equilibrium vapor pressures of ^{3}He and ^{4}He (Table 12.2). Figure 12.6 shows a schematic diagram of a refrigerator built on these principles. The diagram is highly oversimplified. In particular, in actual refrigerators the heat exchanger between the mixing chamber and the still has an elaborate multistage design. An alternate method* to separate the ^{3}He–^{4}He mixture utilizes the superfluidity of ^{4}He below 2.17 K. For a variety of practical reasons it is less commonly used, although its performance is excellent.

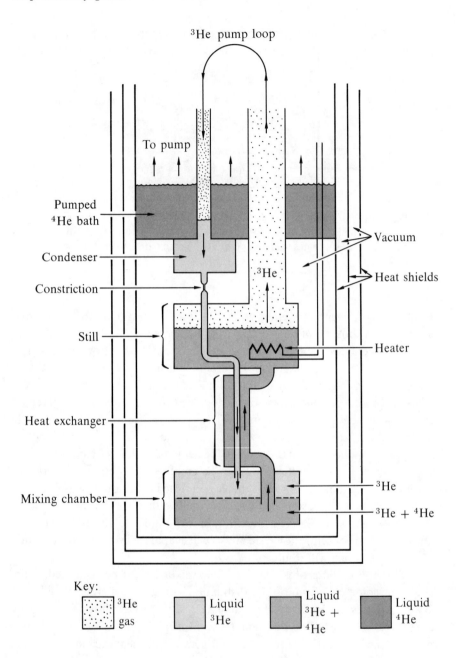

^{3}He pump loop

To pump

Pumped
^{4}He bath

Condenser

Constriction

^{3}He

Still

Heater

Heat exchanger

Mixing chamber

Vacuum

Heat shields

^{3}He

^{3}He + ^{4}He

Key:

^{3}He gas

Liquid ^{3}He

Liquid ^{3}He + ^{4}He

Liquid ^{4}He

Figure 12.6 Helium dilution refrigerator. Precooled liquid ^{3}He enters a mixing chamber at the lower end of the assembly, where cooling takes place by the quasi-evaporation of the ^{3}He atoms into the denser ^{3}He-^{4}He mixed phase underneath. The quasi-gas of ^{3}H atoms dissolved in liquid ^{4}He then diffuses through a counterflow heat exchanger into a still. There the ^{3}He is distilled from the ^{3}He-^{4}He mixture selectively, and is pumped off. To obtain a useful ^{3}He evaporation and circulation rate, heat must be added to the still, to raise its temperature to about 0.7 K, at which temperature the ^{4}He vapor pressure is still much smaller. Thus, the ^{4}He does not circulate to any appreciable extent; the ^{3}He moves through a nearly stationary background of ^{4}He. The pumped-off ^{3}He is returned to the system and is condensed in a condenser that is cooled to about 1 K by contact with a pumped ^{4}He bath. The constriction below the condenser takes up the excess pressure generated by the circulation pump over the pressure in the still. The liquified ^{3}He is cooled further, first in the still, then in the counterflow heat exchanger, before re-entering the mixing chamber.

The helium dilution refrigerator has a low temperature limit. In the conventional evaporation refrigerator this limit arose because of the disappearance of the gas phase, but the quasi-gas phase of ^{3}He persists down to $\tau = 0$. However, the heat of quasi-vaporization of ^{3}He vanishes proportionally to τ^2, and as a result, the heat removal rate from the mixing chamber vanishes as τ^2. The practical low temperature limit is about 10 mK. In one representative device* a temperature of 8.3 mK has been achieved; the same device was capable of removing 40 μW at 80 mK.

Temperatures below 8 mK can be achieved by single shot operation. If, in the design of Figure 12.6, we shut off the ^{3}He supply after some time of operation, there is no need to cool the incoming ^{3}He itself, and the temperature of the mixing chamber drops below its steady state value, until all ^{3}He has been removed from the chamber.

The dilution refrigerator is not the only cooling method in the millikelvin range that utilizes the peculiar properties of ^{3}He. An alternate method, known as Pomeranchuk cooling, utilizes the phase diagram of ^{3}He, as shown in Figure 7.15, with its negative slope of the phase boundary between liquid and solid ^{3}He. The interested reader is referred to the reviews by Huiskamp and Lounasmaa, and by Lounasmaa, cited earlier.

---

* N. H. Pennings, R. de Bruyn Ouboter, K. W. Taconis, Physica B **81**, 101 (1976), and Physica B **84**, 102 (1976).

ISENTROPIC DEMAGNETIZATION:
QUEST FOR ABSOLUTE ZERO

Below 0.01 K the dominant cooling process is the isentropic (adiabatic) demagnetization of a paramagnetic substance. By this process, temperatures of 1 mK have been attained with electronic paramagnetic systems and 1 μK with nuclear paramagnetic systems. The method depends on the fact that at a fixed temperature the entropy of a system of magnetic moments is lowered by application of a magnetic field—essentially because fewer states are accessible to the system when the level splitting is large than when the level splitting is small. Examples of the dependence of the entropy on the magnetic field were given in Chapters 2 and 3.

We first apply a magnetic field B_1 at constant temperature τ_1. The spin excess will attain a value appropriate to the value of B_1/τ_1. If the magnetic field is then reduced to B_2 without changing the entropy of the spin system, the spin excess will remain unchanged, which means that B_2/τ_2 will equal B_1/τ_1. If $B_2 \ll B_1$, then $\tau_2 \ll \tau_1$. When the specimen is demagnetized isentropically, entropy can flow into the spin system only from the system of lattice vibrations, as in Figure 12.7. At the temperatures of interest the entropy of the lattice vibrations is usually negligible; thus the entropy of the spin system will be essentially constant during isentropic demagnetization of the specimen.

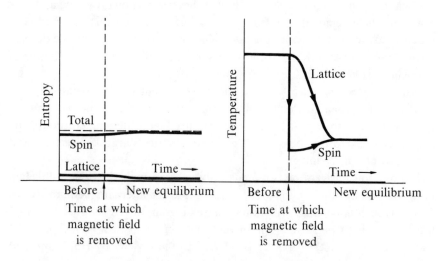

Figure 12.7 During isentropic demagnetization the total entropy of the s specimen is constant. The initial entropy of the lattice should be small in comparison with the entropy of the spin system in order to obtain significant cooling of the lattice.

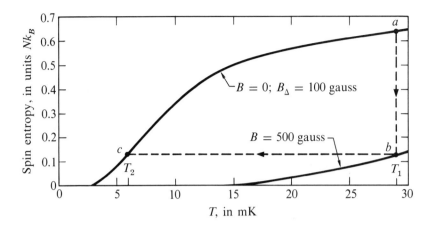

Figure 12.8 Entropy for a spin $\frac{1}{2}$ system as a function of temperature, assuming an internal random magnetic field B_Δ of 100 gauss. The specimen is magnetized isothermally along *ab*, and is then insulated thermally. The external magnetic field is turned off along *bc*. In order to keep the figure on a reasonable scale the initial temperature T_1 and the external magnetic field are lower than would be used in practice.

The steps carried out in the cooling process are shown in Figure 12.8. The field is applied at temperature τ_1 with the specimen in good thermal contact with the surroundings, giving the isothermal path *ab*. The specimen is then insulated ($\Delta\sigma = 0$) and the field removed; the specimen follows the constant entropy path *bc*, ending up at temperature τ_2. The thermal contact at τ_1 is provided by helium gas, and the thermal contact is broken by removing the gas with a pump.

The population of a magnetic sublevel is a function only of mB/τ, where m is the magnetic moment of a spin. The spin-system entropy is a function only of the population distribution; hence the spin entropy is a function only of mB/τ. If B_Δ is the effective field that corresponds to the diverse local interactions among the spins or of the spins with the lattice, the final temperature τ_2 reached in an isentropic demagnetization experiment is

$$\tau_2 = \tau_1(B_\Delta/B), \tag{9}$$

where B is the initial field and τ_1 the initial temperature. Results are shown in Figure 12.9 for the paramagnetic salt known as CMN, which denotes cerous magnesium nitrate.

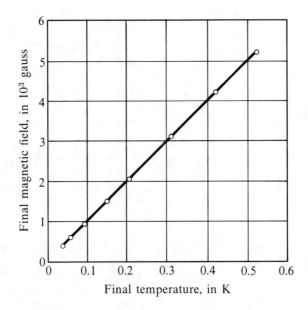

Figure 12.9 Final magnetic field B_f versus final temperature T_f for magnetic cooling of cerous magnesium nitrate. In these experiments the magnetic field was not removed entirely, but only to the indicated values. The initial fields and temperatures were identical in all runs. After unpublished results of J. S. Hill and J. H. Milner, as cited by N. Kurti, Nuovo Cimento (Supplemento) **6**, 1109 (1957).

The process described so far is a single shot process. It is easily converted into a cyclic process by thermally disconnecting, in one way or another, the demagnetized working substance from the load, reconnecting it to the reservoir at τ_1, and repeating the process.*

Nuclear Demagnetization

Because nuclear magnetic moments are weak, nuclear magnetic interactions are much weaker than similar electronic interactions. We expect to reach a temperature 100 times lower with a nuclear paramagnet than with an electron paramagnet. The initial temperature of the nuclear stage in a nuclear spin-

---

* C. V. Heer, C. B. Barnes, and J. G. Daunt, Rev. Sci. Inst. **25**, 1088 (1954); W. P. Pratt, S. S. Rosenblum, W. A. Steyert, and J. A. Barclay, Cryogenics **17**, 381 (1977).

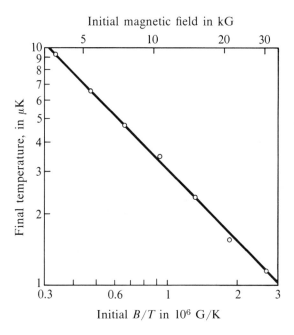

Figure 12.10   Nuclear demagnetizations of copper nuclei in the metal, starting from 0.012 K and various fields. After M. V. Hobden and N. Kurti, Phil. Mag. **4**, 1902 (1959).

cooling experiment must be lower than in an electron spin-cooling experiment. If we start at $B = 50\,\text{kG}$ and $T_1 = 0.01\,\text{K}$, then $mB/k_B T_1 \approx 0.5$, and the entropy decrease on magnetization is over 10 percent of the maximum spin entropy. This is sufficient to overwhelm the lattice and from (9) we estimate a final temperature $T_2 \approx 10^{-7}\,\text{K}$. The first nuclear cooling experiment was carried out by Kurti and coworkers on Cu nuclei in the metal, starting from a first stage at about 0.02 K as attained by electron demagnetization cooling. The lowest temperature reached in this experiment was $1.2 \times 10^{-6}\,\text{K}$. The results in Figure 12.10 fit a line of the form of (9): $T_2 = T_1(3.1/B)$ with B in gauss, so that $B_\Delta = 3.1$ gauss. This is the effective interaction field of the magnetic moments of the Cu nuclei. The motivation for using nuclei in a metal rather than in an insulator is that conduction electrons help ensure rapid thermal contact of lattice and nuclei at the temperature of the first stage.

Temperatures below $1\,\mu\text{K}$ have been achieved in experiments in which the cooling load was the system of nuclear spins itself, particularly in experiments that were combinations of cooling experiments and nuclear magnetic resonance experiments.*

SUMMARY

1. The two dominant principles of the production of low temperatures are the cooling of a gas by letting it do work against a force during an expansion and the isentropic demagnetization of a paramagnetic substance.

2. Joule-Thomson cooling is an irreversible process in which work is done against interatomic attractive forces in a gas. It is used as the last cooling stage in liquefying low-boiling gases.

3. In evaporation cooling the work is also done against the interatomic forces, but starting from the liquid phase rather than the gas phase. Using different working substances, evaporation cooling forms the basis of household cooling devices, automobile air conditioners, and laboratory cooling devices (in the range 4 K down to 10 mK).

4. The helium dilution refrigerator is an evaporation cooling device in which the gas is the virtual gas of ^{3}He atoms dissolved in ^{4}He.

5. Isentropic demagnetization utilizes the lowering of the temperature of a system of magnetic moments, when an external magnetic field is reduced in strength. The magnetic moments may be electronic or nuclear moments. By using nuclear moments, temperatures in the microkelvin range may be achieved.

PROBLEMS

1. Helium as a van der Waals gas. (a) Estimate the liquefaction coefficient λ for helium by treating it as a van der Waals gas. Select the van der Waals coefficients a and b in such a way that for one mole $2Nb$ is the actual molar volume of liquid helium and that $2a/b$ is the actual inversion temperature. Use the data in Table 12.1. Approximate the denominator in (7) by setting

$$H_{\text{out}} - H_{\text{liq}} \simeq \Delta H + \tfrac{5}{2}(\tau_{\text{in}} - \tau_{\text{liq}}) , \tag{10}$$

---

* See, for example, M. Chapellier, M. Goldman, V. H. Chau and A. Abragam, Appl. Phys. **41**, 849 (1970).

where ΔH is the latent heat of vaporization of liquid helium. (Explain how this approximation arises if one treats the expanded gas as an ideal gas). The resulting expression gives λ as a function of the molar volumes V_{in} and V_{out}. Convert to pressures by approximating the V's via the ideal gas law. (b) Insert numerical values for $T = 15\,\mathrm{K}$ and compare with Figure 12.4.

2. *Ideal Carnot liquefier.* (a) Calculate the work W_L that would be required to liquefy one mole of a monatomic ideal gas if the liquefier operated reversibly. Assume that the gas is supplied at room temperature T_0, and under the same pressure p_0 at which the liquefied gas is removed, typically 1 atmosphere. Let T_b be the boiling temperature of the gas at this pressure, and ΔH the latent heat of vaporization. Show that under these conditions

$$W_L = \tfrac{5}{2}RT_0 \times \left(\log \frac{T_0}{T_b} - \frac{T_0 - T_b}{T_0} \right) + \frac{T_0 - T_b}{T_b} \times \Delta H. \qquad (11)$$

To derive (11) assume that the gas is first cooled at fixed pressure p_0 from T_0 to T_b, by means of a reversible refrigerator that operates between the fixed upper temperature $T_h = T_0$ and a variable lower temperature equal to the gas temperature. Initially $T_l = T_0$, and at the end $T_l = T_b$. After reaching T_b the refrigerator extracts the latent heat of vaporization at the fixed lower temperature T_b. (b) Insert $T_0 = 300\,\mathrm{K}$ and values for T_b and ΔH characteristic of helium. Re-express the result as kilowatt-hours per liter of liquid helium. Actual helium liquefiers consume 5 to $10\,\mathrm{kWh/liter}$.

3. *Claude cycle helium liquefier.* Consider a helium liquefier in which $1\,\mathrm{mol\,s^{-1}}$ of gas enters the Linde stage at $T_{in} = 15\,\mathrm{K}$ and at a pressure $p_{in} = 30\,\mathrm{atm}$. (a) Calculate the rate of liquefaction, in liter $\mathrm{hr^{-1}}$. Suppose that all the liquefied helium is withdrawn to cool an external experimental apparatus, releasing the boiled-off helium vapor into the atmosphere. Calculate the cooling load in watts sufficient to evaporate the helium at the rate it is liquefied. Compare this with the cooling load obtainable if the liquefier is operated as a closed-cycle refrigerator by placing the apparatus into the liquid collection vessel of the liquefier, so that the still cold boiled-off helium gas is returned through the heat exchangers. (b) Assume that the heat exchanger between compressor and expansion engine (Figure 12.1) is sufficiently ideal that the expanded return gas that leaves it with pressure p_{out} is at essentially the same temperature T_c as the compressed gas entering it with pressure p_c. Show that under ordinary liquefier operation the expansion engine must extract the work

$$W_e = H(T_c,p_c) - H(T_{in},p_{in})$$

$$- (1 - \lambda)[H(T_c,p_{out}) - H(T_{in},p_{out})] \simeq \tfrac{5}{2}\lambda\, R(T_c - T_{in}) , \qquad (12)$$

per mole of compressed gas. Here T_{in}, p_{in}, p_{out}, and λ have the same meaning as in the Linde cycle section of this chapter. Assume the expansion engine operates isentropically between the pressure-temperature pairs (p_c, T_e) and (p_{in}, T_{in}). From (12) and the given values of (p_{in}, T_{in}), calculate (p_c, T_e). (c) Estimate the minimum compressor power required to operate the liquefier, by assuming that the compression is isothermal from p_{out} to p_c at temperature $T_c = 50°C$. Combine the result with the cooling loads calculated under (a) into a coefficient of refrigerator performance, for both modes of operation. Compare with the Carnot limit.

4. Evaporation cooling limit.   Estimate the lowest temperature T_{min} that can be achieved by evaporation cooling of liquid ^4He if the cooling load is 0.1 W and the vacuum pump has a pump speed $S = 10^2$ liter s^{-1}. Assume that the helium vapor pressure above the boiling helium is equal to the equilibrium vapor pressure corresponding to T_{min}, and assume that the helium gas warms up to room temperature and expands accordingly before it enters the pump. *Note*: The molar volume of an ideal gas at room temperature and atmospheric pressure (760 torr) is about 24 liters. Repeat the calculation for a much smaller heat load (10^{-3} W) and a faster pump (10^3 liter s^{-1}). Pump speed is defined in Chapter 14.

5. Initial temperature for demagnetization cooling.   Consider a paramagnetic salt with a Debye temperature (Chapter 4) of 100 K. A magnetic field of 100 kG or 10 tesla is available in the laboratory. Estimate the temperature to which the salt must be precooled by other means in order that significant magnetic cooling may subsequently be obtained by the isentropic demagnetization process. Take the magnetic moment of a paramagnetic ion to be 1 Bohr magneton. By significant cooling we may understand cooling to 0.1 of the initial temperature.

Semiconductor Statistics

Note: This chapter is written for students with a professional interest in semiconductors. We assume familiarity with conduction and valence bands; electrons and holes; donors and acceptors. The notation is that:

n_e = concentration of conduction electrons;

n_h = concentration of holes;

n_i = value of n_e or n_h for an intrinsic semiconductor;

n_c = effective quantum concentration for conduction electrons;

n_v = effective quantum concentration for holes.

In the semiconductor literature n_c and n_v are called the effective densities of states for the conduction and valence bands. Notice that we use μ for the chemical potential or Fermi level, and we use $\tilde{\mu}$ for carrier mobilities.

ENERGY BANDS; FERMI LEVEL;
ELECTRONS AND HOLES

The application of the Fermi-Dirac distribution to electrons in semiconductors is central to the design and operation of all semiconductor devices, and thus to much of modern electronics. We treat below those aspects of the physics of semiconductors and semiconductor devices that are parts of thermal physics. We assume that the reader is familiar with the basic ideas of the physics of electrons in crystalline solids, as treated in the texts on solid state physics and on semiconductor devices cited in the general references. We assume the concept of energy bands and of conduction by electrons and holes. Our principal aim is to understand the dependence of the all-important concentrations of conduction electrons and of holes upon the impurity concentration and the temperature.

A semiconductor is a system with electron orbitals grouped into two energy bands separated by an energy gap (Figure 13.1). The lower band is the **valence band** and the upper band is the **conduction band**.* In a pure semiconductor at $\tau = 0$ all valence band orbitals are occupied and all conduction band orbitals are empty. A full band cannot carry any current, so that a pure semiconductor at $\tau = 0$ is an insulator. Finite conductivity in a semiconductor follows either from the presence of electrons, called **conduction electrons**, in the conduction band or from unoccupied orbitals in the valence band, called **holes**.

Two different mechanisms give rise to conduction electrons and holes: Thermal excitation of electrons from the valence band to the conduction band, or the presence of impurities that change the balance between the number of orbitals in the valence band and the number of electrons available to fill them.

We denote the energy of the top of the valence band by ε_v, and the energy of the bottom of the conduction band by ε_c. The difference

$$\varepsilon_g = \varepsilon_c - \varepsilon_v, \tag{1}$$

is the **energy gap** of the semiconductor. For typical semiconductors ε_g is between 0.1 and 2.5 electron volts. In silicon, $\varepsilon_g \simeq 1.1$ eV. Because $\tau \simeq 1/40$ eV at room

---

* We treat both bands as single bands; for our purposes it does not matter that both may be groups of bands with additional gaps within each group.

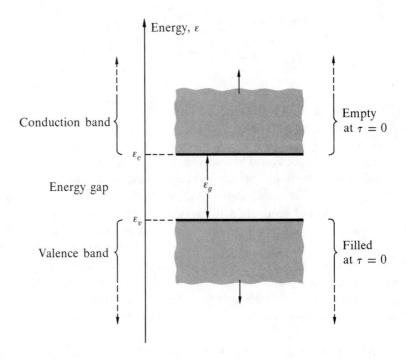

Figure 13.1   Energy band structure of a pure semiconductor or insulator.
ˉThe electron orbitals occur in bands which extend through the crystal.
At $\tau = 0$ all orbitals up to the top of the valence band are filled, and the
conduction band is empty. The energy interval between the bands is called
the energy gap.

temperature, we usually have $\varepsilon_g \gg \tau$. Substances with a gap of more than about
2.5 eV are usually insulators. Table 13.1 gives the energy gaps for selected
semiconductors, together with other properties needed later.

Let n_e denote the concentration of conduction electrons and n_h the concentration of holes. In a pure semiconductor the two will be equal:

$$n_e = n_h, \tag{2}$$

if the crystal is electrically neutral.

Most semiconductors as used in devices have been intentionally doped with
impurities that may become thermally ionized in the semiconductor at room
temperature. Impurities that give an electron to the crystal (and become
positively charged in the process) are called **donors**. Impurities that accept

Table 13.1 Band structure data of some important semiconductors

	Energy gaps at 300 K	Quantum concentrations of electrons and holes at 300 K		Density-of-states effective masses, in units of the free electron mass		Dielectric constants, relative to vacuum
	ε_g, eV	n_c, cm^{-3}	n_v, cm^{-3}	m_e^*/m	m_h^*/m	ϵ/ϵ_0
Si	1.14	2.7×10^{19}	1.1×10^{19}	1.06	0.58	11.7
Ge	0.67	1.0×10^{19}	5.2×10^{18}	0.56	0.35	15.8
GaAs	1.43	4.6×10^{17}	1.5×10^{19}	0.07	0.71	13.13
InP	1.35	4.9×10^{17}	6.9×10^{18}	0.073	0.42	12.37
InSb	0.18	4.6×10^{16}	6.2×10^{18}	0.015	0.39	17.88

an electron from the valence band (and become negatively charged in the process) are called **acceptors**.

Let n_d^+ be the concentration of positively charged donors and n_a^- the concentration of negatively charged acceptors. The difference

$$\Delta n \equiv n_d^+ - n_a^- \tag{3}$$

is called the net ionized donor concentration. The electrical neutrality condition becomes

$$n_e - n_h = \Delta n = n_d^+ - n_a^-, \tag{4}$$

which specifies the difference between electron and hole concentrations.

The electron concentration may be calculated from the Fermi-Dirac distribution function of Chapter 6:

$$f_e(\varepsilon) = \frac{1}{\exp[(\varepsilon - \mu)/\tau] + 1}, \tag{5}$$

where μ is the chemical potential of the electrons. The subscript e refers to electrons. In semiconductor theory the electron chemical potential is always called the **Fermi level**. Further, in semiconductor theory the character μ is almost always reserved for the electron and hole mobilities, and the Fermi level is designated by ε_F or by ζ. To avoid confusion with the Fermi energy of a metal which we designated as ε_F and which stands for the Fermi level in the limit $\tau \to 0$, we shall maintain our previous usage of the letter μ for the chemical potential at any temperature.

Given μ and τ, the number of conduction electrons is obtained by summing the distribution function $f_e(\varepsilon)$ over all conduction band orbitals:

$$N_e = \sum_{\mathrm{CB}} f_e(\varepsilon). \tag{6}$$

The number of holes is

$$N_h = \sum_{\mathrm{VB}} \left[1 - f_e(\varepsilon)\right] = \sum_{\mathrm{VB}} f_h(\varepsilon), \tag{7}$$

where the summation is over all valence band orbitals. Here we have introduced the quantity

$$f_h(\varepsilon) \equiv 1 - f_e(\varepsilon) = \frac{1}{\exp[(\mu - \varepsilon)/\tau] + 1}, \tag{8}$$

which is the probability that an orbital at energy ε is unoccupied. We say that the unoccupied orbital is "occupied by a hole"; then $f_h(\varepsilon)$ is the distribution function for holes just as $f_e(\varepsilon)$ is the distribution function for electrons. Comparison of (8) with (5) shows that the hole occupation probability involves $\mu - \varepsilon$ where the electron occupation probability involves $\varepsilon - \mu$.

The concentrations $n_e = N_e/V$ and $n_h = N_h/V$ depend on the Fermi level. But what is the value of the Fermi level? It is determined by the electrical neutrality requirement (4), now written as $n_e(\mu) - n_h(\mu) = \Delta n$. This is an implicit equation for μ; to solve the equation we must determine the functional dependences $n_e(\mu)$ and $n_h(\mu)$.

Classical Regime

We assume that both electron and hole concentrations are in the classical regime defined by the requirements that $f_e \ll 1$ and $f_h \ll 1$, as in Chapter 6. This will be true if, as in Figure 13.2, the Fermi level lies inside the energy gap and is separated from both band edges by energies large enough that

$$\exp[-(\varepsilon_c - \mu)/\tau] \ll 1; \qquad \exp[-(\mu - \varepsilon_v)/\tau] \ll 1. \tag{9}$$

To satisfy (9) both $(\varepsilon_c - \mu)$ and $(\mu - \varepsilon_v)$ have to be positive and at least a few times larger than τ. Such a semiconductor is called **nondegenerate**. The inequalities (9) place upper limits on the electron and hole concentrations and are satisfied in many applications. With (9) the two occupation probabilities $f_e(\varepsilon)$ and $f_h(\varepsilon)$ reduce to classical distributions:

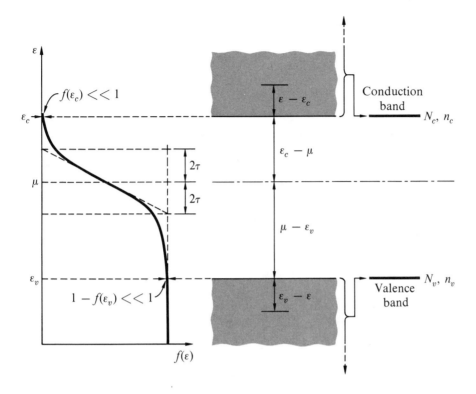

Figure 13.2 Occupancy of orbitals at a finite temperature, according to the Fermi-Dirac distribution function. The conduction and valence bands may be represented in terms of temperature-dependent effective numbers N_c, N_v of degenerate orbitals located at the two band edges ε_c, ε_v. The n_c, n_v are the corresponding quantum concentrations.

$$f_e(\varepsilon) \simeq \exp[-(\varepsilon - \mu)/\tau]; \qquad f_h(\varepsilon) \simeq \exp[-(\mu - \varepsilon)/\tau]. \tag{10}$$

We use (6) and (10) to write the total number of conduction electrons in the form

$$N_e = \sum_{\mathrm{CB}} \exp[-(\varepsilon - \mu)/\tau] = \exp[-(\varepsilon_c - \mu)/\tau]\left\{\sum_{\mathrm{CB}} \exp[-(\varepsilon - \varepsilon_c)/\tau]\right\},$$

or

$$N_e = N_c \exp[-(\varepsilon_c - \mu)/\tau], \tag{11}$$

where we define

$$N_c \equiv \sum_{CB} \exp[-(\varepsilon - \varepsilon_c)/\tau]. \qquad (12)$$

Here $\varepsilon - \varepsilon_c$ is the energy of a conduction electron referred to the conduction band edge ε_c as origin.

The expression for N_c has the mathematical form of a partition function for one electron in the conduction band. In Chapter 3 we evaluated a similar sum denoted there by Z_1, and we can adapt that result to the present problem with an approximate modification for band structure effects. Because of the rapid decrease of $\exp[-(\varepsilon - \varepsilon_c)/\tau]$ as ε increases above its minimum value at ε_c, only the distribution of orbitals within a range of a few τ above ε_c really matters in the evaluation of the sum in (12). The orbitals high in the band make a negligible contribution. The important point is that near the band edge the electrons behave very much like free particles. Not only are the electrons mobile, which causes the conductivity of the semiconductor, but the energy distribution of the orbitals near the band edge usually differs from that of free particles only by a proportionality factor in the energy and eventually in the sum for Z_1.

We can arrange for a suitable proportionality factor by use of a device called the **density-of-states effective mass**. For free particles we calculated the partition function Z_1 in (3.62), but for zero spin. For particles of spin $\frac{1}{2}$ the result is larger by a factor of 2, so that (12) becomes

$$N_c = Z_1 = 2n_Q V = 2(m\tau/2\pi\hbar^2)^{3/2}V. \qquad (13)$$

Numerically, this gives

$$N_c/V \simeq 2.509 \times 10^{19} \times (T/300\,\text{K})^{3/2}\,\text{cm}^{-3} , \qquad (14)$$

where T is in kelvin.

The quantity N_c for actual semiconductors exhibits the same temperature dependence as (13), but differs in magnitude by a proportionality factor. We express this formally by writing, in analogy to (13),

$$N_c = 2(m_e^*\tau/2\pi\hbar^2)^{3/2}V , \qquad (15)$$

where m_e^* is called the density-of-states effective mass for electrons. Experimental values are given in Table 13.1. The introduction of effective masses is more than a formality. In the theory of electrons in crystals it is shown that the dynamical behavior of electrons and holes, under the influence of external forces such as electric fields, is that of particles with effective masses different from the free electron mass. The dynamical masses usually are different from the density-of-states masses, however.

We define the **quantum concentration** n_c for conduction electrons as

$$n_c \equiv N_c/V = 2(m_e * \tau/2\pi\hbar^2)^{3/2}. \tag{16}$$

By (11) the conduction electron concentration $n_e \equiv N_e/V$ becomes

$$n_e = n_c \exp[-(\varepsilon_c - \mu)/\tau]. \tag{17}$$

The earlier assumption (9) is equivalent to the assumption that $n_e \ll n_c$, so that the conduction electrons act as an ideal gas. As an aid to memory, we may think of N_e as arising from N_c orbitals at ε_c, with the Fermi level at μ. *Warning*: In the semiconductor literature n_c is invariably called the **effective density of states** of the conduction band.

Similar reasoning gives the number of holes in the valence band:

$$N_h = \sum_{VB} \exp[-(\mu - \varepsilon)/\tau] = N_v \exp[-(\mu - \varepsilon_v)/\tau], \tag{18}$$

with the definition

$$N_v \equiv \sum_{VB} \exp[-(\varepsilon_v - \varepsilon)/\tau]. \tag{19}$$

We define the quantum concentration n_v for holes as

$$n_v \equiv N_v/V \equiv 2(m_h * \tau/2\pi\hbar^2)^{3/2}, \tag{20}$$

where m_h* is the density-of-states effective mass for holes. By (18) the hole concentration $n_h \equiv N_h/V$ is

$$n_h = n_v \exp[-(\mu - \varepsilon_v)/\tau]. \tag{21}$$

Like (17), this gives the carrier concentration in terms of the quantum concentration and the position of the Fermi level relative to the valence band edge. In the semiconductor literature n_v is called the **effective density of states of the valence band**.

Law of Mass Action

The product $n_e n_h$ is independent of the Fermi level so long as the concentrations are in the classical regime. Then

$$n_e n_h = n_c n_v \exp[-(\varepsilon_c - \varepsilon_v)/\tau] = n_c n_v \exp(-\varepsilon_g/\tau), \qquad (22a)$$

where the energy gap $\varepsilon_g \equiv \varepsilon_c - \varepsilon_v$. In a pure semiconductor we have $n_e = n_h$, and the common value of the two concentrations is called the **intrinsic carrier concentration** n_i of the semiconductor. By (22a),

$$n_i = (n_c n_v)^{1/2} \exp(-\varepsilon_g/2\tau). \qquad (22b)$$

The Fermi level independence of the product $n_e n_h$ means that this product retains its value even when $n_e \neq n_h$, as in the presence of electrically charged impurity atoms, provided both concentrations remain in the classical regime. We may then write (22a) as

$$n_e n_h = n_i^2. \qquad (22c)$$

The value of the product depends only on the temperature. This result is the mass action law of semiconductors, similar to the chemical mass action law (Chapter 9).

Intrinsic Fermi Level

For an intrinsic semiconductor $n_e = n_i$, and we may equate the right-hand sides of (17) and (22b):

$$n_c \exp[-(\varepsilon_c - \mu)/\tau] = (n_c n_v)^{1/2} \exp(-\varepsilon_g/2\tau). \qquad (23)$$

Insert $\varepsilon_g = \varepsilon_c - \varepsilon_v$ and divide by $n_c \exp(-\varepsilon_c/\tau)$:

$$\exp(\mu/\tau) = (n_v/n_c)^{1/2} \exp[(\varepsilon_c + \varepsilon_v)/2\tau].$$

We take logarithms to obtain

$$\mu = \tfrac{1}{2}(\varepsilon_c + \varepsilon_v) + \tfrac{1}{2}\tau \log(n_v/n_c) = \tfrac{1}{2}(\varepsilon_c + \varepsilon_v) + \tfrac{3}{4}\tau \log(m_h^*/m_e^*), \qquad (24)$$

by use of (16) and (20). The Fermi level for an intrinsic semiconductor lies near the middle of the forbidden gap, but displaced from the exact middle by an amount that is usually small.

n-TYPE AND *p*-TYPE SEMICONDUCTORS

Donors and Acceptors

Pure semiconductors are an idealization of little practical interest. Semiconductors used in devices usually have impurities intentionally added in order to increase the concentration of either conduction electrons or holes. A semiconductor with more conduction electrons than holes is called *n*-type; a semiconductor with more holes than electrons is called *p*-type. The letters *n* and *p* signify negative and positive majority carriers. Consider a silicon crystal in which some of the Si atoms have been substituted by phosphorus atoms. Phosphorus is just to the right of Si in the periodic table, hence each P has exactly one electron more than the Si it replaces. These extra electrons do not fit into the filled valence band; hence a Si crystal with some P atoms will contain more conduction electrons and, by the law of mass action, fewer holes than a pure Si crystal. Next consider aluminum atoms. Aluminum is just to the left of Si in the periodic table, hence Al has exactly one electron fewer than the Si it replaces. As a result, Al atoms increase the number of holes and decrease the number of conduction electrons.

Most impurities in the same columns of the periodic table as P and Al will behave in Si just as P and Al behave. What matters is the number of valence electrons relative to Si and not the total number of electrons on the atom. Impurities from other columns of the periodic table will not behave so simply. Similar reasoning can be applied to other semiconductors, for example GaAs. For the present we assume that each donor atom contributes one electron which may enter the conduction band or fill one hole in the valence band. We also assume that each acceptor atom removes one electron, either from the valence band or from the conduction band. These assumptions are called the approximation of fully ionized impurities: all impurities when ionized are either positively charged donors D^+ or negatively charged acceptors A^-.

The electrical neutrality condition (4) told us that

$$\Delta n = n_e - n_h = n_d{}^+ - n_a{}^-. \qquad (25)$$

Because $n_h = n_i^2/n_e$ from the mass action law, we see that (25) leads to a quadratic equation for n_e:

$$n_e^2 - n_e \Delta n = n_i^2. \tag{26}$$

The positive root is

$$n_e = \tfrac{1}{2}\{[(\Delta n)^2 + 4n_i^2]^{1/2} + \Delta n\}, \tag{27a}$$

and because $n_h = n_e - \Delta n$ we have

$$n_h = \tfrac{1}{2}\{[(\Delta n)^2 + 4n_i^2]^{1/2} - \Delta n\}. \tag{27b}$$

Most often the doping concentration is large compared to the intrinsic concentration, so that either n_e or n_h is much larger than n_i:

$$|\Delta n| \gg n_i. \tag{28}$$

This condition defines an **extrinsic semiconductor**. The square roots in (27) can then be expanded:

$$[(\Delta n)^2 + 4n_i^2]^{1/2} = |\Delta n|[1 + (2n_i/\Delta n)^2]^{1/2}$$

$$\simeq |\Delta n| + 2n_i^2/|\Delta n|. \tag{29}$$

In an *n*-type semiconductor Δn is positive and (27) becomes

$$n_e \simeq \Delta n + n_i^2/\Delta n \simeq \Delta n; \qquad n_h \simeq n_i^2/\Delta n \ll n_i. \tag{30}$$

In a *p*-type semiconductor Δn is negative and (27) becomes

$$n_e \simeq n_i^2/|\Delta n| \ll n_i; \qquad n_h \simeq |\Delta n| + n_i^2/|\Delta n| \simeq |\Delta n|. \tag{31}$$

The majority carrier concentration in the extrinsic limit (28) is nearly equal to the magnitude of Δn, while the minority carrier concentration is inversely proportional to $|\Delta n|$.

Fermi Level in Extrinsic Semiconductor

By use of the mass action law we calculated the carrier concentrations without having to calculate the Fermi level first. The Fermi level is obtained from n_e or n_h by solving (17) or (21) for μ:

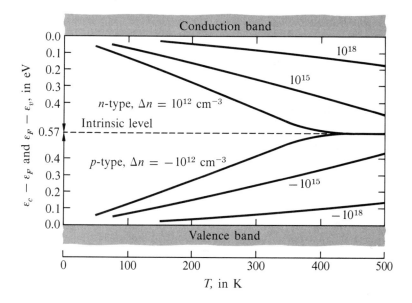

Figure 13.3 The Fermi level in silicon as a function of temperature, for various doping concentrations. The Fermi levels are expressed relative to the band edges. A small decrease of the energy gap with temperature has been neglected.

$$\mu = \varepsilon_c - \tau \log(n_c/n_e) = \varepsilon_v + \tau \log(n_v/n_h). \tag{32}$$

We may now use (27) to find μ as a function of temperature and doping level Δn. Figure 13.3 gives numerical results for Si. With decreasing temperature the Fermi level in an extrinsic semiconductor approaches either the conduction or the valence band edge.

Degenerate Semiconductors

When one of the carrier concentrations is increased and approaches the quantum concentration, we may no longer use the classical distribution (10) for that carrier. The calculation of the carrier concentration now follows the treatment of the Fermi gas in Chapter 7. The sum over all occupied orbitals, which is equal to the number of electrons, is written as an integral over the density of states times the distribution function:

$$N = \int d\varepsilon \, \mathfrak{D}(\varepsilon) f(\varepsilon) , \tag{33}$$

where for free particles of mass m the density of states is

$$\mathfrak{D}(\varepsilon) = \frac{V}{2\pi^2}\left(\frac{2m}{\hbar^2}\right)^{3/2}\varepsilon^{1/2}. \tag{34}$$

That is, $\mathfrak{D}(\varepsilon)d\varepsilon$ is the number of orbitals in the energy interval $(\varepsilon, \varepsilon + d\varepsilon)$. To make the transition to conduction electrons in semiconductors we replace N by $n_e V$; m by $m_e{}^*$; and ε by $\varepsilon - \varepsilon_c$. We obtain

$$n_e = \frac{1}{2\pi^2}\left(\frac{2m_e{}^*}{\hbar^2}\right)^{3/2}\int_{\varepsilon_c}^{\infty}\frac{d\varepsilon(\varepsilon - \varepsilon_c)^{1/2}}{1 + \exp[(\varepsilon - \mu)/\tau]}. \tag{35}$$

Let $x \equiv (\varepsilon - \varepsilon_c)/\tau$ and $\eta \equiv (\mu - \varepsilon_c)/\tau$. We use the definition (16) of n_c to obtain

$$n_e/n_c = I(\eta) = \frac{2}{\sqrt{\pi}}\int_0^{\infty}\frac{dx\,x^{1/2}}{1 + \exp(x - \eta)}. \tag{36}$$

The integral $I(\eta)$ in (36) is known as the **Fermi-Dirac integral**.

When $\varepsilon_c - \mu \gg \tau$ we have $-\eta \gg 1$, so that $\exp(x - \eta) \gg 1$. In this limit

$$n_e/n_c \simeq \frac{2}{\sqrt{\pi}}\,e^{\eta}\int_0^{\infty}dx\,e^{-x}x^{1/2} = \frac{2}{\sqrt{\pi}}\,\Gamma(3/2)e^{\eta} = \exp[(\mu - \varepsilon_c)/\tau], \tag{37}$$

the familiar result for the ideal gas.

In semiconductors the electron concentration rarely exceeds several times the quantum concentration n_c. The deviation between the value of μ from (35) and the approximation (37) then can be expanded into a rapidly converging power series of the ratio $r = n_e/n_c$, called the Joyce–Dixon approximation:*

$$\eta - \log r \simeq \frac{1}{\sqrt{8}}r - \left(\frac{3}{16} - \frac{\sqrt{3}}{9}\right)r^2 + \cdots; \qquad r = n_e/n_c, \tag{38}$$

Figure 13.4 compares the exact relation (36) with the approximations (37) and (38).

---

* W. B. Joyce and R. W. Dixon, Appl. Phys. Lett. **31**, 354 (1977). If the right side of (38) is written as $\sum A_n r^n$, the first four coefficients are $A_1 = 3.53553 \times 10^{-1}$; $A_2 = -4.95009 \times 10^{-3}$; $A_3 = 1.48386 \times 10^{-4}$; $A_4 = -4.42563 \times 10^{-6}$.

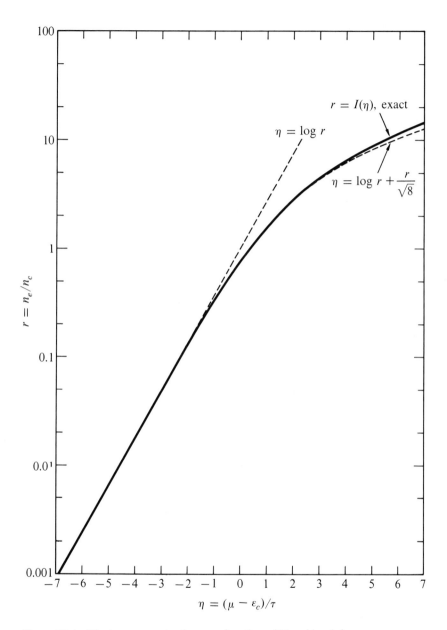

Figure 13.4 Electron concentration as a function of Fermi level, for μ near or above conduction band edge ε_c. The dashed curve represents the first term of the Joyce-Dixon approximation (38).

When n_e is no longer small compared to n_c, the expression of the mass action law must be modified. In Problem 4 we ask the reader to show that

$$n_e n_h = n_i^2 \exp[-n_e/\sqrt{8n_c} + \cdots]. \tag{39}$$

If the gap itself depends on the carrier concentrations, the value of n_i to be used here will depend on concentration.

Impurity Levels

The addition of impurities to a semiconductor moves some orbitals from the conduction or valence band into the energy gap, where the orbitals now appear as localized bound states. We consider phosphorous in a silicon crystal. If the P atom has released its extra electron to the Si conduction band, the atom appears as a positively charged ion. The positive ion attracts the electrons in the conduction band, and the ion can bind an electron just as a proton can bind an electron in a hydrogen atom. However, the binding energy in the semiconductor is several orders of magnitude lower, mostly because the binding energy is to be divided by the square of the static dielectric constant, and partly because of mass effects. Table 13.2 gives the ionization energies for column V donors in Si and Ge. The lowest orbital of an electron bound to a donor corresponds to an energy level $\Delta\varepsilon_d = \varepsilon_c - \varepsilon_d$ below the edge of the conduction band (Figure 13.5). There is one set of bound orbitals for every donor.

A parallel argument applies to holes and acceptors. Orbitals are split off from the valence band, as in Figure 13.5. For each acceptor atom there is one set of bound orbitals with an ionization energy $\Delta\varepsilon_a = \varepsilon_a - \varepsilon_v$, of the same order as $\Delta\varepsilon_d$. Ionization energies for column III acceptors in Si are listed in Table 13.2.

In GaAs the ionization energies for all column VI donors except oxygen are close to 6 meV. For zinc, the most important acceptor, $\Delta\varepsilon_a = 24$ meV. Some

Table 13.2   Ionization energies of column V donors and column III acceptors in Si and Ge, in meV

	Donors			Acceptors			
	P	As	Sb	B	Al	Ga	In
Si	45	49	39	45	57	65	16
Ge	12.0	12.7	9.6	10.4	10.2	10.8	11.2

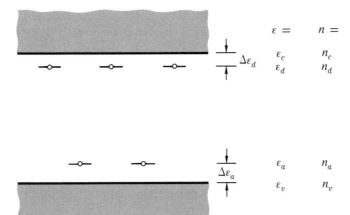

Figure 13.5   Donor and acceptor impurity levels in the energy gap of a semiconductor.

impurities generate orbitals deep inside the forbidden gap, sometimes with multiple orbitals corresponding to different ionization states.

Occupation of Donor Levels

A donor level can be occupied by an electron with either spin up or spin down. Hence there are two different orbitals with the same energy. However, the occupations of these two orbitals are not independent of each other: Once the level is occupied by one electron, the donor cannot bind a second electron with opposite spin. As a result, the occupation probability for a donor level is not given by the simple Fermi-Dirac distribution function, but by a function treated in Chapter 5. We write the probability that the donor orbital is vacant, so that the donor is ionized, in a form slightly different from (5.73):

$$f(\mathrm{D}^+) = \frac{1}{1 + 2\exp[(\mu - \varepsilon_d)/\tau]}. \tag{40}$$

Here ε_d is the energy of a singly occupied donor orbital relative to the origin of the energy. The probability that the donor orbital is occupied by an electron, so that the donor is neutral, is given by (5.74):

$$f(\mathrm{D}) = \frac{1}{1 + \frac{1}{2}\exp[(\varepsilon_d - \mu)/\tau]}. \tag{41}$$

Acceptors require extra thought. In the ionized condition A^- of the acceptor, each of the chemical bonds between the acceptor atom and the surrounding semiconductor atoms contains a pair of electrons with antiparallel spins. There is only one such state, hence the ionized condition contributes only one term, $\exp[(\mu - \varepsilon_a)/\tau]$, to the Gibbs sum for the acceptor. In the neutral condition A of the acceptor, one electron is missing from the surrounding bonds. Because the missing electron may have either spin up or spin down, the neutral condition is represented twice in the Gibbs sum for the acceptor, by a term $2 \times 1 = 2$.

Hence the thermal average occupancy is

$$f(A^-) = \frac{\exp[(\mu - \varepsilon_a)/\tau]}{2 + \exp[(\mu - \varepsilon_a)/\tau]} = \frac{1}{1 + 2\exp[(\varepsilon_a - \mu)/\tau]}. \tag{42}$$

The neutral condition A, with the acceptor orbital unoccupied, occurs with probability

$$f(A) = \frac{2}{2 + \exp[(\mu - \varepsilon_a)/\tau]} = \frac{1}{1 + \frac{1}{2}\exp[(\mu - \varepsilon_a)/\tau]}. \tag{43}$$

The value of $\Delta n \equiv n_d{}^+ - n_a{}^-$ is the difference of concentrations of D^+ and A^-. From (40) or (42) we have

$$n_d{}^+ = n_d f(D^+) = \frac{n_d}{1 + 2\exp[(\mu - \varepsilon_d)/\tau]}, \tag{44}$$

$$n_a{}^- = n_a f(A^-) = \frac{n_a}{1 + 2\exp[(\varepsilon_a - \mu)/\tau]}. \tag{45}$$

The neutrality condition (4) may be rewritten as

$$n^- \equiv n_e + n_a{}^- = n_h + n_d{}^+ \equiv n^+. \tag{46}$$

This expression may be visualized by a logarithmic plot of n^- and n^+ as functions of the position of the Fermi level (Figure 13.6). The four dashed lines represent the four terms in (46); the two solid lines represent the sum of all positive and all negative charges. The actual Fermi level occurs where the total positive charges equal the total negative charges.

For $n_d{}^+ - n_a{}^- \gg n_i$, as in Figure 13.6, the holes can be neglected; for $n_a{}^- - n_d{}^+ \gg n_i$ the electrons can be neglected. If one of the two impurity species can be neglected, the majority carrier concentration can be calculated in closed form. Consider an *n*-type semiconductor with no acceptors. The

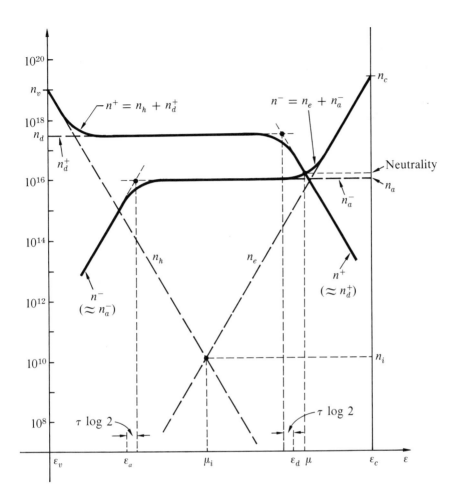

Figure 13.6   Graphical determination of Fermi level and electron concentration in an *n*-type semiconductor containing both donors and acceptors.

neutrality point in Figure 13.6 is now given by the intersection point of the n^+ curve with the n_e curve. If the donor concentration is not too high, the intersection will be on the straight portion of the n_e curve, along which the approximation (17) holds. We rewrite this as

$$\exp(\mu/\tau) = (n_e/n_c)\exp(\varepsilon_c/\tau); \qquad (47)$$

$$\exp[(\mu - \varepsilon_d)/\tau] = (n_e/n_c)\exp[(\varepsilon_c - \varepsilon_d)/\tau] = n_e/n_e^* , \qquad (48)$$

where

$$n_e{}^* \equiv n_c \exp[-(\varepsilon_c - \varepsilon_d)/\tau] = n_c \exp(-\Delta\varepsilon_d/\tau) \tag{49}$$

is the electron concentration that would be present in the conduction band if the Fermi level coincided with the donor level. Here $\Delta\varepsilon_d \equiv \varepsilon_c - \varepsilon_d$ is the donor ionization energy.

We insert (48) into (44) and set $n_e = n_d{}^+$ to obtain

$$n_e = \frac{n_d}{1 + (2n_e/n_e{}^*)}; \tag{50}$$

$$n_e{}^2 + \tfrac{1}{2}n_e n_e{}^* = \tfrac{1}{2}n_d n_e{}^*. \tag{51}$$

This is a quadratic equation in n_e; the positive solution is

$$n_e = \tfrac{1}{4}n_e{}^*\{[1 + (8n_d/n_e{}^*)]^{1/2} - 1\}. \tag{52}$$

For shallow donor levels, $n_e{}^*$ is large and close to n_c. If the doping is sufficiently weak that $8n_d \ll n_e{}^*$, the square root may be expanded by use of

$$(1 + x)^{1/2} \simeq 1 + \tfrac{1}{2}x - \tfrac{1}{8}x^2 + \cdots , \tag{53}$$

for $x \ll 1$. With $x = 8n_d/n_e{}^*$ we obtain

$$n_e \simeq n_d - 2n_d{}^2/n_e{}^* = n_d(1 - 2n_d/n_e{}^*). \tag{54}$$

The second term in the parentheses gives the first order departure from complete ionization. For example, for P in Si at 300 K, we have $\Delta\varepsilon_d \simeq 1.74\tau$ from Table 13.2, so that $n_e{}^* \simeq 0.175n_c$ from (49). If $n_d = 0.01n_c$, Eq. (54) predicts that 11.4 pct of the donors remain un-ionized. The limit of weak ionization is the subject of Problem 6.

Example: Semi-insulating gallium arsenide. Could pure GaAs be prepared, it would have an intrinsic carrier concentration at room temperature of $n_i \lesssim 10^7 \, \text{cm}^{-3}$. With such a low concentration of carriers, 10^{-15} less than a metal, the conductivity would be closer to an insulator than to a conventional semiconductor. Intrinsic GaAs would be useful as an insulating substrate on which to prepare thin layers of doped GaAs as needed for devices. There does not exist a technology to purify any substance to 10^7 impurities per cm^3.

However, it is possible to achieve near intrinsic carrier concentrations in GaAs by doping with high concentrations (10^{15}–10^{17} cm^{-3}) of oxygen and chromium together, two impurities that have their impurity levels near the middle of the energy gap. Oxygen enters an As site and is a donor in GaAs, as expected from the position of O in the periodic table relative to As; the energy level* is about 0.7 eV below ε_c. Chromium is an acceptor with an energy level about 0.84 eV below ε_c.

Consider a GaAs crystal doped with both oxygen and chromium. The ratio of the two concentrations is not critical; anything with an O:Cr ratio between about 1:10 and 10:1 will do. If the concentrations of all other impurities are small compared with those of O and Cr, the position of the Fermi level will be governed by the equilibrium between electrons on O and holes on Cr. The construction of Figure 13.6 applied to this system shows that over the indicated concentration ratio range the Fermi level is pinned to a range between 1.5τ above the O level and 1.5τ below the Cr level. With the Fermi level pinned near the middle of the energy gap, the crystal must act as nearly intrinsic.

Gallium arsenide doped in this way is called semi-insulating GaAs and is used extensively as a high-resistivity (10^8 to 10^{10} Ω cm) substrate for GaAs devices. A similar doping procedure is possible in InP, with iron taking the place of chromium.

p–n JUNCTIONS

Semiconductors used in devices are almost never uniformly doped. An understanding of devices requires an understanding of nonuniformly doped semiconductors, particularly of structures called *p–n* junctions in which the doping changes with position from *p*-type to *n*-type within the same crystal. We consider a semiconductor crystal inside which the doping changes abruptly at $x = 0$ from a uniform donor concentration n_d to a uniform acceptor concentration n_a, as in Figure 13.7a. This is an example of a *p–n* junction. More complicated device structures are made up from simple junctions: a bipolar transistor has two closely spaced *p–n* junctions, of the sequence *p–n–p* or *n–p–n*.

p–n junctions contain a built-in electrostatic potential step V_{bi}, even in the absence of an externally applied voltage (Figure 13.7b). With no externally applied voltage, the electrons on the two sides of the junction are in diffusive equilibrium, which means that the chemical potentials (Fermi levels) of the two sides are the same. Because the position of the Fermi level within the band structure depends on the local doping, constancy of the Fermi level forces a shift in the electron energy bands in crossing the junction (Figure 13.7c). The shift is eV_{bi}. The potential step of height eV_{bi} is an example of the potential step required to equalize the total chemical potential of two systems when the intrinsic chemical potentials are unequal, as discussed in Chapter 5.

---

* R. Zucca, J. Appl. Phys. **48**, 1987 (1977). The energy assignment is somewhat uncertain.

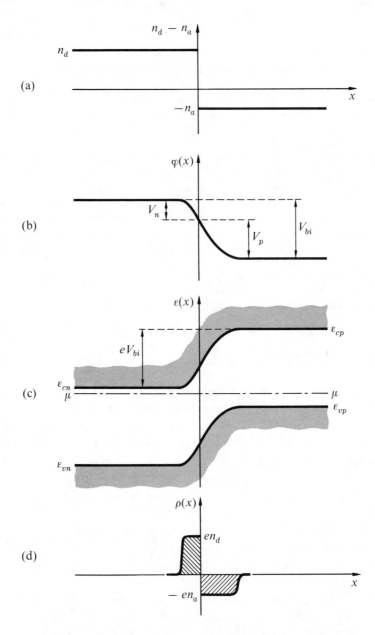

Figure 13.7 A *p-n* junction. (a) Doping distribution. It is assumed that the doping changes abruptly from *n*-type to *p*-type. The two doping levels are usually different. (b) Electrostatic potential. The built-in voltage V_{bi} establishes diffusive equilibrium between the two sides with different electron concentrations as well as hole concentrations. (c) Energy bands. Because the Fermi level must be constant throughout the structure, the bands on the two sides are shifted relative to each other. (d) Space charge dipole required to generate the built-in voltage and to shift the energy bands.

We assume that the two doping concentrations n_d, n_a lie in the extrinsic but nondegenerate range, as defined by

$$n_i \ll n_d \ll n_c; \qquad n_i \ll n_a \ll n_v. \tag{55}$$

If the donors are fully ionized on the n side and the acceptors fully ionized on the p side, then the electron and hole concentrations satisfy

$$n_e \simeq n_d; \qquad n_h \simeq n_a, \tag{56}$$

one on the n side and the other on the p side. (We have dropped the superscripts $\pm$ from n_d, n_a.) The conduction band energies on the n and p sides follow from (17):

$$\varepsilon_{cn} = \mu - \tau \log(n_d/n_c); \tag{57}$$

$$\varepsilon_{cp} = \mu - \tau \log(n_{ep}/n_c) = \mu - \tau \log(n_i^2/n_a n_c), \tag{58}$$

by (22c). Hence

$$eV_{bi} = \varepsilon_{cp} - \varepsilon_{cn} = \tau \log(n_a n_d/n_i^2), \tag{59}$$

or

$$eV_{bi} = \varepsilon_g - \tau \log(n_c n_v/n_d n_a). \tag{60}$$

For doping concentrations $n_d \simeq 0.01 n_c$ and $n_a \simeq 0.01 n_v$, we find $eV_{bi} \simeq \varepsilon_g - 9.2\tau$, which is 0.91 eV in silicon at room temperature.

A step in electrostatic potential is required to shift the band edge energies on the two sides of the junction relative to each other. The electrostatic potential $\varphi(x)$ must satisfy the Poisson equation

(SI)
$$\frac{d^2\varphi}{dx^2} = -\frac{\rho}{\epsilon}, \tag{61}$$

where ρ is the space charge density and ϵ the permittivity of the semiconductor. Space charge must be present whenever φ varies. In the vicinity of the junction the charge carriers no longer neutralize the impurities as in the bulk material. The space charge must be positive on the n side and negative on the p side (Figure 13.7d). Positive space charge on the n side means that the electron concentration is less than the donor concentration. Indeed, as the conduction band edge is raised relative to the fixed Fermi level, (17) predicts an exponential decrease of the electron concentration n_e.

Take the origin of the electrostatic potential at $x = -\infty$, so that $\varphi(-\infty) = 0$. Then $\varepsilon_c(x) = \varepsilon_c(-\infty) - e\varphi(x)$, and (17) becomes

$$n_e(x) = n_d \exp[e\varphi(x)/\tau]. \tag{62}$$

The Poisson equation (61) is

$$\frac{d^2\varphi}{dx^2} = -\frac{\rho}{\epsilon} = -\frac{e}{\epsilon}[n_d - n_e(x)] = -\frac{en_d}{\epsilon}[1 - \exp(e\varphi/\tau)]. \tag{63}$$

Multiply by $2d\varphi/dx$ to obtain

$$2\frac{d\varphi}{dx}\frac{d^2\varphi}{dx^2} = \frac{d}{dx}\left(\frac{d\varphi}{dx}\right)^2 = -\frac{2en_d}{\epsilon}\frac{d}{dx}\left\{\varphi - \frac{\tau}{e}\exp(e\varphi/\tau)\right\}; \tag{64}$$

Integrate with the initial condition $\varphi(-\infty) = 0$:

$$\left(\frac{d\varphi}{dx}\right)^2 = -\frac{2en_d}{\epsilon}\left[\varphi + \frac{\tau}{e} - \frac{\tau}{e}\exp(e\varphi/\tau)\right]. \tag{65}$$

At the interface $x = 0$ we assume that

$$-\varphi(0) \equiv V_n \gg \tau/e\,, \tag{66}$$

where V_n is that part of the built-in electrostatic potential drop that occurs on the n side. The exponential on the right-hand side of (65) can be neglected, and we obtain

$$E = [(2en_d/\epsilon)(V_n - \tau/e)]^{1/2} \tag{67}$$

for the x component of the electric field $E = -d\varphi/dx$ at the interface. Similarly,

$$E = [(2en_a/\epsilon)(V_p - \tau/e)]^{1/2}\,, \tag{68}$$

where V_p is that part of the built-in electrostatic potential drop that occurs on the p side. The two E fields must be the same; from this and from $V_n + V_p = V_{bi}$ we find

$$E = \left(\frac{2e}{\epsilon}\frac{n_a n_d}{n_a + n_d}(V_{bi} - 2\tau/e)\right)^{1/2}. \tag{69}$$

The field E is the same as if on the n-type side all electrons had been depleted from the junction to a distance

$$w_n = \frac{\epsilon E}{e n_d} = \left(\frac{2\epsilon}{e n_d}(V_n - \tau/e)\right)^{1/2} = \left(\frac{2\epsilon}{e}\frac{n_a}{n_d(n_a + n_d)}(V_{bi} - 2\tau/e)\right)^{1/2}, \quad (70)$$

with no depletion at $|x| > w_n$. The distance w_n is used in semiconductor device theory as a measure of the depth of penetration of the space charge transition layer into the n side.

Similarly, on the p side,

$$w_p = \frac{\epsilon E}{e n_a} = \left(\frac{2\epsilon}{e n_a}(V_p - \tau/e)\right)^{1/2} = \left(\frac{2\epsilon}{e}\frac{n_d}{n_a(n_a + n_d)}(V_{bi} - 2\tau/e)\right)^{1/2}. \quad (71)$$

The total depletion width $w_n + w_p$ is

$$w = \left(\frac{2\epsilon}{e}\frac{n_a + n_d}{n_a n_d}(V_{bi} - 2\tau/e)\right)^{1/2} = \frac{2(V_{bi} - 2\tau/e)}{E}. \quad (72)$$

If we assume $n_a = n_d = 10^{15}\,\text{cm}^{-3}$; $\epsilon = 10\epsilon_0$; and $V_{bi} - 2\tau/e = 1$ volt, we find $E = 4.25 \times 10^4\,\text{V cm}^{-1}$ and $w = 4.70 \times 10^{-5}\,\text{cm}$.

Reverse-Biased Abrupt *p–n* Junction

Let a voltage V be applied to a p–n junction, of such sign that the p side is at a negative voltage relative to the n side, which means that V raises the potential energy of the electrons on the p side. This voltage will drive conduction electrons from the p side to the n side, and holes from the n side to the p side. But the p side in bulk contains a very low concentration of conduction electrons, and the n side contains a very low concentration of holes, consistent with the mass action law. As a result, very little current flows. The distributions of electrons, holes, and potential are approximately the same as if the built-in voltage were increased by the applied voltage, Figure 13.8. The field at the interface is now given by

$$E = \left(\frac{2e}{\epsilon}\frac{n_a n_d}{n_a + n_d}[(|V| + V_{bi}) - 2\tau/e]\right)^{1/2}, \quad (73)$$

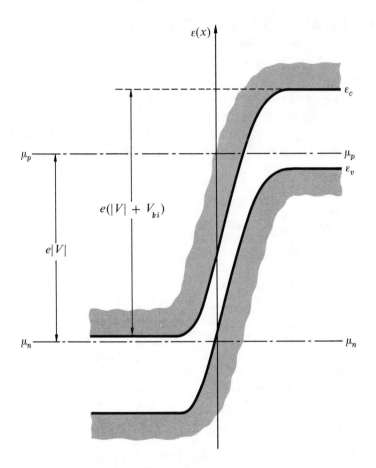

Figure 13.8   Reverse-biased *p-n* junction, showing the quasi-Fermi levels μ_n and μ_P.

and the junction thickness is given by

$$w = \left(\frac{2\epsilon}{e} \frac{n_a + n_d}{n_a n_d} \left[|V| + V_{bi} - 2\tau/e \right] \right)^{1/2} = \frac{2\left[|V| + V_{bi} - 2\tau/e \right]}{E}. \qquad (74)$$

In the semiconductor device literature we often find (73) and (74) without the term $2\tau/e$, because certain approximations have been made about the space charge and field distribution; we have solved the Poisson equation with the correct electron distribution (62).

NONEQUILIBRIUM SEMICONDUCTORS

Quasi-Fermi Levels

When a semiconductor is illuminated with light of quantum energy greater than the energy gap, electrons are raised from the valence band to the conduction band. The electron and the hole concentrations created by illumination are larger than their equilibrium concentrations. Similar nonequilibrium concentrations arise when a forward-biased *p–n* junction injects electrons into a *p*-type semiconductor or holes into an *n*-type semiconductor. The electric charge associated with the injected carrier type attracts oppositely charged carriers from the external electrodes of the semiconductor so that both carrier concentrations increase.

The excess carriers eventually recombine with each other. The recombination times vary greatly with the semiconductor, from less than 10^{-9} s to longer than 10^{-3} s. Recombination times in high purity Si are near 10^{-3} s. Even the shortest recombination times are much longer than the times ($\sim 10^{-12}$ s) required at room temperature for the conduction electrons to reach thermal equilibrium with each other in the conduction band, and for the holes to reach thermal equilibrium with each other in the valence band. Thus the orbital occupancy distributions of electrons and of holes are very close to equilibrium Fermi-Dirac distributions in each band separately, but the total number of holes is not in equilibrium with the total number of electrons.

We can express this steady state or quasi-equilibrium condition by saying that there are different Fermi levels μ_c and μ_v for the two bands, called **quasi-Fermi levels**:

$$f_c(\varepsilon,\tau) = \frac{1}{1 + \exp[(\varepsilon - \mu_c)/\tau]};$$

$$f_v(\varepsilon,\tau) = \frac{1}{1 + \exp[(\varepsilon - \mu_v)/\tau]}. \tag{75}$$

Quasi-Fermi levels are used extensively in the analysis of semiconductor devices.

Current Flow: Drift and Diffusion

If the conduction band quasi-Fermi level is at a constant energy throughout a semiconductor crystal, the conduction electrons throughout the crystal are in thermal and diffusive equilibrium, and no electron current will flow. Any conduction electron flow in a semiconductor at a uniform temperature must be caused by a position-dependence of the conduction band quasi-Fermi level.

If the gradient of this level is sufficiently weak, we may assume that the contribution of conduction electrons to the total electrical current density is proportional to this gradient:

$$\mathbf{J}_e \propto \operatorname{grad} \mu_c. \tag{76}$$

Here $\mathbf{J}_e$ is an electrical current density, not a particle flux density. Because each electron carries the charge $-e$, we have

$$\mathbf{J}_e = (-e) \times (\text{electron flux density}), \tag{77}$$

where the electron flux density is defined as the number of conduction electrons crossing unit area in unit time. The close connection of (76) to Ohm's law is treated in Chapter 14. Because the flow of particles is from high to low chemical potential, the conduction electron flux is opposite to $\operatorname{grad} \mu_c$, but because electrons carry a negative charge, the associated electrical current density is in the direction of $\operatorname{grad} \mu_c$. We view $\operatorname{grad} \mu_c$ as the driving force for this current. For a given driving force, the current density is proportional to the concentration n_e of conduction electrons. Thus we write

$$\mathbf{J}_e = \tilde{\mu}_e n_e \operatorname{grad} \mu_c , \tag{78}$$

where the proportionality constant $\tilde{\mu}_e$ is the **electron mobility**. The symbol $\tilde{\mu}_e$ should not be confused with the conduction band quasi-Fermi level, μ_c.

If the electron concentration is in the extrinsic but nondegenerate range,

$$n_i \ll n_e \ll n_c , \tag{79}$$

the conduction band quasi-Fermi level is given by (15), which can be written in terms of the electron concentration as

$$\mu_c = \varepsilon_c + \tau \log(n_e/n_c). \tag{80}$$

Thus (78) becomes

$$\mathbf{J}_e = \tilde{\mu}_e n_e \operatorname{grad} \varepsilon_c + \tilde{\mu}_e \tau \operatorname{grad} n_e. \tag{81}$$

A gradient in the conduction band edge arises from a gradient in the electrostatic potential and thus from an electric field:

$$\operatorname{grad} \varepsilon_c = -e \operatorname{grad} \varphi = e\mathbf{E}. \tag{82}$$

We introduce an electron diffusion coefficient D_e by the Einstein relation

$$D_e = \tilde{\mu}_e \tau / e \,,$$

(83)

discussed in Chapter 14. We now write (78) or (81) in the final form

$$\mathbf{J}_e = e\tilde{\mu}_e n_e \mathbf{E} + eD_e \operatorname{grad} n_e.$$

(84)

There are two different contributions to the current: one caused by an electric field and one caused by a concentration gradient.

Analogous results apply to holes, with one difference. The valence band quasi-Fermi level is not the chemical potential for holes, but is the chemical potential for the electrons in the valence band. Holes are missing electrons; a hole current to the right is really an electron current to the left. But holes carry a positive rather than a negative charge. The two sign reversals cancel, and we may view grad μ_v as the driving force for the contribution $\mathbf{J}_h$ of holes to the total electrical current density. We write, analogously to (78),

$$\mathbf{J}_h = \tilde{\mu}_h n_h \operatorname{grad} \mu_v.$$

(85)

Carrying through the rest of the argument leads to

$$\mathbf{J}_h = +e\tilde{\mu}_h n_h \mathbf{E} - eD_h \operatorname{grad} n_h$$

(86)

as the analog of (84), with the Einstein relation $D_h = \tilde{\mu}_h \tau / e$. Note the different sign in the diffusion term: Holes, like electrons, diffuse from high to low concentrations, but hole diffusion makes the opposite contribution to the electric current, because holes carry the opposite charge.

Example: Injection laser. The highest nonequilibrium carrier concentrations in semiconductors occur in injection lasers. When by electron injection the occupation $f_e(\varepsilon_c)$ of the lowest conduction band orbital becomes higher than the occupation $f_e(\varepsilon_v)$ of the highest valence band orbital, the population is said to be inverted. Laser theory tells us that light with a quantum energy $\varepsilon_c - \varepsilon_v = \varepsilon_g$ can then be amplified by stimulated emission. The condition for population inversion is that

$$f_c(\varepsilon_c) > f_v(\varepsilon_v).$$

(87)

With the quasi-Fermi distributions (75) this condition is expressed as

$$\mu_c - \mu_v > \varepsilon_c - \varepsilon_v = \varepsilon_g.$$

(88)

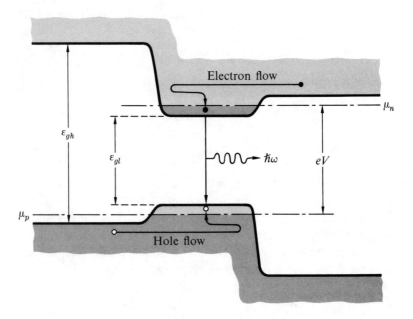

Figure 13.9 Double-heterostructure injection laser. Electrons flow from the right into the active layer, where they form a degenerate electron gas. The potential barrier provided by the wide energy gap on the *p* side prevents the electrons from escaping to the left. Holes flow from the left into the active layer, but cannot escape to the right. When (88) is attained, laser action becomes possible.

For laser action the quasi-Fermi levels must be separated by more than the energy gap. The condition (88) requires that at least one of the quasi-Fermi levels lie inside the band to which it refers. This is a necessary, but not a sufficient condition for laser operation. An important additional condition is that the energy gap is a direct gap rather than an indirect gap. The distinction is treated in solid state physics texts. The most important semiconductors with a direct gap are GaAs and InP.

The population inversion is most easily achieved in the double heterostructure of Figure 13.9; here the lasing semiconductor is embedded between two wider-gap semiconductor regions of opposite doping. An example is GaAs embedded in AlAs. In such a structure there is a potential barrier that prevents the outflow of electrons to the *p*-type region, and an opposite potential barrier that prevents the outflow of holes to the· *n*-type region. Except for the current caused by the recombination itself, the electrons in the active layer are in diffusive equilibrium with the electrons in the *n* contact, and the electron quasi-Fermi level in the active layer lines up with the Fermi level in the *n* contact. Similarly, the valence band quasi-Fermi level lines up with the Fermi level in the *p* contact. Inversion can be achieved if we apply a bias voltage larger than the voltage equivalent of the active layer energy gap. Most injection lasers utilize this double heterostructure principle.

Example: Carrier recombination through an impurity level. Electrons and holes can re-combine either by an electron falling directly into a hole with the emission of a photon, or they can recombine through an impurity level in the energy gap. The impurity process is dominant in silicon. We discuss the process as an instructive example of quasi-equilibrium semiconductor statistics. Consider an impurity recombination orbital at energy ε_r in Figure 13.10. Four transition processes are indicated in the figure. We assume that the rate R_{cr} at which conduction electrons fall into the recombination orbitals is described by a law of the form

$$R_{cr} = (1 - f_r)n_e/t_e , \tag{89}$$

where f_r is the fraction of recombination orbitals already occupied by an electron (and hence not available), and t_e is a characteristic time constant for the capture process. We assume the reverse process proceeds at the rate

$$R_{rc} = f_r n_c/t_e' , \tag{90}$$

where t_e' is the time constant for the reverse process. We take R_{rc} independent of the con-centration of conduction electrons, because we assume that $n_e \ll n_c$. The time constants t_e

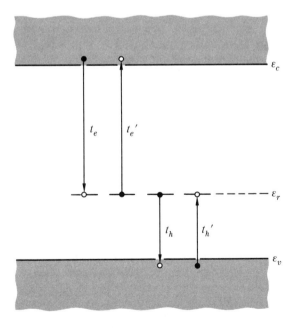

Figure 13.10 Electron-hole recombination through impurity recombination orbitals at ε_r inside the energy gap.

and t_e' are related, because in equilibrium the two rates R_{rc} and R_{cr} must cancel. Thus

$$\frac{n_c}{t_e'} = \left(\frac{1 - f_r}{f_r}\frac{n_e}{t_e}\right)_{eq} , \tag{91}$$

with f_r and n_e evaluated in thermal equilibrium, which means we use (17) for n_e. We ignore the spin multiplicity of the recombination levels. With the equilibrium Fermi-Dirac distribution for f_r, we have

$$(1 - f_r)/f_r = \exp[-(\mu - \varepsilon_r)/\tau]. \tag{92}$$

Thus (91) becomes

$$\frac{n_c}{t_e'} = \frac{n_c}{t_e}\exp[-(\varepsilon_c - \varepsilon_r)/\tau] = \frac{n_e^*}{t_e} , \tag{93}$$

where n_e^* is defined as the conduction electron concentration that would be present if the equilibrium Fermi level μ in (17) coincided with the recombination level. If (92) and (93) are inserted into (89) and (90), the net electron recombination rate becomes

$$R_e = R_{cr} - R_{rc} = \frac{1}{t_e}[(1 - f_r)n_e - f_r n_e^*]. \tag{94}$$

The analogous recombination rate for holes is obtained by the substitutions

$$R_e, n_e, n_e^*, t_e \rightarrow R_h, n_h, n_h^*, t_h;$$

and

$$f_r \rightarrow 1 - f_r; \qquad 1 - f_r \rightarrow f_r. \tag{95}$$

Here t_h is the lifetime of holes in the limit that all recombination centers are occupied by electrons, and n_h^* is, by definition,

$$n_h^* \equiv n_v \exp[-(\varepsilon_r - \varepsilon_v)/\tau] = n_i^2/n_e^*. \tag{96}$$

With these substitutions the net hole recombination rate is

$$R_h = [f_r n_h - (1 - f_r)n_h^*]/t_h. \tag{97}$$

In steady state the two recombination rates must be equal: $R_e = R_h = R$. Equations (94) and (97) are two equations for the two unknowns f_r and R. We eliminate f_r to find

$$R = \frac{n_e n_h - n_i^2}{(n_e^* + n_e)t_h + (n_h^* + n_h)t_e}. \tag{98}$$

This is the basic result of the Hall-Shockley-Read recombination theory.* Applications are developed in Problems 10 and 11.

SUMMARY

1. In semiconductors the electron orbitals are grouped into a valence band (completely occupied at $\tau = 0$ in a pure semiconductor) and a conduction band (completely empty at $\tau = 0$ in a pure semiconductor), separated by an energy gap. Electrons in the conduction band are called conduction electrons; empty orbitals in the valence band are called holes.

2. The probability of occupancy of a band orbital with energy ε is governed by the Fermi-Dirac distribution function

$$f(\varepsilon) = \frac{1}{1 + \exp[(\varepsilon - \mu)/\tau]}.$$

   Here μ is the chemical potential of the electrons, called the Fermi level.

3. The energetic location of the Fermi level in an electrically neutral semiconductor is governed by the neutrality condition

$$n_e - n_h = \Delta n.$$

   Here n_e and n_h are the concentrations of conduction electrons and holes, and Δn is the excess concentration of positively charged impurities over negatively charged impurities.

4. A semiconductor is said to be in the classical regime when $n_e \ll n_c$ and $n_h \ll n_v$. Here

$$n_{c,v} = 2(m_{e,h}{}^*\tau/2\pi\hbar^2)^{3/2}$$

   are the quantum concentrations for electrons and holes; $m_e{}^*$ and $m_h{}^*$ are effective masses for electrons and holes. In the semiconductor literature, n_c and n_v are called the effective densitites of states for the conduction and valence bands.

---

* R. N. Hall, Phys. Rev. **87**, 387 (1952); W. Shockley and W. T. Read, Jr., Phys. Rev. **87**, 835 (1952).

5. In the classical regime

$$n_e = n_c \exp[-(\varepsilon_c - \mu)/\tau] \, ,$$

$$n_h = n_v \exp[-(\mu - \varepsilon_v)/\tau] \, ,$$

where ε_c and ε_v are the energies of the edges of the conduction and valence bands.

6. The mass action law states that in the classical regime the product

$$n_e n_h = n_i^2 = n_c n_v \exp(-\varepsilon_g/\tau)$$

is independent of the impurity concentration. The intrinsic concentration n_i is the common value of n_e and n_h in an intrinsic ($=$ pure) semiconductor. The quantity

$$\varepsilon_g = \varepsilon_c - \varepsilon_v$$

is the energy gap.

7. A semiconductor is called n-type when negative charge carriers ($=$ conduction electrons) dominate; it is called p-type when positive charge carriers ($=$ holes) dominate. The sign of the dominant charge carriers is opposite to the sign of the dominant ionized impurities.

8. A p–n junction is a rectifying semiconductor structure with an internal transition from p-type to n-type. A p–n junction contains internal electric fields even in the absence of an applied voltage. For an abrupt junction the field at the p–n interface is

$$E = \left(\frac{2e}{\epsilon} \frac{n_a n_d}{n_a + n_d} \left[(|V| + V_{bi}) - 2\tau/e \right] \right)^{1/2}.$$

Here ϵ is the permittivity, n_a and n_d are ionized acceptor and donor concentrations, and $|V|$ and V_{bi} are the applied and the built-in reverse bias.

9. The electric current densities due to electron and hole flow are given by

$$\mathbf{J}_n = e\tilde{\mu}_e n_e \mathbf{E} + eD_e \operatorname{grad} n_e \, ,$$

$$\mathbf{J}_p = e\tilde{\mu}_h n_h \mathbf{E} - eD_h \operatorname{grad} n_h.$$

Here $\tilde{\mu}_e$ and $\tilde{\mu}_h$ are the electron and hole mobilities, and

$$D_e = \tilde{\mu}_e \tau / e, \qquad D_h = \tilde{\mu}_h \tau / e$$

are the electron and hole diffusion coefficients.

PROBLEMS

1. Weakly doped semiconductor. Calculate the electron and hole concentrations when the net donor concentration is small compared to the intrinsic concentration, $|\Delta n| \ll n_i$.

2. Intrinsic conductivity and minimum conductivity. The electrical conductivity is

$$\sigma = e(n_e \tilde{\mu}_e + n_h \tilde{\mu}_h) , \qquad (99)$$

where $\tilde{\mu}_e$ and $\tilde{\mu}_h$ are the electron and hole mobilities. For most semiconductors $\tilde{\mu}_e > \tilde{\mu}_h$. (a) Find the net ionized impurity concentration $\Delta n = n_d^+ - n_a^-$ for which the conductivity is a minimum. Give a mathematical expression for this minimum conductivity. (b) By what factor is it lower than the conductivity of an intrinsic semiconductor? (c) Give numerical values at 300 K for Si for which the mobilities are $\tilde{\mu}_e = 1350$ and $\tilde{\mu}_h = 480 \, \mathrm{cm^2 \, V^{-1} \, s^{-1}}$, and for InSb, for which the mobilities are $\tilde{\mu}_e = 77\,000$ and $\tilde{\mu}_h = 750 \, \mathrm{cm^2 \, V^{-1} \, s^{-1}}$. Calculate missing data from Table 13.1.

3. Resistivity and impurity concentration. A manufacturer specifies the resistivity $\rho = 1/\sigma$ of a Ge crystal as 20 ohm cm. Take $\tilde{\mu}_e = 3900 \, \mathrm{cm^2 \, V^{-1} \, s^{-1}}$ and $\tilde{\mu}_h = 1900 \, \mathrm{cm^2 \, V^{-1} \, s^{-1}}$. What is the net impurity concentration a) if the crystal is n-type; b) if the crystal is p-type?

4. Mass action law for high electron concentrations. Derive (39), which is the form of the law of mass action when n_e is no longer small compared to n_c.

5. Electron and hole concentrations in InSb. Calculate n_e, n_h, and $\mu - \varepsilon_c$ for n-type InSb at 300 K, assuming $n_d^+ = 4.6 \times 10^{16} \, \mathrm{cm^{-3}} = n_c$. Because of the high ratio n_v / n_c and the narrow energy gap, the hole concentration is not negligible under these conditions, nor is the nondegenerate approximation $n_e \ll n_c$ applicable. Use the generalized mass action law (39). Solve the transcendental equation for n_e by iteration or graphically.

6. Incomplete ionization of deep impurities. Find the fraction of ionized donor impurities if the donor ionization energy is large enough that $\Delta \varepsilon_d$ is larger than $\tau \log(n_c/8 n_d)$ by several times τ. The result explains why substances with large impurity ionization energies remain insulators, even if impure.

7. Built-in field for exponential doping profile.   Suppose that in a p-type semiconductor the ionized acceptor concentration at $x = x_1$ is $n_a^- = n_1 \ll n_v$ and falls off exponentially to a value $n_a^- = n_2 \gg n_i$ at $x = x_2$. What is the built-in electric field in the interval (x_1, x_2)? Give numerical values for $n_1/n_2 = 10^3$ and $x_2 - x_1 = 10^{-5}$ cm. Assume $T = 300$ K. Impurity distributions such as this occur in the base region of many n–p–n transistors. The built-in field aids in driving the injected electrons across the base.

8. Einstein relation for high electron concentrations.   Use the Joyce-Dixon approximation (38) to give a series expansion of the ratio $D_e/\tilde{\mu}_e$ for electron concentrations approaching or exceeding n_c.

9. Injection laser.   Use the Joyce-Dixon approximation to calculate at $T = 300$ K the electron-hole pair concentration in GaAs that satisfies the inversion condition (88), assuming no ionized impurities.

10. Minority carrier lifetime.   Assume both electron and hole concentrations in a semiconductor are raised by δn above their equilibrium values. Define a net minority carrier lifetime t by $R = \delta n/t$. Give expressions for t in terms of the carrier concentrations n_e and n_h; the energy of the recombination level, as expressed by n_e^* and n_h^*; and the time constants t_e and t_h, in the limits of very small and very large values of δn. Under what doping conditions is t independent of δn?

11. Electron-hole pair generation.   Inside a reverse biased p–n junction both electrons and holes have been swept out. (a) Calculate the electron-hole pair generation rate under these conditions, assuming $n_e^* = n_h^*$ and $t_e = t_h = t$. (b) Find the factor by which this generation rate is higher than the generation rate in an n-type semiconductor from which the holes have been swept out, but in which the electron concentration remains equal to $n_d^+ \gg n_i$. (c) Give a numerical value for this ratio for Si with $n_d^+ = 10^{16}$ cm^{-3}.

Kinetic Theory

I am conscious of being only an individual struggling weakly against the stream of time. But it still remains in my power to contribute in such a way that, when the theory of gases is again revived, not too much will have to be rediscovered.

                                                    L. Boltzmann

In this chapter we give a kinetic derivation of the ideal gas law, the distribution of velocities of gas molecules, and transport processes in gases: diffusion, thermal conductivity, and viscosity. The Boltzmann transport equation is discussed. We also treat gases at very low pressures, with reference to vacuum pumps. The chapter is essentially classical physics because the quantum theory of transport is difficult.

KINETIC THEORY OF THE IDEAL GAS LAW

We apply the kinetic method to obtain an elementary derivation of the ideal gas law, $pV = N\tau$. Consider molecules that strike a unit area of the wall of a container. Let v_z denote the velocity component normal to the plane of the wall, as in Figure 14.1. If a molecule of mass M is reflected specularly (mirror-like) from the wall, the change of momentum of the molecule is

$$-2M|v_z|. \tag{1}$$

This gives an impulse $2M|v_z|$ to the wall, by Newton's second law of motion. The pressure on the wall is

$$p = \left(\begin{array}{c}\text{momentum change}\\ \text{per molecule}\end{array}\right)\left(\begin{array}{c}\text{number of molecules striking}\\ \text{unit area per unit time}\end{array}\right). \tag{2}$$

Let $a(v_z)dv_z$ be the number of molecules per unit volume with the z component of the velocity between v_z and $v_z + dv_z$. Here $\int a(v_z)dv_z = N/V = n$. The number in this velocity range that strike a unit area of the wall in unit time is $a(v_z)v_z\,dv_z$. The momentum change of these molecules is $-2Mv_za(v_z)v_z\,dv_z$, so that the total pressure is

$$p = \int_0^\infty 2Mv_z^2a(v_z)dv_z = M\int_{-\infty}^\infty v_z^2a(v_z)dv_z. \tag{3}$$

The integral on the right is the thermal average of v_z^2 times the concentration, so that $p = Mn\langle v_z^2\rangle$. The average value of $\frac{1}{2}Mv_z^2$ is $\frac{1}{2}\tau$, by equipartition of

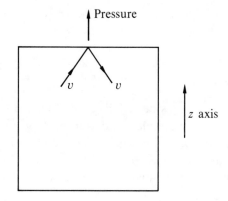

Figure 14.1   The change of momentum of a molecule of velocity v which is reflected from the wall of the container is $-2M|v_z|$.

energy (Chapter 3). Thus the pressure is

$$p = nM\langle v_z^2 \rangle = n\tau = (N/V)\tau; \qquad pV = N\tau. \tag{4}$$

This is the ideal gas law.

The assumption of specular reflection is convenient, but it is immaterial to the result. What comes into the surface must go back, with the same distribution, if thermal equilibrium is to be maintained.

Maxwell Distribution of Velocities

We now transform the energy distribution function of an ideal gas into a classical velocity distribution function. Often when we mean "speed" we shall say "velocity", as this is the tradition in physics when no confusion is caused. In Chapter 6 we found the distribution function of an ideal gas to be

$$f(\varepsilon_n) = \lambda \exp(-\varepsilon_n/\tau) , \tag{5}$$

where $f(\varepsilon_n)$ is the probability of occupancy of an orbital of energy

$$\varepsilon_n = \frac{\hbar^2}{2M}\left(\frac{\pi n}{L}\right)^2 \tag{6}$$

in a cube of volume $V = L^3$. The average number of atoms with quantum number between n and $n + dn$ is (the number of orbitals in this range) × (the probability such an orbital is occupied). The number of orbitals in the positive octant of a spherical shell of thickness dn is $\frac{1}{8}(4\pi n^2)dn$, whence the desired product is

$$(\tfrac{1}{2}\pi n^2\,dn)f(\varepsilon_n) = \tfrac{1}{2}\pi\lambda n^2\exp(-\varepsilon_n/\tau)dn. \tag{7}$$

We take the spin of the atom as zero.

To obtain the probability distribution of the classical velocity, we must find a connection between the quantum number **n** and the classical velocity of a particle in the orbital ε_n. The classical kinetic energy $\frac{1}{2}Mv^2$ is related to the quantum energy (6) by

$$\tfrac{1}{2}Mv^2 = \frac{\hbar^2}{2M}\left(\frac{\pi n}{L}\right)^2; \qquad v = \frac{\hbar\pi}{ML}n; \qquad n = \frac{ML}{\hbar\pi}v. \tag{8}$$

We consider a system of N particles in volume V. Let $NP(v)dv$ be the number of atoms with velocity magnitude, or speed, in the range dv at v. This is evaluated from (7) and (8) by setting $dn = (dn/dv)dv = (ML/\hbar\pi)dv$. We have

$$NP(v)dv = \tfrac{1}{2}\pi\lambda n^2\exp(-\varepsilon_n/\tau)\frac{dn}{dv}\,dv$$

$$= \tfrac{1}{2}\pi\lambda\left(\frac{ML}{\hbar\pi}\right)^3 v^2\exp(-Mv^2/2\tau)dv. \tag{9}$$

From Chapter 6 we know that $\lambda = n/n_Q = (N/L^3)(2\pi\hbar^2/M\tau)^{3/2}$, so that the factor standing to the left of v^2 becomes

$$\frac{\pi N(2\pi)^{3/2}\hbar^3 M^3 L^3}{2L^3 M^{3/2}\tau^{3/2}\hbar^3\pi^3} = 4\pi N\left(\frac{M}{2\pi\tau}\right)^{3/2}. \tag{10}$$

Thus

$$P(v) = 4\pi(M/2\pi\tau)^{3/2}v^2\exp(-Mv^2/2\tau). \tag{11}$$

This is the **Maxwell velocity distribution** (Figure 14.2). The quantity $P(v)dv$ is the probability that a particle has its speed in dv at v. Numerical values of the root mean square thermal velocity and the mean speed are given in Table 14.1, using the results $v_{\text{rms}} = (3\tau/M)^{1/2}$ and $\bar{c} = (8\tau/\pi M)^{1/2}$ from Problem 1.

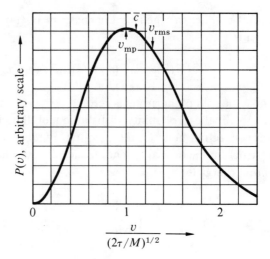

Figure 14.2 Maxwell velocity distribution as a function of the speed in units of the most probable speed $v_{mp} = (2\,\tau/M)^{1/2}$. Also shown are the mean speed $\bar{c}$ and the root mean square velocity v_{rms}.

Table 14.1   Molecular velocities at 273 K, in $10^4\,cm\,s^{-1}$

Gas	v_{rms}	$\bar{c}$	Gas	v_{rms}	$\bar{c}$
H_2	18.4	16.9	O_2	4.6	4.2
He	13.1	12.1	Ar	4.3	4.0
H_2O	6.2	5.7	Kr	2.86	2.63
Ne	5.8	5.3	Xe	2.27	2.09
N_2	4.9	4.5	Free electron	1100.	1013.

Experimental verification.   The velocity distribution of atoms of potassium which exit from the slit of an oven has been studied by Marcus and McFee.* The curve in Figure 14.3 compares the experimental results with the prediction of (12) below; the agreement is excellent. We need an expression for the velocity distribution of atoms that exit from a small hole[†] in an oven. This distribution is different from the velocity distribution within the oven, because the flux through the hole involves an extra factor, the velocity component normal to the wall. The exit beam is weighted in favor of atoms of high velocity at the expense of those at low velocity. In proportion to their concentration in the oven, fast atoms

---

* P. M. Marcus and J. H. McFee, *Recent research in molecular beams*, ed. I. Esterman, Academic Press, 1959.
[†] In such experiments a round hole is said to be small if the diameter is less than a mean free path of an atom in the oven. If the hole is not small in this sense, the flow of gas from it will be governed by the laws of hydrodynamic flow and not by gas kinetics.

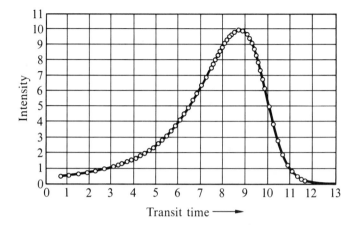

Figure 14.3   Measured transmission points and calculated Maxwell transmission curve for potassium atoms that exit from an oven at a temperature 157°C. The horizontal axis is the transit time of the atoms transmitted. The intensity is in arbitrary units; the curve and the points are normalized to the same maximum value. After Marcus and McFee.

strike the walls more often than slow atoms strike the walls. The weight factor is the velocity component $v \cos \theta$ normal to the plane of the hole. The average of $\cos \theta$ over the forward hemisphere is just a numerical factor, namely $\frac{1}{2}$. The probability that an atom which leaves the hole will have a velocity between v and $v + dv$ defines the quantity $P_{\text{beam}}(v)dv$, where

$$P_{\text{beam}}(v) \propto vP_{\text{Maxwell}} \propto v^3 \exp(-Mv^2/2\tau) , \qquad (12)$$

with P_{Maxwell} given by (11). The distribution (12) of the transmission through a hole is called the Maxwell transmission distribution.

Collision Cross Sections and Mean Free Paths

We can estimate the collision rates of gas atoms viewed as rigid spheres. Two atoms of diameter d will collide if their centers pass within the distance d of each other. From Figure 14.4 we see that one collision will occur when an atom has traversed an average distance

$$l = 1/n\pi d^2 , \qquad (13)$$

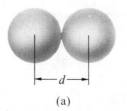

(a)

Figure 14.4 (a) Two rigid spheres will collide if their centers pass within a distance d of each each other. (b) An atom of diameter d which travels a long distance L will sweep out a volume $\pi d^2 L$, in the sense that it will collide with any atom whose center lies within the volume. If n is the concentration of atoms, the average number of atoms in this volume is $n\pi d^2 L$. This is the number of collisions. The average distance between collisions is

$$l = \frac{L}{n\pi d^2 L} = \frac{1}{n\pi d^2}.$$

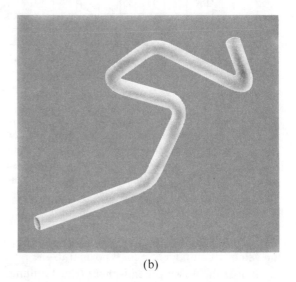

(b)

where n is the number of atoms per unit volume. The length l is called the **mean free path**: it is the average distance traveled by an atom between collisions. Our result neglects the velocity of the target atoms.

We estimate the order of magnitude of the mean free path. If the atomic diameter d is 2.2 Å as for helium, then the collision cross section σ_c is

$$\sigma_c = \pi d^2 = (3.14)(2.2 \times 10^{-8}\,\text{cm})^2 = 15.2 \times 10^{-16}\,\text{cm}^2. \qquad (14)$$

The concentration of molecules of an ideal gas at 0°C and 1 atm is given by the **Loschmidt number**

$$n_0 = 2.69 \times 10^{19}\,\text{atoms cm}^{-3}, \qquad (15)$$

defined as the Avogadro number divided by the molar volume at 0°C and 1 atm. The Avogadro number is the number of molecules in one mole; the molar volume is the volume occupied by one mole. We combine (14) and (15) to obtain the mean free path under standard conditions:

$$l = \frac{1}{\pi d^2 n_0} = \frac{1}{(15.2 \times 10^{-16}\,\text{cm}^2)(2.69 \times 10^{19}\,\text{cm}^{-3})} = 2.44 \times 10^{-5}\,\text{cm.} \quad (16a)$$

This length is about 1000 times larger than the diameter of an atom. The associated collision rate is

$$\frac{v_{\text{rms}}}{l} \approx \frac{10^5\,\text{cm s}^{-1}}{10^{-5}\,\text{cm}} \approx 10^{10}\,\text{s}^{-1}. \quad (16b)$$

At a pressure of 10^{-6} atm or 1 dyne cm^{-2}, the concentration of atoms is reduced by 10^{-6} and the mean free path is increased to 25 cm. At 10^{-6} atm the mean free path may not be small in comparison with the dimensions of any particular experimental apparatus. Then we are in what is called the **high vacuum region**, also called the **Knudsen region**. We assume below that the mean free path is small in comparison with the relevant dimension of the apparatus, except in the section on laws of rarefied gases.

TRANSPORT PROCESSES

Consider a system not in thermal equilibrium, but in a nonequilibrium steady state with a constant flow from one end of the system to the other. For example, we may create a steady state nonequilibrium condition in a system by placing opposite ends in thermal contact with large reservoirs at two different temperatures. If reservoir 1 is at the higher temperature, energy will flow through the system from reservoir 1 to reservoir 2. Energy flow in this direction will increase the total entropy of reservoir 1 + reservoir 2 + system. The temperature gradient in the system is the driving force; the physical quantity that is transported through the specimen in this process is energy.

There is a linear region in most transport processes in which the flux is directly proportional to the driving force:

$$\text{flux} = (\text{coefficient}) \times (\text{driving force}), \quad (17)$$

provided the force is not too large. Such a relation is called a linear phenomenological law, such as Ohm's law for the conduction of electricity. The definition of the flux density of a quantity A is:

$$\mathbf{J}_A = \text{flux density of } A = \text{net quantity of } A \text{ transported across}$$
$$\text{unit area in unit time.} \quad (18)$$

Table 14.2 Summary of phenomenological transport laws

Effect	Flux of particle property	Gradient	Coefficient	Law	Name of law	Approximate expression for coefficient
Diffusion	Number	$\dfrac{dn}{dz}$	Diffusivity D	$\mathbf{J}_n = -D\,\mathrm{grad}\,n$	Fick's law	$D = \tfrac{1}{3}\bar{c}l$
Viscosity	Transverse momentum	$M\dfrac{dv_x}{dz}$	Viscosity η	$\dfrac{F_x}{A} = J_{\mathbf{p}}{}^x = -\eta\dfrac{dv_x}{dz}$	Newtonian viscosity	$\eta = \tfrac{1}{3}\rho\bar{c}l$
Thermal conductivity	Energy	$\dfrac{d\rho_u}{dz} = \hat{C}_V\dfrac{dT}{dz}$	Thermal conductivity K	$\mathbf{J}_u = -K\,\mathrm{grad}\,\tau$	Fourier's law	$K = \tfrac{1}{3}\hat{C}_V\bar{c}l$
Electrical conductivity	Charge	$-\dfrac{d\varphi}{dz} = E_z$	Conductivity σ	$\mathbf{J}_q = \sigma\mathbf{E}$	Ohm's law	$\sigma = \dfrac{nq^2l}{M\bar{c}}$

SYMBOLS:
n = number of particles per unit volume
$\bar{c}$ = mean thermal speed = $\langle |v| \rangle$
l = mean free path
$\hat{C}_V$ = heat capacity per unit volume
ρ_u = thermal energy per unit volume
F_x/A = shear force per unit area

φ = electrostatic potential
$\mathbf{E}$ = electric field intensity
q = electric charge
M = mass of particle
ρ = mass per unit volume
$\mathbf{p}$ = momentum

The net transport is the transport in one direction minus the transport in the opposite direction. Various transport laws are summarized in Table 14.2.

Particle Diffusion

In Figure 14.5 we consider a system with one end in diffusive contact with a reservoir at chemical potential μ_1; the other end is in diffusive contact with a reservoir at chemical potential μ_2. The temperature is constant. If reservoir 1 is at the higher chemical potential, then particles will flow through the system from reservoir 1 to reservoir 2. Particle flow in this direction will increase the total entropy of reservoir 1 + reservoir 2 + system.

Consider particle diffusion, first when the difference of chemical potential is caused by a difference in particle concentration. The flux density $\mathbf{J}_n$ is the number of particles passing through a unit area in unit time. The driving force of isothermal diffusion is usually taken as the gradient of the particle concentration along the system:

$$\mathbf{J}_n = -D \operatorname{grad} n. \tag{19}$$

The relation is called **Fick's law**; here D is the particle diffusion constant or **diffusivity**.

Particles travel freely over distances of the order of the mean free path l before they collide. We assume that in a collision at position z the particles come into a local equilibrium condition at the local chemical potential $\mu(z)$ and local concentration $n(z)$. Let l_z be the z component of the mean free path. Across the plane at z there is a particle flux density in the positive z direction equal to $\frac{1}{2}n(z - l_z)\overline{c}_z$ and a flux density in the negative z direction equal to $-\frac{1}{2}n(z + l_z)\overline{c}_z$. Here $n(z - l_z)$ means the particle concentration at $z - l_z$. The net particle

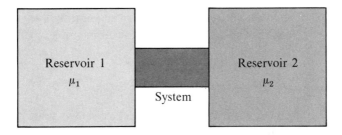

Figure 14.5 Opposite ends of the system are in diffusive contact with reservoirs at chemical potentials μ_1 and μ_2. The temperature is constant everywhere.

flux density is the average over all directions on a hemisphere of

$$J_n{}^z = \tfrac{1}{2}[n(z - l_z) - n(z + l_z)]\bar{c}_z = -\frac{dn}{dz}\bar{c}_z l_z. \tag{20}$$

We want to express the average value of $c_z l_z$ in terms of $\bar{c}l$. Here $l_z = l\cos\theta$ is the projection of the mean free path, and $\bar{c}_z = \bar{c}\cos\theta$ is the projection of the speed on the z axis. The average is taken over the surface of a hemisphere, because all forward directions are equally likely. The element of surface area is $2\pi\sin\theta\,d\theta$. Thus

$$\langle \bar{c}_z l_z \rangle = \bar{c}l\,\frac{2\pi\int_0^{\frac{1}{2}\pi}\cos^2\theta\sin\theta\,d\theta}{2\pi} = \tfrac{1}{3}\bar{c}l\;, \tag{21}$$

so that

$$J_n{}^z = -\tfrac{1}{3}\bar{c}l\,\frac{dn}{dz}. \tag{22}$$

On comparison with (19) we see that the diffusivity is given by

$$\boxed{D = \tfrac{1}{3}\bar{c}l.} \tag{23}$$

The particle diffusion problem is the model for other transport problems. In particle diffusion we are concerned with the transport of particles; in thermal conductivity with the transport of energy by particles; in viscosity with the transport of momentum by particles; and in electrical conductivity with the transport of charge by particles. The linear transport coefficients that describe the processes are proportional to the particle diffusivity D.

Let ρ_A denote the concentration of the physical quantity A. If A is a quantity like charge or mass that has the same value for all the molecules, then the flux density of A in the z direction is

$$J_A{}^z = \rho_A\langle v_z\rangle\;, \tag{24}$$

where $\langle v_z\rangle$ is the mean drift velocity of the particles in the z direction. The drift velocity is zero in thermal equilibrium.

If A is a quantity like energy or momentum that depends on the velocity of a molecule, then we always find a similar expression:

$$J_A{}^z = f_A\rho_A\langle v_z\rangle\;, \tag{25}$$

where f_A is a factor with magnitude of the order of unity. The exact value of f_A depends on the velocity dependence of A and may be calculated by the method of the Boltzmann transport equation treated at the end of this chapter. For simplicity we set $f_A = 1$ in this discussion. By analogy with (19) for particle diffusion, the phenomenological law for the transport of A is

$$\mathbf{J}_A = -D \operatorname{grad} \rho_A , \tag{26}$$

with the particle diffusivity D given by (22).

Thermal Conductivity

Fourier's law

$$\mathbf{J}_u = -K \operatorname{grad} \tau \tag{27}$$

describes the energy flux density $\mathbf{J}_u$ in terms of the **thermal conductivity** K and the temperature gradient (Figure 14.6). This form assumes that there is a net transport of energy, but not of particles. Another term must be added if additional energy is transported by means of particle flow, as when electrons flow under the influence of an electric field.

The energy flux density in the z direction is

$$J_u^z \simeq \rho_u \langle v_z \rangle , \tag{28}$$

where $\langle v_z \rangle$ is the mean drift velocity; ρ_u is the energy density. This result is valid within a factor of the order of unity, as discussed. By analogy with the diffusion equation, the right-hand side is equal to

$$-D \, d\rho_u/dx = -D(\partial \rho_u/\partial \tau)(d\tau/dx). \tag{29}$$

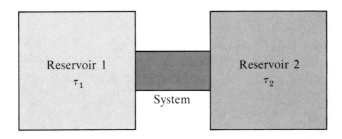

Figure 14.6 Opposite ends of the system are in thermal contact with reservoirs at temperatures τ_1 and τ_2.

This describes the diffusion of energy. Now $\partial \rho_u / \partial \tau$ is just the heat capacity per unit volume, denoted by $\hat{C}_V$. Thus

$$\mathbf{J}_u = -D\hat{C}_V \operatorname{grad} \tau; \tag{30}$$

on comparison with (27) the thermal conductivity is

$$K = D\hat{C}_V = \tfrac{1}{3}\hat{C}_V \bar{c} l. \tag{31}$$

The thermal conductivity of a gas is independent of pressure until at very low pressures the mean free path becomes limited by the dimensions of the apparatus, rather than by intermolecular collisions. Until very low pressures are attained there is no advantage to evacuating a Dewar vessel, because the heat losses are independent of pressure as long as (31) applies.

Viscosity

Viscosity is a measure of the diffusion of momentum parallel to the flow velocity and transverse to the gradient of the flow velocity. Consider a gas with flow velocity in the x direction, with the flow velocity gradient in the z direction. The **viscosity coefficient** η is defined by

$$X_z = -\eta \frac{dv_x}{dz} = J_z(p_x). \tag{32}$$

Here v_x is the x component of the flow velocity of the gas; p_x denotes the x component of momentum; and X_z is the x component of the shear force exerted by the gas on a unit area of the xy plane normal to the z direction. By Newton's second law of motion a shear stress X_z acts on the xy plane if the plane receives a net flux density of x momentum $J_z(p_x)$, because this flux density measures the rate of change of the momentum of the plane, per unit area.

In diffusion the particle flux density in the z direction is given by the number density n times the mean drift velocity $\langle v_z \rangle$ in the z direction, so that $J_n{}^z = n\langle v_z \rangle = -D\, dn/dz$. In the viscosity equation the transverse momentum density is nMv_x; its flux density in the z direction is $(nMv_x)\langle v_z \rangle$. By analogy with (26) this flux density equals $-D\, d(nMv_x)/dz$, within a factor of the order of unity. With $\rho = nM$ as the mass density,

$$J_z(p_x) = \rho v_x \langle v_z \rangle = -D\rho\, dv_x/dz = -\eta\, dv_x/dz. \tag{33}$$

Thus, with D given by (23),

$$\boxed{\eta = D\rho = \tfrac{1}{3}\rho\bar{c}l}\tag{34}$$

gives the coefficient of viscosity. The CGS unit of viscosity is called the **poise**.

The mean free path is $l = 1/\pi d^2 n$ from (13), where d is the molecular diameter and n is the concentration. Thus the viscosity may be expressed as

$$\eta = M\bar{c}/3\pi d^2 \;,\tag{35}$$

which is independent of the gas pressure. The independence fails at very high pressures when the molecules are nearly always in contact or at very low pressures when the mean free path is longer than the dimensions of the apparatus.

Robert Boyle in 1660 reported an early observation on the pressure independence of the damping of a pendulum in air:

Experiment 26 . . . . We observ'd also that when the Receiver was full of Air, the included Pendulum continu'd its Recursions about fifteen minutes (or a quarter of an hour) before it left off swinging; and that after the exsuction of the Air, the Vibration of the same Pendulum (being fresh put into motion) appear'd not (by a minutes Watch) to last sensibly longer. So that the event of this experiment being other than we expected, scarce afforded us any other satisfaction, than that of our not having omitted to try it.

Although at first glance implausible, this result is readily understood. With decreasing pressure the rate of momentum-transfer collisions decreases, but each colliding particle comes from farther away. The larger the distance, the larger the momentum difference; the increasing momentum transfer per collision cancels the decreasing collision rate.

It is easier to measure the viscosity than the diffusivity. If $D = \eta/\rho$ as predicted by (34), then K is related to η by

$$K = \eta\hat{C}_V/\rho.\tag{36}$$

The observed values of the ratio $K\rho/\eta\hat{C}_V$ given in Table 14.3 are somewhat higher than the value unity predicted by our approximate calculations. Improved calculations of the kinetic coefficients K, D, η take account of minor,

Table 14.3   Experimental values of K, D, η, and $K\rho/\eta C_V$ at 0°C and 1 atm

Gas	K, in mW cm^{-1} K^{-1}	D, in cm^2 s^{-1}	η, in μpoise	$K\rho/\eta \hat{C}_V$
He	1.50	—	186.	2.40
Ar	0.18	0.158	210.	2.49
H_2	1.82	1.28	84.	1.91
N_2	0.26	—	167.	1.91
O_2	0.27	—	189.	1.90

NOTE: Values of the thermal conductivity are at 300 K.

but difficult, effects we have neglected; see the works cited in the general references.

Comment.   The diffusivity of gas atoms is directly proportional to their viscosity. The diffusivity of a particle suspended in a liquid or gas is a different problem: the viscosity of the solvent opposes the diffusion of the suspended particle. We find $D \propto 1/\eta$, where D refers to the particles and η refers to the liquid. The **Stokes-Einstein relation** for suspended particles is $D = \tau/6\pi\eta R$, where R is the radius of the sphere in suspension.

Comment.   The quantity $v = \eta/\rho$ is called the **kinematic viscosity**; if (34) holds, v should be equal to the diffusivity D. The ratio η/ρ enters into hydrodynamic theory and into the Reynold's number criterion for laminar flow.

Generalized Forces

The transfer of entropy from one part of a system to another is a consequence of any transport process. We can relate the rate of change of entropy to the flux density of particles and of energy. By analogy with the thermodynamic identity at constant volume,

$$d\sigma = \frac{1}{\tau} dU - \frac{\mu}{\tau} dN \; , \qquad (37)$$

we write the entropy current density J_σ as

$$\mathbf{J}_\sigma = \frac{1}{\tau} \mathbf{J}_u - \frac{\mu}{\tau} \mathbf{J}_n. \qquad (38)$$

Let $\hat{\sigma}$ denote the entropy density; let $\partial\hat{\sigma}/\partial t$ denote the net rate of change of entropy density at a fixed position $\mathbf{r}$. Then, by the equation of continuity,

$$\partial\hat{\sigma}/\partial t = g_\sigma - \operatorname{div} \mathbf{J}_\sigma. \tag{39}$$

In a unit volume element the net rate of appearance of entropy is equal to the rate of production g_σ minus the loss $-\operatorname{div} \mathbf{J}_\sigma$ attributed to the transport current.

In a transfer process U and N are conserved. The equation of continuity for the energy density u is

$$\frac{\partial u}{\partial t} = -\operatorname{div} \mathbf{J}_u; \tag{40}$$

the equation of continuity for the particle concentration n is

$$\frac{\partial n}{\partial t} = -\operatorname{div} \mathbf{J}_n. \tag{41}$$

Let us take the divergence of $\mathbf{J}_\sigma$ in (38):

$$\operatorname{div} \mathbf{J}_\sigma = \frac{1}{\tau} \operatorname{div} \mathbf{J}_u + \mathbf{J}_u \cdot \operatorname{grad}(1/\tau)$$

$$- (\mu/\tau) \operatorname{div} \mathbf{J}_n - \mathbf{J}_n \cdot \operatorname{grad}(\mu/\tau). \tag{42}$$

Let (37) refer to unit volume; we take a partial derivative with respect to time to obtain the net rate of entropy change:

$$\frac{\partial\hat{\sigma}}{\partial t} = \frac{1}{\tau}\frac{\partial u}{\partial t} - \frac{\mu}{\tau}\frac{\partial n}{\partial t}. \tag{43}$$

We use (40)–(43) to rearrange (39) in a form suggestive of the ohmic power dissipation:

$$g_\sigma = \mathbf{J}_u \cdot \operatorname{grad}(1/\tau) + \mathbf{J}_n \cdot \operatorname{grad}(-\mu/\tau) , \tag{44}$$

or

$$g_\sigma = \mathbf{J}_u \cdot \mathbf{F}_u + \mathbf{J}_n \cdot \mathbf{F}_n. \tag{45}$$

Here $\mathbf{F}_u$ and $\mathbf{F}_n$ are **generalized forces** defined by

$$\mathbf{F}_u \equiv \operatorname{grad}(1/\tau); \qquad \mathbf{F}_n \equiv \operatorname{grad}(-\mu/\tau). \tag{46}$$

Einstein relation

In an isothermal process $\mathbf{F}_n$ in (46) may be written as $\mathbf{F}_n = (-1/\tau)\operatorname{grad}\mu$ or, in terms of the internal and external parts of the chemical potential, as

$$\mathbf{F}_n = -(1/\tau)[\operatorname{grad}\mu_{\text{int}} + \operatorname{grad}\mu_{\text{ext}}] \tag{47}$$

For an ideal gas $\mu_{\text{int}} = \tau\log(n/n_Q)$, so that $\operatorname{grad}\mu_{\text{int}} = (\tau/n)\operatorname{grad} n$; for an electrostatic potential $\operatorname{grad}\mu_{\text{ext}} = q\operatorname{grad}\varphi = -q\mathbf{E}$. Thus

$$\mathbf{F}_n = -(1/\tau)[\tau n^{-1}\operatorname{grad} n - q\mathbf{E}] \tag{48}$$

Now the particle flux density also has two terms, written as

$$\mathbf{J}_n = -D_n\operatorname{grad} n + n\tilde{\mu}\mathbf{E} , \tag{49}$$

where D_n is the diffusivity and $\tilde{\mu}$ is the mobility, which is the drift velocity per unit electric field. The ratio of the coefficients of $\operatorname{grad} n$ to $\mathbf{E}$ is $D_n/n\tilde{\mu}$ in (49) and τ/nq in (48). These ratios must be equal, so that

$$\boxed{D_n = \tau\tilde{\mu}/q ,} \tag{50}$$

which is called the Einstein relation between the diffusivity and the mobility for a classical gas.

Comment. We gain an advantage, for reasons related to the thermodynamics of irreversible processes, if we use F_u and F_n in (46) as the driving forces for the linear transport processes. We write

$$\mathbf{J}_u = L_{11}\mathbf{F}_u + L_{12}\mathbf{F}_n; \qquad \mathbf{J}_n = L_{21}\mathbf{F}_u + L_{22}\mathbf{F}_n , \tag{51}$$

The **Onsager relation** of irreversible thermodynamics is that

$$L_{ij}(\mathbf{B}) = L_{ji}(-\mathbf{B}) \tag{52}$$

where $\mathbf{B}$ is the magnetic field intensity. If $\mathbf{B} = 0$, then $L_{ij} = L_{ji}$ always. For (52) to hold, the driving forces $\mathbf{F}$ must be defined as in (46). Other definitions of the forces are perfectly valid, such as the pair $\operatorname{grad}\tau$ and $\operatorname{grad} n$, but do not necessarily lead to coefficients L that satisfy the Onsager relation. For a derivation see the book by Landau and Lifshitz cited in the general references.

KINETICS OF DETAILED BALANCE

Consider a system with two states, one at energy Δ and one at energy $-\Delta$. In an ensemble of N such systems, N^+ are at Δ and N^- are at $-\Delta$, with $N = N^+ + N^-$. To establish thermal equilibrium there must exist some mechanism whereby systems can pass between the two states. Consider the rate equation for transitions into and out of the upper state:

$$dN^+/dt = \alpha N^- - \beta N^+ , \qquad (53)$$

where α, β may be functions of the temperature. The transition rate from $-$ to $+$ is directly proportional to the number of systems in the $-$ state. The transition rate from $+$ to $-$ is directly proportional to the number of systems in the $+$ state.

In thermal equilibrium $\langle dN^+/dt \rangle = 0$, which can be satisfied only if

$$\alpha/\beta = \langle N^+ \rangle/\langle N^- \rangle = \exp(-2\Delta/\tau) , \qquad (54)$$

the Boltzmann factor. This result expresses a relation between $\alpha(\tau)$ and $\beta(\tau)$ that must be satisfied by any and every mechanism that assists in the transitions. As an example, suppose that the transition $+ \rightarrow -$ proceeds with the excitation of a harmonic oscillator from a state of energy $s\varepsilon$ to a state of energy $(s + 1)\varepsilon$; in the inverse process $- \rightarrow +$ the oscillator goes from $s\varepsilon$ to $(s - 1)\varepsilon$. In the quantum mechanical theory of the oscillator it is shown that

$$\frac{\beta}{\alpha} = \frac{\text{Prob}(s \rightarrow s + 1)}{\text{Prob}(s \rightarrow s - 1)} = \frac{s + 1}{s} ,$$

for the excitation and de-excitation of the oscillator, a result derived in most texts on quantum theory. The value of $\langle s \rangle$ is found from the Planck distribution:

$$\langle s \rangle = \frac{1}{\exp(\varepsilon/\tau) - 1}; \qquad \langle s \rangle + 1 = \frac{\exp(\varepsilon/\tau)}{\exp(\varepsilon/\tau) - 1} ,$$

so that, with $\varepsilon = 2\Delta$ to conserve energy,

$$\alpha/\beta = \langle s \rangle/\langle s + 1 \rangle = \exp(-2\Delta/\tau). \qquad (55)$$

This satisfies the condition (54).

The principle of detailed balance emerges as a generalization of this argument: in thermal equilibrium the rate of any process that leads to a given state must equal exactly the rate of the inverse process that leads from the state. One common application of the principle is to the Kirchhoff law for the absorption and emission of radiation by a solid, already discussed in Chapter 4: radiation of a wavelength that is absorbed strongly by a solid will also be emitted strongly—otherwise the specimen would heat up because it could not come into thermal equilibrium with the radiation.

ADVANCED TREATMENT: BOLTZMANN TRANSPORT EQUATION

The classical theory of transport processes is based on the Boltzmann transport equation. We work in the six-dimensional space of Cartesian coordinates $\mathbf{r}$ and velocity $\mathbf{v}$. The classical distribution function $f(\mathbf{r},\mathbf{v})$ is defined by the relation

$$f(\mathbf{r},\mathbf{v})d\mathbf{r}\,d\mathbf{v} = \text{number of particles in } d\mathbf{r}\,d\mathbf{v}. \tag{56}$$

The Boltzmann equation is derived by the following argument. We consider the effect of a time displacement dt on the distribution function. The Liouville theorem of classical mechanics tells us that if we follow a volume element along a flowline the distribution is conserved:

$$f(t + dt,\mathbf{r} + d\mathbf{r},\mathbf{v} + d\mathbf{v}) = f(t,\mathbf{r},\mathbf{v}) , \tag{57}$$

in the absence of collisions. With collisions

$$f(t + dt,\mathbf{r} + d\mathbf{r},\mathbf{v} + d\mathbf{v}) - f(t,\mathbf{r},\mathbf{v}) = dt(\partial f/\partial t)_{\text{collisions}}. \tag{58}$$

Thus

$$dt(\partial f/\partial t) + d\mathbf{r} \cdot \text{grad}_r\ f + d\mathbf{v} \cdot \text{grad}_v\ f = dt(\partial f/\partial t)_{\text{coll}}. \tag{59}$$

Let $\boldsymbol{\alpha}$ denote the acceleration $d\mathbf{v}/dt$; then

$$\boxed{\partial f/\partial t + \mathbf{v} \cdot \text{grad}_r\ f + \boldsymbol{\alpha} \cdot \text{grad}_v\ f = (\partial f/\partial t)_{\text{coll}}.} \tag{60}$$

This is the **Boltzmann transport equation**.

In many problems the collision term $(\partial f/\partial t)_{\text{coll}}$ may be treated by the introduction of a relaxation time $\tau_c(\mathbf{r},\mathbf{v})$, defined by the equation

$$(\partial f/\partial t)_{\text{coll}} = -(f - f_0)/\tau_c. \tag{61}$$

Here f_0 is the distribution function in thermal equilibrium. Do not confuse τ_c for relaxation time with τ for temperature. Suppose that a nonequilibrium distribution of velocities is set up by external forces which are suddenly removed. The decay of the distribution towards equilibrium is then obtained from (61) as

$$\frac{\partial(f - f_0)}{\partial t} = -\frac{f - f_0}{\tau_c} ,$$

(62)

if we note that $\partial f_0/\partial t = 0$ by definition of the equilibrium distribution. This equation has the solution

$$(f - f_0)_t = (f - f_0)_{t=0} \exp(-t/\tau_c).$$

(63)

It is not excluded that τ_c may be a function of $\mathbf{r}$ and $\mathbf{v}$.

We combine (56), (60), and (61) to obtain the Boltzmann transport equation in the relaxation time approximation:

$$\boxed{\frac{\partial f}{\partial t} + \boldsymbol{\alpha} \cdot \mathrm{grad}_\mathbf{v}\, f + \mathbf{v} \cdot \mathrm{grad}_\mathbf{r}\, f = -\frac{f - f_0}{\tau_c}.}$$

(64)

In the steady state $\partial f/\partial t = 0$ by definition.

Particle Diffusion

Consider an isothermal system with a gradient of the particle concentration. The steady-state Boltzmann transport equation in the relaxation time approximation becomes

$$v_x\, df/dx = -(f - f_0)/\tau_c ,$$

(65)

where the nonequilibrium distribution function f varies along the x direction. We may write (65) to first order as

$$f_1 \simeq f_0 - v_x \tau_c\, df_0/dx ,$$

(66)

where we have replaced $\partial f/\partial x$ by df_0/dx. We can iterate to obtain higher order solutions when desired. Thus the second order solution is

$$f_2 = f_0 - v_x \tau_c\, df_1/dx = f_0 - v_x \tau_c\, df_0/dx + v_x^2 \tau_c^2\, d^2f_0/dx^2.$$

(67)

The iteration is necessary for the treatment of nonlinear effects.

Classical Distribution

Let f_0 be the distribution function in the classical limit:

$$f_0 = \exp[(\mu - \varepsilon)/\tau] , \qquad (68)$$

as in Chapter 6. We are at liberty to take whatever normalization for the distribution function is most convenient because the transport equation is linear in f and f_0. We can take the normalization as in (68) rather than as in (56). Then

$$df_0/dx = (df_0/d\mu)(d\mu/dx) = (f_0/\tau)(d\mu/dx) , \qquad (69)$$

and the first order solution (66) for the nonequilibrium distribution becomes

$$f = f_0 - (v_x \tau_c f_0/\tau)(d\mu/dx). \qquad (70)$$

The particle flux density in the x direction is

$$J_n{}^x = \int v_x f \mathfrak{D}(\varepsilon) d\varepsilon , \qquad (71)$$

where $\mathfrak{D}(\varepsilon)$ is the density of orbitals per unit volume per unit energy range:

$$\mathfrak{D}(\varepsilon) = \frac{1}{4\pi^2}\left(\frac{2M}{\hbar^2}\right)^{3/2} \varepsilon^{1/2} , \qquad (72)$$

as in (7.65) for a particle of spin zero. Thus

$$J_n{}^x = \int v_x f_0 \mathfrak{D}(\varepsilon) d\varepsilon - (d\mu/dx) \int (v_x{}^2 \tau_c f_0/\tau)\mathfrak{D}(\varepsilon) d\varepsilon \qquad (73)$$

The first integral vanishes because v_x is an odd function and f_0 is an even function of v_x. This confirms that the net particle flux vanishes for the equilibrium distribution f_0. The second integral will not vanish.

Before evaluating the second integral, we have an opportunity to make use of what we may know about the velocity dependence of the relaxation time τ_c. Only for the sake of example we assume that τ_c is constant, independent of velocity; τ_c may then be taken out of the integral:

$$J_n{}^x = -(d\mu/dx)(\tau_c/\tau) \int v_x{}^2 f_0 \mathfrak{D}(\varepsilon) d\varepsilon. \qquad (74)$$

The integral may be written as

$$\tfrac{1}{3} \int v^2 f_0 \mathfrak{D}(\varepsilon) d\varepsilon = \frac{2}{3M} \int (\tfrac{1}{2} M v^2) f_0 \mathfrak{D}(\varepsilon) d\varepsilon = n\tau/M \ , \tag{75}$$

because the integral is just the kinetic energy density $\tfrac{3}{2} n\tau$ of the particles. Here $\int f_0 \mathfrak{D}(\varepsilon) d\varepsilon = n$ is the concentration. The particle flux density is

$$J_n{}^x = -(n\tau_c/M)(d\mu/dx) = -(\tau_c \tau/M)(dn/dx) \ , \tag{76}$$

because $\mu = \tau \log n + \text{constant}$. The result (76) is of the form of the diffusion equation with the diffusivity

$$D = \tau_c \tau/M = \tfrac{1}{3} \langle v^2 \rangle \tau_c. \tag{77}$$

Another possible assumption about the relaxation time is that it is inversely proportional to the velocity, as in $\tau_c = l/v$, where the mean free path l is constant. Instead of (74) we have

$$J_n{}^x = -(d\mu/dx)(l/\tau) \int (v_x{}^2/v) f_0 \mathfrak{D}(\varepsilon) d\varepsilon \ , \tag{78}$$

and now the integral may be written as

$$\tfrac{1}{3} \int v f_0 \mathfrak{D}(\varepsilon) d\varepsilon = \tfrac{1}{3} n \bar{c} \ , \tag{79}$$

where $\bar{c}$ is the average speed. Thus

$$J_n{}^x = -\tfrac{1}{3} (l \bar{c} n/\tau)(d\mu/dx) = -\tfrac{1}{3} l \bar{c} (dn/dx) \ , \tag{80}$$

and the diffusivity is

$$D = \tfrac{1}{3} l \bar{c}. \tag{81}$$

Fermi-Dirac Distribution

The distribution function is

$$f_0 = \frac{1}{\exp[(\varepsilon - \mu)/\tau] + 1}. \tag{82}$$

To form df_0/dx as in (69) we need the derivative $df_0/d\mu$. We argue below that

$$df_0/d\mu \simeq \delta(\varepsilon - \mu) ,\tag{83}$$

at low temperatures $\tau \ll \mu$. Here δ is the Dirac delta function, which has the property for a general function $F(\varepsilon)$ that

$$\int_{-\infty}^{\infty} F(\varepsilon)\delta(\varepsilon - \mu)d\varepsilon = F(\mu).\tag{84}$$

Now consider the integral $\int_0^\infty F(\varepsilon)(df_0/d\mu)d\varepsilon$. At low temperatures $df_0/d\mu$ is very large for $\varepsilon \simeq \mu$ and is small elsewhere. Unless the function $F(\varepsilon)$ is very rapidly varying near μ we may take $F(\varepsilon)$ outside the integral, with the value $F(\mu)$:

$$\int_0^\infty F(\varepsilon)(df_0/d\mu)d\varepsilon \simeq F(\mu) \int_0^\infty (df_0/d\mu)d\varepsilon = -F(\mu) \int_0^\infty (df_0/d\varepsilon)d\varepsilon$$
$$= -F(\mu)[f_0(\varepsilon)]_0^\infty = F(\mu)f_0(0) ,\tag{85}$$

where we have used $df_0/d\mu = -df_0/d\varepsilon$. We have also used $f_0 = 0$ for $\varepsilon = \infty$. At low temperatures $f(0) \simeq 1$; thus the right-hand side of (85) is just $F(\mu)$, consistent with the delta function approximation. Thus

$$df_0/dx = \delta(\varepsilon - \mu)d\mu/dx.\tag{86}$$

The particle flux density is, from (71),

$$J_n{}^x = -(d\mu/dx)\tau_c \int v_x{}^2\delta(\varepsilon - \mu)\mathfrak{D}(\varepsilon)d\varepsilon ,\tag{87}$$

where τ_c is the relaxation time at the surface $\varepsilon = \mu$ of the Fermi sphere. The integral has the value

$$\tfrac{1}{3}v_F{}^2(3n/2\varepsilon_F) = n/m ,\tag{88}$$

by use of $\mathfrak{D}(\mu) = 3n/2\varepsilon_F$ at absolute zero, from (7.17), where $\varepsilon_F \equiv \tfrac{1}{2}mv_F{}^2$ defines the velocity v_F on the Fermi surface. Thus

$$J_n{}^x = -(n\tau_c/m)d\mu/dx.\tag{89}$$

At absolute zero $\mu(0) = (\hbar^2/2m)(3\pi^2n)^{2/3}$, whence

$$d\mu/dx = \{\tfrac{2}{3}(\hbar^2/2m)(3\pi^2)^{2/3}/n^{1/3}\}dn/dx$$
$$= \tfrac{2}{3}(\varepsilon_F/n)dn/dx ,\tag{90}$$

so that (87) becomes

$$J_n{}^x = -(2\tau_c/3m)\varepsilon_F \, dn/dx = -\tfrac{1}{3}v_F{}^2\tau_c \, dn/dx. \tag{91}$$

The diffusivity is the coefficient of dn/dx:

$$D = \tfrac{1}{3}v_F{}^2\tau_c \;, \tag{92}$$

closely similar in form to the result (77) for the classical distribution of velocities. In (92) the relaxation time is to be taken at the Fermi energy.

We see we can solve transport problems where the Fermi-Dirac distribution applies, as in metals, as easily as where the classical approximation applies.

Electrical Conductivity

The isothermal electrical conductivity σ follows from the result for the particle diffusivity when we multiply the particle flux density by the particle charge q and replace the gradient $d\mu/dx$ of the chemical potential by the gradient $qd\varphi/dx = -qE_x$ of the external potential, where E_x is the x component of the electric field intensity. The electric current density follows from (76):

$$\mathbf{J}_q = (nq^2\tau_c/m)\mathbf{E}; \qquad \sigma = nq^2\tau_c/m \;, \tag{93}$$

for a classical gas with relaxation time τ_c. For the Fermi-Dirac distribution, from (89),

$$\mathbf{J}_q = (nq^2\tau_c/m)\mathbf{E}; \qquad \sigma = nq^2\tau_c/m. \tag{94}$$

LAWS OF RAREFIED GASES

Thus far in this chapter the discussion of transport has assumed that the molecular mean free path is short in comparison with the dimensions of the apparatus. At a gas pressure of 10^{-6} atm at room temperature, the mean free path of a molecule is of the order of 25 cm. The diameter of a laboratory vacuum system connection may be of the order of 25 cm, thus of the order of the mean free path. We may usefully draw a line here and denote pressures lower than 1×10^{-6} atm as high vacuum. This pressure is approximately $0.1 \, \mathrm{N\,m^{-2}}$ or $1 \times 10^{-6} \, \mathrm{kg\,cm^{-2}}$ or $7.6 \times 10^{-4} \, \mathrm{mm\,Hg}$ or 7.6×10^{-4} torr. The Knudsen region of pressures is understood to be the region in which the mean free path is much greater than the dimensions of the apparatus. A knowledge of the behavior of gases in this pressure region is important in the use of high vacuum pumps and allied equipment.

The terminology recommended by the American Vacuum Society is expressed in terms of torr, where 1 torr $\equiv$ 1 mm Hg $= 1.333 \times 10^{-3}$ bar $= 133.3$ N m$^{-2} =$ 1333 dyne cm^{-2}; here 1 bar $\equiv 10^6$ dyne cm$^{-2} = 0.987$ standard atmospheres. Then:

high vacuum	10^{-3}–10^{-6} torr
very high vacuum	10^{-6}–10^{-9} torr
ultra high vacuum	below 10^{-9} torr.

Flow of Molecules Through a Hole

In the Knudsen regime we do not need to solve a hydrodynamic flow problem in order to get the rate of efflux of gas molecules through a hole, because the molecules do not see each other. We have merely to calculate the rate J_n at which molecules strike unit area of surface per unit time. We find for the flux density

$$\boxed{J_n = \tfrac{1}{4} n \bar{c} \; ,}$$

$$(95)$$

where n is the concentration and $\bar{c}$ is the mean speed of a gas molecule. To prove (95), consider a unit cube containing n molecules. Each molecule strikes the $+z$ face of the cube $\tfrac{1}{2}\bar{c}_z$ times per unit time, so that in unit time $\tfrac{1}{2} n \bar{c}_z$ molecules strike unit area.

We solve for $\bar{c}_z$ in terms of $\bar{c}$. Because $c_z = c \cos \theta$, we require the average of $\cos \theta$ over a hemisphere:

$$\langle \cos \theta \rangle = \frac{2\pi \int_0^{\pi/2} \cos \theta \sin d\theta}{2\pi \int_0^{\pi/2} \sin \theta \, d\theta} = \frac{1}{2}.$$

$$(96)$$

Therefore $\bar{c}_z = \tfrac{1}{2}\bar{c}$, and (95) is obtained. The expression (95) for the flux forms the basis for many calculations of gas flow in vacuum physics in the Knudsen regime.

If A is the area of the hole, the total particle flux, which is the number of molecules per unit time, is

$$\Phi = \tfrac{1}{4} A n \bar{c} = nS \; ,$$

$$(97)$$

where

$$S = \tfrac{1}{4}A\bar{c}. \tag{98}$$

The **conductance** S of the hole is defined as the volume of gas per unit time flowing through the hole, with the volume taken at the actual pressure p of the gas. The conductance is usually expressed in liters per second. For the average air molecule at $T = 300\,\mathrm{K}$ we have $\bar{c} \approx 4.7 \times 10^4\,\mathrm{cm\,s^{-1}}$; for a circular hole of 10 cm diameter, (98) leads to a conductance of 917 liter/sec, roughly 1000 liter/sec.

For a hole with a given conductance the total particle flux is proportional to the concentration n or, because $p = n\tau$, to the pressure p:

$$\Phi = \frac{p}{\tau}S = \frac{1}{\tau}Q. \tag{99}$$

Here we have defined the quantity

$$Q = pS , \tag{100}$$

sometimes called the **throughput**, which is widely used by vacuum physicists as a convenient measure of the flow. The quantity Q is numerically (not dimensionally) equal to the gas volume flowing per unit time, but referred to the volume at unit pressure, in whatever units are used to express pressure. Vacuum physicists like to express pressure in torr, hence they usually express flux in torr-liters per second. From the ideal gas law one finds that 1 torr-liter at 300 K is equivalent to 5.35×10^{-5} mole or 3.22×10^{19} molecules.

Our calculations have expressed the flow of gas through a hole into a perfect vacuum. With gas on both sides, the net flux from side 1 to side 2 will be

$$\Delta\Phi = \tfrac{1}{4}A(n_1\bar{c}_1 - n_2\bar{c}_2) = \tfrac{1}{4}A\left(\frac{p_1\bar{c}_1}{\tau_1} - \frac{p_2\bar{c}_2}{\tau_2}\right). \tag{101}$$

The condition for zero net flux is

$$\frac{p_1}{p_2} = \frac{\tau_1}{\tau_2}\frac{\bar{c}_2}{\bar{c}_1} = \left(\frac{\tau_1}{\tau_2}\right)^{1/2} , \tag{102}$$

using the proportionality of $\bar{c}$ to $\tau^{1/2}$. In the Knudsen regime equal pressures do not imply zero net flux if the temperatures on the two sides are different. At equal pressures gas will flow from the cold side to the hot side; zero gas flow requires a higher pressure on the hot side.

If $\tau_1 = \tau_2$, Eq. (101) can be written

$$\Delta\Phi = (n_1 - n_2)S = \frac{1}{\tau}(p_1 - p_2)S = \frac{1}{\tau}\Delta Q \ , \tag{103}$$

where

$$\Delta Q = p_1 S - p_2 S = Q_1 - Q_2. \tag{104}$$

Example: Flow through a long tube. We assume that the molecules which strike the inner wall of the tube are re-emitted in all directions; that is, the reflection at the surface is assumed to be diffuse. Thus when there is a net flow there is a net momentum transfer to the tube, and we must provide a pressure head to supply the momentum transfer. Let u be the velocity component of the gas molecules parallel to the wall before striking the wall. We estimate the momentum transfer to the wall on the assumption that every collision with the wall transfers momentum $M\langle u \rangle$. The rate of flow down the tube is $nA\langle u \rangle$, where A is the area of the opening. The rate at which molecules strike the wall is, from (95)

$$\tfrac{1}{4}\pi L\, d\, n\bar{c} \ , \tag{105}$$

where d is the diameter and L the length of the tube. The momentum transfer to the tube must equal the force due to the pressure differential Δp:

$$\tfrac{1}{4}\pi L\, dn\, \bar{c}M\langle u \rangle = A\,\Delta p. \tag{106}$$

We solve for the flow velocity $\langle u \rangle$ to obtain

$$\langle u \rangle = \frac{\Delta p}{n}\frac{1}{M\bar{c}}\frac{4A}{\pi Ld} = \frac{\Delta p}{n}\frac{1}{M\bar{c}}\frac{d}{L}. \tag{107}$$

The net flux is

$$\Delta\Phi = n\langle u \rangle A = \Delta p\,\frac{Ad}{M\bar{c}L} = \frac{\Delta p}{\tau}S \ , \tag{108}$$

where

$$S = \tau\,\Delta\Phi/\Delta p = \frac{\tau}{M\bar{c}}\frac{Ad}{L} \tag{109}$$

is the conductance of the tube, defined analogously to the conductance of a hole, Eq. (97).

A more detailed calculation, with averages over the velocity distribution taken more carefully, leads to a conductance differing from (109) by a factor $8/3\pi$:

$$S = \frac{8}{3\pi} \frac{\tau A d}{M\bar{c}L} = \frac{2\tau d^3}{3M\bar{c}L}. \tag{110}$$

The conductance of a tube cannot be larger than that of a hole with the same area. From (98), (110), and (121) below,

$$S_{\text{tube}}/S_{\text{hole}} = \frac{32}{3\pi} \frac{\tau d}{M\bar{c}^2 L} = \frac{4d}{3L}. \tag{111}$$

This ratio will be larger than unity for $3L < 4d$. In writing (106) we assumed implicitly that every molecule hits the tube wall. This will not be true for a short tube. For our result to be valid we must suppose that the tube is long enough to make the ratio (111) be small compared to unity, which means

$$L \gg \tfrac{4}{3}d. \tag{112}$$

Using our earlier example for the conductance of a hole, we find that the conductance of a tube 1 meter long and 10 cm in diameter is about 122 liter/sec, for air at 300 K.

Speed of a Pump

The **speed** of a pump is defined similarly to the conductance of a hole or of a tube; it is defined as the volume pumped per unit time, with the volume taken at the intake pressure of the pump. The same symbol S is used as for conductance; sometimes the conductance of an aperture or a tube is referred to as its speed. The product $Q = pS$ for a pump is often called the throughput of the pump.

If a pump of speed S_p evacuates a vacuum system through a tube of conductance S_t, the effective pumping speed S_{eff} of the combination is given by

$$\frac{1}{S_{\text{eff}}} = \frac{1}{S_p} + \frac{1}{S_t}, \tag{113}$$

just as for the conductance of two electrical conductors in series.

Proof: Let p_1 denote the pressure at the input end of the tube, and let p_2 denote the pump intake pressure at the output end of the tube. Continuity of flux requires that

$$p_1 S_{\text{eff}} = (p_1 - p_2)S_t = p_2 S_p, \tag{114}$$

so that

$$\frac{p_1}{p_2} = \frac{S_p}{S_{\text{eff}}} = \frac{S_t + S_p}{S_t} = S_p\left(\frac{1}{S_p} + \frac{1}{S_t}\right), \tag{115}$$

equivalent to (113).

The relation (113) for S_{eff} explains why in high vacuum systems the connections between the pump and the vessel to be evacuated must be as short and of as large a diameter as possible. A long and narrow connecting tube makes poor use of a high speed pump. Further, the speed of the pump itself cannot be larger than the conductance of its own aperture.

How rapidly does a pump with effective speed S evacuate a volume V? From the ideal gas law $pV = N\tau$, and from the definition of pump speed analogous to Eq. (99) we find

$$\frac{dp}{dt} = \frac{\tau}{V}\frac{dN}{dt} = \frac{Q}{V} = \frac{pS}{V}. \tag{116}$$

If the pump speed is independent of pressure, this differential equation has the solution

$$p(t) = p(0)\exp(-t/t_0); \qquad t_0 = V/S. \tag{117}$$

For a volume of 100 liters connected to a pump with a speed of 100 liter/sec, the pressure should decrease by $1/e$ per second.

Any user of vacuum technology soon discovers that the pumpdown of a vacuum system proceeds much more slowly in the high and ultrahigh vacuum regions than expected on the basis of pumping speed and system volume. The desorption of surface gas predominates—often by many orders of magnitude—over volume gas. The surface emits adsorbed molecules as fast as the pump evacuates molecules from the volume.

SUMMARY

1. The probability that an atom has velocity in dv at v is

$$P(v)dv = 4\pi(M/2\pi\tau)^{3/2}v^2\exp(-Mv^2/2\tau)dv ,$$

the Maxwell velocity distribution.

2. Diffusion is described by

$$\mathbf{J}_n = -D \operatorname{grad} n; \qquad D = \tfrac{1}{3}\bar{c}l \ ,$$

where $\bar{c}$ is the mean speed and l is the mean free path.

3. Thermal conductivity is described by

$$\mathbf{J}_u = -K \operatorname{grad} \tau; \qquad K = \tfrac{1}{3}\hat{C}_V \bar{c}l \ ,$$

where $\hat{C}_V$ refers to unit volume.

4. The coefficient of viscosity is given by

$$\eta = \tfrac{1}{3}\rho \bar{c}l \ ,$$

where ρ is the mass density.

5. According to the principle of detailed balance, in thermal equilibrium the rate of any process that leads to a given state must equal exactly the rate of the inverse process that leads from the state.

6. The Boltzmann transport equation in the relaxation time approximation is

$$\frac{\partial f}{\partial t} + \boldsymbol{\alpha} \cdot \operatorname{grad}_v f + \mathbf{v} \cdot \operatorname{grad}_r f = -\frac{f - f_0}{\tau_c}.$$

7. The electrical conductivity of a Fermi gas is

$$\sigma = nq^2 \tau_c/m \ ,$$

where τ_c is the relaxation time.

PROBLEMS

1. Mean speeds in a Maxwellian distribution. (a) Show that the root mean square velocity v_{rms} is

$$v_{\text{rms}} = \langle v^2 \rangle^{1/2} = (3\tau/M)^{1/2}. \tag{118}$$

Because $\langle v^2 \rangle = \langle v_x^2 \rangle + \langle v_y^2 \rangle + \langle v_z^2 \rangle$ and $\langle v_x^2 \rangle = \langle v_y^2 \rangle = \langle v_z^2 \rangle$, it follows that

$$\langle v_x^2 \rangle^{1/2} = (\tau/M)^{1/2} = v_{\text{rms}}/3^{1/2}. \tag{119}$$

The results can also be obtained directly from the expression in Chapter 3 for the average kinetic energy of an ideal gas. (b) Show that the most probable value of the speed v_{mp} is

$$v_{mp} = (2\tau/M)^{1/2}. \tag{120}$$

By most probable value of the speed we mean the maximum of the Maxwell distribution as a function of v. Notice that $v_{mp} < v_{rms}$. (c) Show that the mean speed $\bar{c}$ is

$$\bar{c} = \int_0^\infty dv\, vP(v) = (8\tau/\pi M)^{1/2}. \tag{121}$$

The mean speed may also be written as $\langle|v|\rangle$. The ratio

$$v_{rms}/\bar{c} = 1.086. \tag{122}$$

(d) Show that $\bar{c}_z$, the mean of the absolute value of the z component of the velocity of an atom, is

$$\bar{c}_z \equiv \langle|v_z|\rangle = \tfrac{1}{2}\bar{c} = (2\tau/\pi M)^{1/2}. \tag{123}$$

2. Mean kinetic energy in a beam.   (a) Find the mean kinetic energy in a beam of molecules that exits from a small hole in an oven at temperature τ. (b) Assume now that the molecules are collimated by a second hole farther down the beam, so that the molecules that pass through the second hole have only a small velocity component normal to the axis of emission. What is the mean kinetic energy? *Comment*: The molecules in the beam do not collide and are not in real thermal equilibrium after they have exited from the oven. The gas left in the oven is depleted with respect to fast molecules, and the residual gas will cool down if it is not reheated by heat flowing in through the walls of the oven.

3. Ratio of thermal to electrical conductivity. Show for a classical gas of particles of charge q that

$$K/\tau\sigma = 3/2q^2\ ,\qquad \text{or}\qquad K/T\sigma = 3k_B^2/2q^2 \tag{124}$$

in conventional units for K and T. This is known as the **Wiedemann-Franz ratio**.

4. Thermal conductivity of metals. The thermal conductivity of copper at room temperature is largely carried by the conduction electrons, one per atom. The mean free path of the electrons at 300 K is of the order of 400×10^{-8} cm.

The conduction electron concentration is 8×10^{22} per cm^3. Estimate (a) the electron contribution to the heat capacity; (b) the electronic contribution to the thermal conductivity; (c) the electrical conductivity. Specify units.

5. Boltzmann equation and thermal conductivity. Consider a medium with temperature gradient $d\tau/dx$. The particle concentration is constant. (a) Employ the Boltzmann transport equation in the relaxation time approximation to find the first order nonequilibrium classical distribution:

$$f \simeq f_0 - v_x \tau_c \left(-\frac{3}{2\tau} + \frac{\varepsilon}{\tau^2} \right) f_0 \frac{d\tau}{dx}. \tag{125}$$

(b) Show that the energy flux in the x direction is

$$J_u = -\left(\frac{d\tau}{dx} \right) \tau_c \int v_x^2 \left(-\frac{3\varepsilon}{2\tau} + \frac{\varepsilon^2}{\tau^2} \right) f_0 \mathfrak{D}(\varepsilon) d\varepsilon , \tag{126}$$

where $v_x^2 = 2\varepsilon/3m$. (c) Evaluate the integral to obtain for the thermal conductivity $K = 5n\tau\tau_c/m$.

6. Flow through a tube. Show that when a liquid flows through a narrow tube under a pressure difference p between the ends, the total volume flowing through the tube in unit time is

$$\dot{V} = (\pi a^4/8\eta L)p , \tag{127}$$

where η is the viscosity; L is the length; a is the radius. Assume that the flow is laminar and that the flow velocity at the walls of the tube is zero.

7. Speed of a tube. Show that for air at 20°C the speed of a tube in liters per second is given by, approximately,

$$S_T \simeq \frac{12d^3}{L + \frac{4}{3}d} , \tag{128}$$

where the length L and diameter d are in centimeters; we have tried to correct for end effects on a tube of finite length by treating the ends as two halves of a hole in series with the tube.

Chapter 15

Propagation

The purpose of this terminal chapter is to bring within the compass of the text the most important problems in the propagation of heat and the propagation of sound, both classical subjects that are part of an education in thermal physics.

HEAT CONDUCTION EQUATION

Consider first the derivation of the diffusion equation, which is found from the Fick law (14.19) for the particle flux density:

$$\mathbf{J}_n = -D_n \operatorname{grad} n \ , \tag{1}$$

where D_n is the particle diffusivity and n the particle concentration. The equation of continuity,

$$\frac{\partial n}{\partial t} + \operatorname{div} \mathbf{J}_n = 0 \ , \tag{2}$$

assures that the number of particles is conserved. Because div grad $\equiv \nabla^2$, substitution of (1) in (2) gives

$$\frac{\partial n}{\partial t} = D_n \nabla^2 n. \tag{3}$$

This partial differential equation describes the time-dependent diffusion of the particle concentration n.

The thermal conductivity equation is derived similarly. By (14.27–14.30) we have in a homogeneous medium

$$\mathbf{J}_u = -K \operatorname{grad} \tau. \tag{4}$$

The equation of continuity for the energy density is

$$\hat{C} \frac{\partial \tau}{\partial t} + \operatorname{div} \mathbf{J}_u = 0 \ , \tag{5}$$

where $\hat{C}$ is the heat capacity per unit volume. We combine (4) and (5) to obtain the heat conduction equation

$$\frac{\partial \tau}{\partial t} = D_\tau \nabla^2 \tau; \qquad D_\tau \equiv K/\hat{C}. \tag{6}$$

This equation describes the time-dependent diffusion of the temperature. The equation is of the form of the particle diffusion equation (3). The quantity D_τ is called the thermal diffusivity; for a gas it is approximately equal to the particle diffusivity, as in (14.23).

Comment.   The eddy current equation of electromagnetic theory* has the same form as (3) and (6). If B is the magnetic field intensity, then

$$\frac{\partial B}{\partial t} = D_B \nabla^2 B. \tag{7}$$

The constant D_B may be called the magnetic diffusivity and in SI is equal to $1/\sigma\mu$; in CGS, $D_B = c^2/4\pi\sigma\mu$. It has the dimensions (length)^2 (time)$^{-1}$ and is directly proportional to the (skin depth)^2 times the frequency. When we have solved one equation, say (3), we have solved three problems.

Dispersion Relation, ω Versus k

We look for solutions of the diffusivity equation

$$D\nabla^2 \theta = \partial\theta/\partial t \tag{8}$$

that have the wavelike form

$$\theta = \theta_0 \exp[i(\mathbf{k} \cdot \mathbf{r} - \omega t)] , \tag{9}$$

with ω as the angular frequency and $\mathbf{k}$ as the wavevector. Plane wave analysis is an excellent approach to this problem, even though it will turn out that the diffusion waves or heat waves are so highly damped that they are hardly waves

---

* See, for example, W. R. Smythe, *Static and dynamic electricity*, McGraw-Hill, 3rd ed., 1968, p. 369. This book has an unusually full treatment of eddy current problems.

at all. Substitute (9) in (8) to obtain the relation between k and ω:

$$Dk^2 = i\omega. \tag{10}$$

A relation $\omega(\mathbf{k})$ for a plane wave is called a **dispersion relation**.

Penetration of Temperature Oscillation

Consider the variation of temperature in the semi-infinite medium $z > 0$ when the temperature of the plane $z = 0$ is varied periodically with time as

$$\theta(0,t) = \theta_0 \cos \omega t , \tag{11}$$

which is the real part of $\theta_0 \exp(-i\omega t)$, for real θ_0. Then in the medium $z > 0$ the temperature is

$$\begin{aligned}
\theta(z,t) &= \theta_0 \operatorname{Re}\{\exp[i(kz - \omega t)]\} \\
&= \theta_0 \operatorname{Re}\{\exp[i^{3/2}(\omega/D)^{1/2}z - i\omega t]\} ,
\end{aligned} \tag{12}$$

where Re denotes real part and $i^{3/2} = (i - 1)/\sqrt{2}$. Thus, with $\delta \equiv (2D/\omega)^{1/2}$,

$$\begin{aligned}
\theta(z,t) &= \theta_0 \operatorname{Re}\{\exp(-z/\delta)\exp[i(z/\delta) - i\omega t]\} \\
&= \theta_0 \exp(-z/\delta)\cos(\omega t - z/\delta).
\end{aligned} \tag{13}$$

The quantity $\delta = (2D/\omega)^{1/2}$ has the dimensions of a length and represents the characteristic penetration depth of the temperature variation: at this depth the amplitude of the oscillations of θ is reduced by e^{-1}. The characteristic depth is called the skin depth if we are dealing with the eddy current equation. The wave is highly damped in the medium—the wave amplitude decreases by e^{-1} in a distance equal to a wavelength/2π.

If the thermal diffusivity of soil is taken as $D \approx 1 \times 10^{-3}\,\mathrm{cm^2\,s^{-1}}$, then the penetration depth of the diurnal cycle of heating of the ground by the sun and cooling of the ground by the night sky ($\omega = 0.73 \times 10^{-4}\,\mathrm{s^{-1}}$) is

$$L(\text{diurnal}) = (2D/\omega)^{1/2} \approx 5\,\mathrm{cm}.$$

For the annual cycle,

$$L(\text{annual}) \approx 1\,\mathrm{m}.$$

A layer of 10 cm of earth on top of a cellar will tend to average out day/night variations of surface temperature, but the summer/winter variation at the top of the cellar requires several meters of earth. Actual values of the thermal diffusivity are sensitive to the composition and condition of the soil or rock. Notice that a figure of merit for cellar construction involves the thermal diffusivity, and not the conductivity alone.

Development of a Pulse

In addition to the wavelike solutions of the form (9), the diffusion equation has several other useful forms of solutions. We confirm by insertion in (8) that

$$\theta(x,t) = (4\pi Dt)^{-1/2} \exp(-x^2/4Dt) \tag{14}$$

is a solution. The proportionality factor has been chosen so that

$$\int_{-\infty}^{+\infty} \theta(x,t)dx = 1. \tag{15}$$

The solution (14) corresponds to the time development of a pulse which at $t = 0$ has the form of a Dirac delta function $\delta(x)$, sharply localized at $x = 0$, and zero elsewhere.

The pulse might be a temperature pulse, as when a pulsed laser or pulsed electron beam heats a surface briefly. Let Q be the quantity of heat deposited on the surface, per unit area. The temperature distribution is then given by

$$\theta(x,t) = (2Q/\hat{C}_V)(4\pi Dt)^{-1/2} \exp(-x^2/4Dt) \,, \tag{16}$$

where $\hat{C}_V$ is the heat capacity per unit volume of the material. The function is plotted in Figure 15.1. The factor 2 arises because all heat is assumed to flow inwards from the surface, while for the solution (14) symmetrical flow was assumed. Another example of the application of (14) is the diffusion of impurities deposited on the surface of a semiconductor, to form a p–n junction inside the semiconductor.

The pulse spreads out with increasing time. The mean square value of x is given by

$$\langle x^2 \rangle = \int x^2 \theta(x,t)dx \Big/ \int \theta(x,t)dx = 2Dt \,, \tag{17}$$

after evaluating the Gaussian integrals. The root mean square value is

$$x_{\text{rms}}(t) = \langle x^2 \rangle^{1/2} = (2Dt)^{1/2}. \tag{18}$$

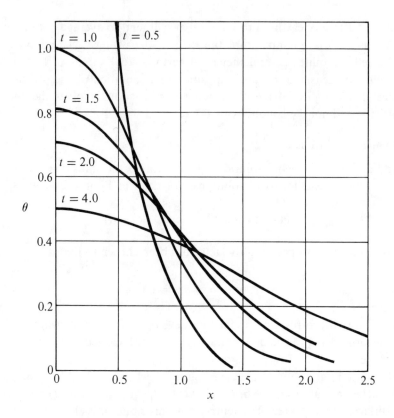

Figure 15.1 Plot of spread of temperature pulse with time, for $4\pi D = 1$, from Eq. (16). At $t = 0$ the pulse is a delta function.

Comment. This result shows that the width of the distribution increases as $t^{1/2}$, which is a general characteristic of diffusion and random walk problems in one dimension. It is quite unlike the motion of a wave pulse in a nondispersive medium, which is a medium for which $\omega = vk$, where v is the constant velocity. The connection with Brownian motion or the random walk problem follows if we let t_0 be the duration of each step of a random walk; then $t = Nt_0$, where N is the number of steps. It follows that

$$x_{\text{rms}}(t) = (2Dt_0)^{1/2}N^{1/2}, \tag{19}$$

so that the rms displacement is proportional to the square root of the number of steps. This is the result observed in studies of the Brownian motion, the random motion of suspensions of small particles in liquids.

Diffusion with a Fixed Boundary Condition at $x = 0$

If a solution of (8) is differentiated or integrated with respect to any of its independent variables, the result may again be a solution. An important example is obtained by integrating (14) with respect to x:

$$\theta(x,t) = (4\pi Dt)^{-\frac{1}{2}} \int_0^x dx' \exp(-x'^2/4Dt)$$

$$= \frac{1}{\sqrt{\pi}} \int_0^u ds \exp(-s^2) = \tfrac{1}{2} \mathrm{erf}\, u \ , \tag{20}$$

where $u \equiv x/(4Dt)^{1/2}$. Here we have introduced the **error function** defined by

$$\mathrm{erf}\, u \equiv \frac{2}{\sqrt{\pi}} \int_0^u ds \exp(-s^2). \tag{21}$$

Tables of the error function are readily available. The error function has the properties

$$\mathrm{erf}(0) = 0; \qquad \lim_{x \to \infty} \mathrm{erf}(x) = 1. \tag{22}$$

Of particular practical interest is the diffusion of heat or of particles into an infinite solid from a surface at $x = 0$, with the fixed boundary condition $\theta = \theta_0$ at $x = 0$ and $\theta = 0$ at $x = \infty$. (For $t < 0$ we assume $\theta = 0$ everywhere.) The solution is

$$\theta(x,t) = \theta_0[1 - \mathrm{erf}(x/(4Dt)^{1/2})]. \tag{23}$$

Again we see that the distance at which $\theta(x,t)$ reaches a specified value is proportional to $(4Dt)^{1/2}$. The application of this solution to the diffusion of impurities into a semiconductor is discussed in a problem.

Time-Independent Distribution

Let us look at a solution of (8) that is independent of the time. The diffusivity equation reduces to the Laplace equation

$$\nabla^2 \theta = 0. \tag{24}$$

Consider a semi-infinite medium bounded by the plane $z = 0$ and extending along the positive z axis. Let the temperature vary sinusoidally in the boundary plane:

$$\theta(x,y,0) = \theta_0 \sin kx. \tag{25}$$

The solution of (24) in the medium is

$$\theta(x,y,z) = \theta_0 \sin kx \exp(-kz). \tag{26}$$

The temperature variation is damped exponentially with the distance from the boundary plane. The temperature distribution in the time-independent problem must be maintained by constant heat sources on the boundary plane $z = 0$.

PROPAGATION OF SOUND WAVES IN GASES

Results developed earlier in this book can be applied to the study of sound waves in gases. Thermal effects are important in this problem. Let $\delta p(x,t)$ denote the pressure associated with the sound wave; the form of the wave may be written as

$$\delta p = \delta p_0 \exp[i(kx - \omega t)] , \tag{27}$$

where k is the wavevector and ω is the angular frequency. The wave propagates in the x direction.

We suppose the equation of state is that of an ideal gas:

$$pV = N\tau , \quad \text{or} \quad p = \rho\tau/M , \tag{28}$$

where $\rho = NM/V$ is the mass density, and M is the mass of a molecule. The force equation referred to unit volume is

$$\rho \frac{\partial u}{\partial t} = -\frac{\partial p}{\partial x} = -\frac{\tau}{M}\frac{\partial \rho}{\partial x} - \frac{\rho}{M}\frac{\partial \tau}{\partial x}. \tag{29}$$

Here u is the x component of the velocity of a volume element. The motion is subject to the equation of continuity

$$\partial\rho/\partial t + \text{div}(\rho\mathbf{v}) = 0 , \tag{30}$$

or, in one dimension,

$$\partial\rho/\partial t + \partial(\rho u)/\partial x = 0. \tag{31}$$

The thermodynamic identity is

$$dU + pdV = \tau d\sigma , \tag{32a}$$

which can also be written

$$\frac{dU}{V} + p\frac{dV}{V} = \tau\frac{d\sigma}{V}. \tag{32b}$$

If we assume (pending discussion below) that there is no entropy exchange during the passage of a sound wave, Eq. (32b) becomes

$$\hat{C}_V(\partial\tau/\partial t) + (p/V)(\partial V/\partial t) = 0 , \tag{33}$$

where $\hat{C}_V$ is the heat capacity at constant volume, per unit volume. We can re-write the second term in terms of $\partial\rho/\partial t$ because $\rho = NM/V$ and $(1/V)(\partial V/\partial t) = -(1/\rho)(\partial\rho/\partial t)$. Now the thermodynamic identity appears as

$$\hat{C}_V(\partial\tau/\partial t) - (p/\rho)(\partial\rho/\partial t) = 0. \tag{34}$$

Let us define the fractional deviations s, θ by

$$\rho = \rho_0(1 + s); \quad \tau = \tau_0(1 + \theta) , \tag{35}$$

where ρ_0, τ_0 are the density and temperature in the absence of the sound wave. We assume that u, s, θ have the form of a traveling wave: $\exp[i(kx - \omega t)]$. The three equations (29), (31), (34) that govern the motion now become

$$-i\omega\rho u + ik[(\tau\rho_0/M)s + (\rho\tau_0/M)\theta] = 0; \tag{36}$$

$$-i\omega\rho_0 s + ik(\rho_0 us + \rho u) = 0; \tag{37}$$

$$-i\omega\tau_0\hat{C}_V\theta + i\omega(p/\rho)\rho_0 s = 0. \tag{38}$$

We assume that at sufficiently small wave amplitudes it is a good approximation to neglect in these equations terms in the squares and cross products of $u, s,$ and θ. For example, $\rho u = \rho_0(1 + s)u$ becomes $\rho_0 u$ if the cross product su is neglected. The equations thus reduce to, with the subscripts dropped from ρ and τ,

$$\omega u - (k\tau/M)s - (k\tau/M)\theta = 0; \tag{39}$$

$$\omega s - ku = 0; \tag{40}$$

$$\tau\hat{C}_V\theta - ps = 0; \quad \text{or } \hat{C}_V\theta - ns = 0 , \tag{41}$$

where n is the concentration. These equations have a solution only if

$$\omega = (\gamma\tau/M)^{1/2}k \ , \tag{42}$$

where $\gamma = (\hat{C}_V + n)/\hat{C}_V = \hat{C}_p/\hat{C}_V$ in our units. The velocity of sound is

$$v_s = \partial\omega/\partial k = (\gamma\tau/M)^{1/2}. \tag{43}$$

This result applies to monatomic gases from the lowest frequencies up to high frequencies limited only by the requirement that the acoustic wavelength should be much larger than the mean free path of the atoms. This requirement is the criterion for the applicability of the hydrodynamic approach embodied in the force equation (29).

Thermal Relaxation

With polyatomic gases (43) is valid at low frequencies, but as the frequency is increased there is a transition frequency region above which the velocity of sound increases. The transition region between low frequency and high frequency propagation is associated with relaxation effects.

Thermal relaxation describes the establishment of thermal equilibrium in a system. Energy dissipation results when all parts of a system are not at the same temperature; the dissipation is strongest when the period of the heating and cooling half-cycle in the sound wave is comparable with the time required for heat exchange between the different degrees of freedom of the system. In polyatomic gases under standard conditions there are time delays of the order of 10^{-5} s in the transfer of energy between the internal vibrational states of a molecule and the external translation states.

Let the heat capacity C_1 and temperature $\tau_1 = \tau_0(1 + \theta_1)$ refer to the internal states, while C_V and $\tau = \tau_0(1 + \theta)$ refer to the translational states. Then (34) becomes

$$C_1(\partial\tau_1/\partial t) + C_V(\partial\tau/\partial t) - (p/\rho)(\partial\rho/\partial t) = 0 \ , \tag{44}$$

or, in place of (38),

$$-i\omega\tau_0 C_1\theta_1 - i\omega\tau_0 C_V\theta + i\omega(p/\rho)\rho_0 s = 0. \tag{45}$$

Suppose that the transfer of energy between the internal and external states has the characteristic time delay t_0 such that

$$\partial\tau_1/\partial t = -(\tau - \tau_1)/t_0 \ , \tag{46}$$

or

$$i\omega\theta_1 = (\theta - \theta_1)/t_0. \tag{47}$$

Here t_0 is called the **relaxation time**. There will be separate relaxation times for the rotational-translational transfer and the vibrational-translational transfer.

We combine (39), (40), (45) and (47) to obtain the dispersion relation

$$k^2 = \omega^2(M/\tau) \frac{C_V + C_1 + i\omega t_0 C_V}{C_p + C_1 + i\omega t_0 C_p}, \tag{48}$$

where C_V, C_p refer to the translational states alone. In the low frequency limit $\omega t_0 \ll 1$ and

$$k^2 = \omega^2(M/\tau) \frac{C_V + C_1}{C_p + C_1} = \omega^2(M/\gamma_0\tau), \tag{49}$$

where γ_0 is the low frequency limit of the total heat capacity ratio $(C_p + C_1)/(C_V + C_1)$. The low frequency limit of the velocity of sound is

$$v_s(0) = (\gamma_0\tau/M)^{1/2}. \tag{50}$$

In the high frequency limit $\omega t_0 \gg 1$ and

$$k^2 = \omega^2(M/\tau)(C_V/C_p) = \omega^2(M/\gamma_\infty\tau). \tag{51}$$

Here γ_∞ refers only to the translational states; at high frequencies the internal states are not excited by the sound wave. The high frequency limit of the velocity of sound is

$$v_s(\infty) = (\gamma_\infty\tau/M)^{1/2}. \tag{52}$$

Values of γ_0 are given in Table 15.1; if no internal states at all are excited, $\gamma_\infty = \frac{5}{3}$.

The wave is attenuated when k is complex; the imaginary part of k gives the pressure attenuation coefficient α. From (48) it is found that the maximum absorption per wavelength occurs when $\omega = 2\pi/t_0$ and is given approximately by

$$(\alpha\lambda)_{\max} \simeq \frac{\pi}{2} \cdot \frac{C_p - C_V}{C_p} \cdot \frac{C_1}{C_V + C_1}. \tag{53}$$

Table 15.1   Ratio $C_p/C_V = \gamma$ for gases

Gas	Temperature, °C	γ
Air	17	1.403
H_2O	100	1.324
H_2	15	1.410
O_2	−181	1.450
	15	1.401
	200	1.396
	2000	1.303
CO_2	15	1.304
Ar	15	1.668
He	−180	1.660

NOTE: For a monatomic ideal gas, $C_p/C_V = 5/3 = 1.667$, as for Ar and He. For a diatomic gas at a temperature high enough to excite the rotational motion, $C_p/C_V = 7/5 = 1.40$, as for O_2 and H_2 at room temperature; at temperatures sufficiently high to excite also the vibrational motion, $C_p/C_V = 9/7 = 1.286$, as for O_2 at 2000°C. The values given are of γ_0, applicable to static processes and to sound waves in the limit of low frequencies. For very high frequency sound waves only the translational motion is excited and $\gamma_\infty = 5/3$ is applicable.

For CO_2 gas at the relaxation frequency of 20 kHz under standard conditions the intensity is observed to decrease by $1/e$ in about 4 wavelengths—a massive absorption, in agreement with theory.

Example: Heat transfer in a sound wave.   Equation (33) expresses the isentropic assumption: the equation neglects the thermal conductivity which gives rise to some transfer of thermal energy within the sound wave between successive warm and cool half cycles. The assumption that $d\sigma = 0$ must be modified to take account of heat flow. The heat conduction equation (6) may be written as

$$K\, \partial^2\tau/\partial x^2 = \tau\, \partial\hat{\sigma}/\partial t \ , \tag{54}$$

where $\hat{\sigma}$ is the entropy density. Then (34) becomes

$$\hat{C}_V(\partial\tau/\partial t) - (p/\rho)(\partial\rho/\partial t) = K(\partial^2\tau/\partial x^2) \ ,$$

or

$$-i\omega\tau\hat{C}_V\theta + i\omega ps = -K\tau k^2\theta. \tag{55}$$

When we use this in place of (41), the dispersion relation $\omega(k)$ becomes

$$k^2 = \omega^2(M/\tau)\left(\frac{\hat{C}_V + iWk^2}{\hat{C}_p + iWk^2}\right),\qquad(56)$$

with $W \equiv K/\omega$. At low frequencies Wk^2 is much smaller than $\hat{C}_V$, so that the sound velocity is equal to the isentropic result $v_s = (\gamma_0\tau/M)^{1/2}$, as before. The condition $Wk^2 \ll \hat{C}_V$ is essentially the condition $l \ll \lambda$, where l is the molecular mean free path and λ is the wavelength of the sound wave. The attenuation of the pressure oscillation is given by the imaginary part of the wavevector k and is denoted by α. The result from (56) in the low frequency region $Wk^2 \ll \hat{C}_V$ is that

$$\alpha = (\gamma_0 - 1)\rho K\omega^2/2v_s^3\hat{C}_p,\qquad(57)$$

where $\hat{C}_p$ refers to unit volume.

SUMMARY

1. The heat conduction equation is the partial differential equation that follows when the phenomenological transport equation (here the Fourier law) is combined with the equation of continuity. We obtain

$$\frac{\partial\tau}{\partial t} = D_\tau\nabla^2\tau;\qquad D_\tau \equiv K/\hat{C}.$$

2. The time-dependent diffusion equation and the eddy current equation have the same form, so that their solutions may be translated from the solutions of the heat conduction equation, these being often more familiar in the literature.

3. Frequently it is useful to construct solutions in the form of superpositions of plane waves of the form

$$\theta = \theta_0\exp[i(\mathbf{k}\cdot\mathbf{r} - \omega t)].$$

The differential equation then gives the relation between ω and $\mathbf{k}$, called the dispersion relation of the problem.

4. The propagation of sound waves in gases depends on the rate of exchange of energy between the translational, rotational, and vibrational motions of a molecule. A low frequency sound wave is described by isentropic, and not isothermal, parameters—a result that seems paradoxical at first sight.

PROBLEMS

1. Fourier analysis of pulse. Consider a distribution that at the initial time $t = 0$ has the form of a Dirac delta function $\delta(x)$. A delta function can be represented by a Fourier integral:

$$\theta(x,0) = \delta(x) = \frac{1}{2\pi} \int_{-\infty}^{\infty} dk \exp(ikx). \tag{58}$$

At later times the pulse becomes

$$\theta(x,t) = \frac{1}{2\pi} \int_{-\infty}^{\infty} dk \exp[i(kx - \omega t)] , \tag{59}$$

or, by use of (10),

$$\theta(x,t) = \frac{1}{2\pi} \int_{-\infty}^{\infty} dk \exp(ikx - Dk^2 t). \tag{60}$$

Evaluate the integral to obtain the result (14). The method can be extended to describe the time development of any distribution given at $t = 0$. If the distribution is $f(x,0)$, then by the definition of the delta function

$$f(x,0) = \int dx' f(x',0)\delta(x - x'). \tag{61}$$

The time development of $\delta(x - x')$ is

$$\theta(x - x',t) = (4\pi Dt)^{-\frac{1}{2}} \exp[-(x - x')^2/4Dt] , \tag{62}$$

by (14). Thus at time t the distribution $f(x,0)$ has evolved to

$$f(x,t) = (4\pi Dt)^{-\frac{1}{2}} \int dx' f(x',0) \exp[-(x - x')^2/4Dt]. \tag{63}$$

This is a powerful general solution.

2. Diffusion in two and three dimensions. (a) Show that the diffusion equation in two dimensions admits the solution

$$\theta_2(t) = (C_2/t) \exp(-r^2/4Dt) \tag{64}$$

and in three dimensions

$$\theta_3(t) = (C_3/t^{3/2}) \exp(-r^2/4Dt). \tag{65}$$

(b) Evaluate the constants C_2 and C_3. These solutions are analogous to (14) and describe the evolution of a delta function at $t = 0$.

3. Temperature variations in soil. Consider a hypothetical climate in which both the daily and the annual variations of the temperature are purely sinusoidal, with amplitudes $\theta_d = 10°C$. The mean annual temperature $\theta_0 = 10°C$. Take the thermal diffusivity of the soil to be 1×10^{-3} cm^2 s^{-1}. What is the minimum depth at which water pipes should be buried in this climate?

4. Cooling of a slab. Suppose a hot slab of thickness $2a$ and initial uniform temperature θ_1 is suddenly immersed into water of temperature $\theta_0 < \theta_1$, thereby reducing the temperature at the surface of the slab abruptly to θ_0 and keeping it there. Expand the temperature in the slab in a Fourier series. After some time all but the longest wavelength Fourier component of the temperature will have decayed, and then the temperature distribution becomes sinusoidal. After what time will the temperature difference between the center of the slab and its surface decay to 0.01 of the initial difference $\theta_1 - \theta_0$?

5. p–n junction: diffusion from a fixed surface concentration. Suppose a silicon crystal is p-type doped with a concentration of $n_a = 10^{16}$ cm^{-3} of boron atoms. If the crystal slab is heated in an atmosphere containing phosphorus atoms, the latter will diffuse as donors with a concentration $n_d(x)$ into the semiconductor. They will form a p–n junction at that depth at which $n_a = n_d$. Assume that the diffusion conditions are such that the phosphorus concentration at the surface is maintained at $n_d(0) = 10^{17}$ cm^{-3}. Take the diffusion coefficient of donors to be $D = 10^{-13}$ cm^2 s^{-1}. What is the value of the constant C in the equation $x = Ct^{1/2}$, where x is the depth of the p–n junction and t is the time?

6. Heat diffusion with internal sources. When internal heat sources are present, the continuity equation (5) must be modified to read

$$\hat{C}\frac{\partial \tau}{\partial t} + \text{div } J_u = g_u , \tag{66}$$

where g_u is the heat generation rate per unit volume. Examples include Joule heat generated in a wire; heat from the radioactive decay of trace elements inside the Earth or the Moon. Give an expression for the temperature rise at the center of (a) a cylindrical wire and (b) the spherical Earth, on the assumption that g_u is independent of position and is constant with time.

7. Critical size of nuclear reactor. Extend the considerations of the preceding problem to particle diffusion, and assume that there is a net particle generation rate g_n that is proportional to the local particle concentration, $g_n = n/t_0$, where t_0 is a characteristic time constant. Such behavior describes the neutron generation in a nuclear reactor. The value of t_0 depends on the concentration of ^{235}U nuclei; if no surface losses took place, the neutron concentration would grow as $\exp(t/t_0)$. Consider a reactor in the shape of a cube of volume L^3 and assume that surface losses pin the neutron surface concentration at zero. Show that Eq. (3), if augmented by a generation term $g_n = n/t_0$, has solutions of the form

$$n(x,y,z,t) \propto \exp(t/t_1)\cos(k_x x)\cos(k_y y)\cos(k_z z) , \tag{67}$$

where $k_x L$, $k_y L$ and $k_z L$ are integer multiples of π. Give the functional dependence of the net time constant t_1 on k_x, k_y, k_z and t_0, and show that for at least one of the solutions of the form (67) the neutron concentration grows with time if L exceeds a critical value L_{crit}. Express L_{crit} as a function of D_n and t_0. In actual nuclear reactors this increase is ultimately halted because the neutron generation rate g_n decreases with increasing temperature.

Appendix A
Some Integrals Containing
Exponentials

THE GAUSS INTEGRAL

Let

$$I_0 = \int_{-\infty}^{+\infty} \exp(-x^2)\,dx = 2\int_0^{\infty} \exp(-x^2)\,dx. \tag{1}$$

The following trick is used to evaluate I_0. Write (1) in terms of a different integration variable:

$$I_0 = \int_{-\infty}^{+\infty} \exp(-y^2)\,dy. \tag{2}$$

Multiply (1) and (2) and convert the result to a double integral:

$$I_0^2 = \int_{-\infty}^{+\infty} \exp(-x^2)\,dx \int_{-\infty}^{+\infty} \exp(-y^2)\,dy = \int_{-\infty}^{+\infty}\int_{-\infty}^{+\infty} \exp - (x^2 + y^2)\,dx\,dy. \tag{3}$$

This is an integral over the entire x–y plane. Convert to polar coordinates r and φ, as shown in Figure A.1. Then, $x^2 + y^2 = r^2$, and the area element $dA = dx\,dy$ becomes $dA = r\,dr\,d\varphi$:

$$I_0^2 = \int_0^{2\pi}\left[\int_{-0}^{\infty} \exp(-r^2)\,r\,dr\right]d\varphi = 2\pi\int_0^{\infty} \exp(-r^2)\,r\,dr.$$

Because of $d[\exp(-r^2)] = -2\exp(-r^2)r\,dr$, the integral over r is elementary:

$$I_0^2 = -\pi\int d[\exp(-r^2)] = -\left[\pi\exp(-r^2)\right]_{r=0}^{r=\infty} = \pi.$$

Thus

$$\boxed{I_0 = \int_{-\infty}^{+\infty} \exp(-x^2)\,dx = 2\int_0^{\infty} \exp(-x^2)\,dx = \pi^{1/2}.} \tag{4}$$

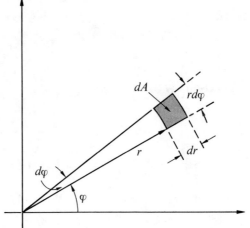

Figure A.1   The area element $dA = rdrd\varphi$.

GENERALIZED GAUSS INTEGRALS, AND GAMMA FUNCTION INTEGRALS

Integrals of the form

$$I_m = 2 \int_0^\infty x^m \exp(-x^2)\,dx, \qquad (m > -1) , \tag{5}$$

where m need not be an integer, may be reduced to the widely tabulated gamma function $\Gamma(z)$, by the substitutions $x^2 = y$, $2dx = y^{-\frac{1}{2}}\,dy$:

$$I_m = \int_0^\infty y^n e^{-y}\,dy = \Gamma(n + 1), \qquad n = (m - 1)/2. \tag{6}$$

The integral in (6) may be viewed as the definition of $\Gamma(z)$ for noninteger positive values of z.

The gamma function satisfies the recursion relation

$$\Gamma(n + 1) = n\Gamma(n) . \tag{7}$$

It is easily obtained for $n > 0$ from (6) by integration by parts, and it is used to extend the definition (6) of $\Gamma(z)$ to negative values of z. By using (7) repeatedly it is always possible to reduce $\Gamma(z)$ for arbitrary argument to a value in the interval $0 < z \le 1$.

For $m = 0$, $n = -\frac{1}{2}$; from (4):

$$I_0 = \int_0^\infty y^{-\frac{1}{2}} e^{-y}\, dy = \Gamma(\tfrac{1}{2}) = \pi^{1/2}. \tag{8}$$

If m is an even integer, $m = 2l > 0$, n is a half-integer, and $n = l - \frac{1}{2}$, then we find by repeated application of (7), with the aid of (8), that

$$I_{2l} = 2 \int_0^\infty x^{2l} \exp(-x^2)\, dx = \int_0^\infty y^{l-\frac{1}{2}} e^{-y}\, dy$$
$$= \Gamma(l + \tfrac{1}{2}) = (l - \tfrac{1}{2}) \times (l - \tfrac{3}{2}) \times \cdots \times \tfrac{3}{2} \times \tfrac{1}{2} \times \pi^{1/2}. \tag{9}$$

For $m = 1$, $n = 0$:

$$I_1 = 2 \int_0^\infty x \exp(-x^2)\, dx = \int_0^\infty e^{-y}\, dy = \Gamma(1) = 1. \tag{10}$$

If m is an odd integer, $m = 2l + 1 \geq 1$, n is an integer, $n = l \geq 0$, and we find similarly, with the aid of (10),

$$I_{2l+1} = 2 \int_0^\infty x^{2l+1} \exp(-x^2)\, dx = \int_0^\infty y^l e^{-y}\, dy$$
$$= \Gamma(l + 1) = l \times (l - 1) \times \cdots \times 2 \times 1 = l! \tag{11}$$

The gamma function for positive integer argument is simply the factorial of the integer preceding the argument.

THE STIRLING APPROXIMATION

For large values of n, $n!$ can be approximated by

$$n! \simeq (2\pi n)^{1/2} n^n \exp\left[-n + \frac{1}{12n} + 0\!\left(\frac{1}{n^2}\right) \right] \tag{12a}$$

or

$$\log n! \simeq \tfrac{1}{2} \log 2\pi + (n + \tfrac{1}{2}) \log n - n + \frac{1}{12n} + 0\!\left(\frac{1}{n^2}\right). \tag{12b}$$

Here the term $1/12n$ is the first term of an expansion by powers of $1/n$, and $0(1/n^2)$ stands for omitted higher order terms in this expansion, of order $1/n^2$

or higher. In practice, even the term $1/12n$ is usually omitted. Its principal role is to check on the accuracy of the approximation. If the effect of the $1/12n$-correction introduces only a change below the desired accuracy, the entire expression has the desired accuracy.

To derive (12) we write, in accordance with (11),

$$n! = \int_0^\infty x^n e^{-x} \, dx = \int_0^\infty \exp[f(x)] \, dx \, , \tag{13}$$

where

$$f(x) = n \log x - x. \tag{14}$$

We make the substitution

$$x = n + yn^{\frac{1}{2}} = n(1 + yn^{-\frac{1}{2}}) \, , \qquad dx = n^{\frac{1}{2}} \, dy. \tag{15}$$

Then

$$f(x) = n \log n - n + g(y) \, , \tag{16}$$

where

$$g(y) = n[\log(1 + yn^{-\frac{1}{2}}) - yn^{-\frac{1}{2}}]. \tag{17}$$

With these,

$$\exp[f(x)] = n^n e^{-n} \exp[g(y)] \, , \tag{18}$$

$$n! = n^{1/2} n^n e^{-n} \int_{-n^{1/2}}^\infty \exp[g(y)] \, dy. \tag{19}$$

The function $g(y)$ has its maximum at $y = 0$: $g(0) = 0$. Using the Taylor expansion of the logarithm,

$$\log(1 + s) = s - \tfrac{1}{2}s^2 + \tfrac{1}{3}s^3 - \tfrac{1}{4}s^4 + \cdots \, , \tag{20}$$

with $s = (y^2/n)^{1/2}$, we expand $g(y)$:

$$g(y) = n\left[-\frac{1}{2}\frac{y^2}{n} + \frac{1}{3}\left(\frac{y^2}{n}\right)^{3/2} - \cdots \right]$$

$$= -\tfrac{1}{2}y^2 + O(s^3). \tag{21}$$

In the limit $n \to \infty$, $s \to 0$, and all but the first term in (21) vanish, and the integral in (19) becomes

$$\int_{-n^{1/2}}^{\infty} \exp[g(y)]dy = \int_{-\infty}^{+\infty} \exp(-y^2/2)dy = (2\pi)^{1/2} , \qquad (22)$$

with the aid of (4). If (22) is inserted into (19) the result is identical to (12a) except for the correction term $1/12n$. Its derivation is a bit tedious. We work with $\log n!$ and write

$$\log n! = \tfrac{1}{2}\log 2\pi + (n + \tfrac{1}{2})\log n - n + \frac{A}{n} + 0\!\left(\frac{1}{n^2}\right). \qquad (23a)$$

If we replace n by $n - 1$,

$$\log(n - 1)! = \tfrac{1}{2}\log 2\pi + (n - \tfrac{1}{2})\log(n - 1) - (n - 1)$$

$$+ \frac{A}{n - 1} + 0\!\left(\frac{1}{n^2}\right). \qquad (23b)$$

We subtract (23b) from (23a):

$$\log n! - \log(n - 1)! = \log\frac{n!}{(n - 1)!} = \log n$$

$$= (n + \tfrac{1}{2})\log n - (n - \tfrac{1}{2})\log(n - 1) - 1$$

$$+ \frac{A}{n} - \frac{A}{n - 1} + 0\!\left(\frac{1}{n^3}\right) , \qquad (24)$$

where all omitted terms are now at least of order $1/n^3$. The two terms in A can be combined:

$$\frac{A}{n} - \frac{A}{n - 1} = -\frac{A}{n(n - 1)} = -\frac{A}{n^2} + 0\!\left(\frac{1}{n^3}\right). \qquad (25)$$

If this is inserted into (24), we find

$$\frac{A}{n^2} = (n - \tfrac{1}{2})\log\frac{n}{n - 1} - 1 + 0\!\left(\frac{1}{n^3}\right). \qquad (26)$$

For large n the logarithm may be expanded according to (20), with $s = -1/n$:

$$\log\frac{n}{n-1} = -\log\left(1 - \frac{1}{n}\right) = \frac{1}{n} + \frac{1}{2n^2} + \frac{1}{3n^3} + 0\left(\frac{1}{n^4}\right) \tag{27}$$

$$(n - \tfrac{1}{2})\log\frac{n}{n-1} = 1 + \frac{1}{2n} - \frac{1}{2n} + \frac{1}{3n^2} - \frac{1}{4n^2} + 0\left(\frac{1}{n^3}\right)$$

$$= 1 + \frac{1}{12n^2} + 0\left(\frac{1}{n^3}\right). \tag{28}$$

If this is inserted in (26) we see that $A = 1/12$.

We are often interested not in $n!$ but only in $\log n!$, and only to an accuracy such that the relative error between an approximate value of $\log n!$ and the true value decreases with increasing n. Such an approximation is obtained by neglecting all terms in (12b) that increase less rapidly than linearly with n:

$$\boxed{\log n! \simeq n\log n - n.} \tag{29}$$

Appendix B
Temperature Scales

DEFINITION OF THE KELVIN SCALE

Numerical values of temperature*,[†] are not expressed in practice in terms of the fundamental temperature τ, whose unit is the unit of energy, but on the (absolute) thermodynamic temperature scale T, the **Kelvin scale**, whose unit is the kelvin, symbol K. The kelvin was defined in 1954 by international agreement as the fraction $1/273.16$ of the temperature T_t of the triple point of pure water. Hence, by this definition, $T_t \equiv 273.16$ K, exactly. This temperature is 0.01 K above the atmospheric-pressure freezing point of water (the ice point), $T_0 = 273.15$ K. The triple point is more easily and accurately reproducible than the ice point. The triple point establishes itself automatically in any clean evacuated vessel that is partially backfilled with pure water and cooled until part but not all of the water is frozen, leading to an equilibrium between solid ice, liquid water, and the water vapor above the ice-water mixture.

The **Celsius temperature scale** t is defined in terms of the Kelvin scale, by

$$t \equiv T - 273.15 \text{ K}. \tag{1}$$

Temperatures on this scale are expressed in degrees Celsius, symbol °C. Temperature differences have the same value on both Kelvin and Celsius scales.

The conversion factor k_B between the fundamental temperature τ and the Kelvin temperature,

$$\tau = k_B T \, , \tag{2}$$

---

* We appreciate the assistance of Norman E. Phillips in the preparation of this appendix.

[†] The ultimate survey of the state of development of precise temperature measurements is the proceedings of an international symposium taking place every few years under the title *Temperature, its measurement and control in science and industry*. The proceedings are published under the same title. Vol. 1: C. O. Fairchild, editor; Reinhold, 1940. This volume is largely obsolete. Vol. 2: E. C. Wolfe, editor; Reinhold, 1955. Although no longer reflecting the state of the art, this volume is still useful for its thorough introductory discussions of principles of various methods of temperature measurements, not all of which are repeated in the later volumes. Vol. 3 (3 parts): C. M. Herzfeld, editor; Reinhold, 1962. Perhaps the most useful volume because of its introductory discussion of principles of various methods of temperature measurements. Vol. 4 (3 parts): H. H. Plumb, editor; Instrument Society of America, Pittsburgh, 1972. Most useful for representing the state of the art of various methods of temperature measurements; contains less introductory review material than Vol. 3.

is called the **Boltzmann constant**. Its numerical value must be determined experimentally; the best current value* is

$$k_B = (1.380662 \pm 0.000044) \times 10^{-16} \, \text{erg} \, \text{K}^{-1}. \tag{3}$$

The value of the Boltzmann constant is determined with the aid of certain model systems whose structures are sufficiently simple that one can calculate the energy distribution of the quantum states, and from it the entropy as a function of the energy, $\sigma = \sigma(U)$. The fundamental temperature as a function of the energy is $\tau(U) = (\partial\sigma/\partial U)^{-1}$. Examples of model systems used in the determination of k_B are the following:

(*a*) *Ideal gas.*   In the limit of low particle concentration all gases behave as ideal gases, satisfying $pV = Nk_BT$. One obtains k_B by measuring the pV product of a known amount of gas at a known kelvin temperature T, extrapolated to vanishing pressure. The determination of the number of particles N invariably involves the Avogadro constant N_A, independently known.

(*b*) *Black body radiation.*   We can obtain k_B by fitting the measured spectral distribution of a black body of known Kelvin temperature T to the Planck radiation law (Chapter 4). Because this law involves τ through the ratio $\hbar\omega/\tau = \hbar\omega/k_BT$, this determination requires the independent knowledge of Planck's constant.

(*c*) *Spin paramagnetism.*   In the limit of vanishing interaction the magnetic moment M of a system of N spins in a magnetic field B, at temperature τ, is given by Eq. (3.46). Various paramagnetic salts, such as cerous magnesium nitrate (CMN) are good approximations to noninteracting spin systems if the temperature is not too low. By fitting measured values of M as a function of B/T to (3.46) we can determine the ratio m/k_B, where m is the intrinsic magnetic moment of the electron, known independently. Usually only the low-field portion is used, in which case the number of spins must also be known, which involves again N_A. Precision results require correction for weak residual spin interactions, similar to corrections for particle interactions in a gas.

The k_B value given in (3) is a weighted average of several determinations. With an uncertainty of about 32 parts per million, it is one of the least accurately known fundamental constants. Most of this uncertainty is due to the difficulty of the measurements and to the nonideality of the systems used for these measurements. About 5 parts per million are due to the limited accuracy with which $\hbar$ and N_A are known.

---

* E. R. Cohen and B. N. Taylor, *J. Phys. Chem. Reference Data* **2**, No. 4 (1973).

When expressing temperature as conventional Kelvin temperature T rather than fundamental temperature τ, it is customary to absorb the Boltzmann constant into the definition of a conventional entropy S,

$$S \equiv k_B \sigma. \tag{4}$$

The relation $dQ = \tau d\sigma$ between reversible heat transfer and entropy transfer then becomes

$$dQ = T dS. \tag{5}$$

PRIMARY AND SECONDARY THERMOMETERS

Any accurately measurable physical property X whose value is an accurately known function of the temperature, $X = X(T)$, may be used as a thermometric parameter to measure the temperature of the system possessing the property X and of any system in thermal equilibrium with it. Used in this way, the system with the property X is a thermometer. The principles underlying the most commonly used thermometers are listed in Tables B.1 and B.2. The thermometers listed in Table B.2 are called secondary thermometers, defined as thermometers whose temperature dependence $X(T)$ must be calibrated empirically, by comparison with another thermometer whose calibration is already known. The calibration of all secondary thermometers must ultimately be traceable to a primary thermometer. But once calibrated, secondary thermometers are easier to use and are more reproducible than the primary thermometers available at the same temperature.

Any calculable model system that can be used to determine the value of the Boltzmann constant k_B can be used as a primary thermometer, and the three model systems discussed above are the most important primary thermometers (Table B.1).

The precision and accuracy of thermometers vary greatly. Precision is expressed by the variation ΔT observed when the same temperature is measured at different times with the same instrument. Accuracy is expressed by the uncertainty ΔT with which the thermometer reproduces the true Kelvin scale. Secondary thermometers based on electrical resistance measurements may achieve a precision of 1 part in 10^5. The precision of thermometers based on mechanical pressure measurements is much poorer, particularly at low pressures. For example, helium vapor pressure thermometers at the lower end of their useful range have a precision of about 1 part in 10^3. The accuracy of secondary thermometers is limited by the accuracy of the primary thermometers

Table B.1 Principles of the most important primary thermometers

Model system and property utilized	Defining equation		Typical working substances	Temperature range used [K]
Ideal gas				
a) Static pressure at constant volume	$pV = N\tau$;	(3.73)	^{4}He, N_2	4–1400
b) Speed of sound	$v_s = (\gamma\tau/M)^{1/2}$;	(15.43)	^{3}He, ^{4}He	2–20
Magnetic susceptibility of noninteracting spin system	$\chi = nm^2/\tau$;	(3.46)*		
a) Electronic spins			CMN†	0.001–4
b) Nuclear spins			Metals: Cu, Tl, Pt	<0.001–0.05
Black body radiation	$u_\omega \propto \dfrac{(\hbar\omega/\tau)^3}{\exp(\hbar\omega/\tau) - 1}$;	(4.22)		>1300

NOTE: The column "Defining equation" indicates which equation underlies the basic idea, plus text reference. In practice, corrections for various nonidealities may be needed. The temperature ranges quoted are the ranges used in practice for primary thermometry, including those temperatures at which the principle is used mainly to establish a desirable overlap with other methods. Some of the principles can actually be used over much wider ranges.

* Obtained from (3.46) by differentiation for low fields: $\mu_M \approx nm^2 B/\tau$, $\chi = (\partial M/\partial B)_\tau$

† Cerous magnesium nitrate, and diluted CMN in lanthanum magnesium nitrate.

Table B.2   Principles of the most important secondary
thermometers

Physical property	Useful range in K
Thermoelectric voltage of thermocouples*	400–1400
Thermal expansion of liquid in glass	200–400
Electrical resistance	
metals*	14–700
semiconductors (germanium)[†]	0.05–77
commercial carbon resistors[†]	0.05–20
Vapor pressure of liquefied gas	
^{4}He	1–5.2
^{3}He	0.3–3.2

NOTE: The temperature ranges are approximate ranges of wide utility,
not ultimate limits.
* Used as interpolating instrument in the IPTS.
[†] Widely used as cryogenic laboratory thermometer; each specimen
must be individually calibrated to deliver usable accuracy.

used to calibrate them. The accuracy of primary thermometers is limited by
their relatively poor precision and by residual variations between different
thermometers. As a rough estimate, the present-day accuracy of primary
thermometers is about 1 part in 10^4 above 100 K, about 1 part in 10^3 around
1 K, and about 1 part in 10^2 near 0.01 K.

THERMODYNAMIC THERMOMETRY

It is possible to perform primary thermometry without relying on the theoretically known properties of simple model systems, by somehow utilizing the
relation (5). We give three examples.

(a) Carnot cycle.   Consider a Carnot cycle operating between a known temperature T_1 and the unknown temperature T_2. Because of entropy conservation
the heat transfers at the two temperatures satisfy $Q_1/T_1 = Q_2/T_2$. The unknown temperature can be determined by measuring the ratio of the two heat
transfers. The method is not very practical.

(b) Magnetic calorimetry.   Suppose a paramagnetic substance is initially at a
known temperature T_1, in a magnetic field B_1. Let the substance be cooled by
isentropic demagnetization to the unknown temperature T_2. If now a known

small amount $đQ$ of heat is added to the substance, its entropy is raised by $dS = đQ/T_2$. The substance is then isentropically re-magnetized, and the magnetic field B_2 is determined at which the temperature has returned exactly to T_1. The field B_2 will be found slightly different than $B_1: B_2 = B_1 + dB$. Entropy conservation requires

$$dS = đQ/T_2 = S(T_1,B_2) - S(T_1,B_1) = (\partial S/\partial B)_T \, dB. \qquad (6)$$

From the thermodynamic identity for the Helmholtz free energy for a magnetizable substance,

$$dF = -SdT - MdB , \qquad (7)$$

one obtains, by the usual cross-differentiation, the Maxwell relation

$$(\partial S/\partial B)_T = (\partial M/\partial T)_B. \qquad (8)$$

We insert (8) into (6) to find the expression for the unknown temperature:

$$T_2 = (đQ/dB)(\partial M/\partial T)_B. \qquad (9)$$

The quantities $đQ$ at $T = T_2$ and dB at $T = T_1$ are known, and the temperature derivative of M at $T = T_1$ and $B = B_1$ is easily measured. The method makes no assumptions about the ideality of the paramagnetic substance, and it has therefore been used extensively at low temperatures.

(c) *Clausius-Clapeyron thermometry.* The melting temperature T_m of a substance varies with pressure p according to the Clausius-Clapeyron equation of Chapter 10:

$$dT_m/dp = T_m \Delta V/\Delta H , \qquad (10)$$

where ΔV is the volume change during melting, and ΔH the latent heat of fusion. If both quantities have been measured as functions of pressure, (10) can be integrated:

$$T_2/T_1 = \exp \int_{p_1}^{p_2}(\Delta V/\Delta H)dp. \qquad (11)$$

If T_1 and p_1 are known, a measurement of the pressure p_2 at which the unknown temperature T_2 is the equilibrium melting temperature permits calculation of

T_2 from (11). By utilization of the strong temperature dependence of the solidification pressure of liquid ^{3}He, the method has been used as an alternative to magnetic thermometry at low temperatures.

INTERNATIONAL PRACTICAL TEMPERATURE SCALE (IPTS)

Many known phase equilibrium temperatures can be reproduced far more precisely than the accuracy with which their exact location on the Kelvin scale can be determined by primary thermometry. To facilitate practical thermometry, a number of easily reproducible phase equilibrium temperatures have been determined as accurately as possible and have been assigned best values to define an International Practical Temperature Scale (IPTS). On the IPTS the selected equilibrium points are treated as if their temperatures were known to be exactly equal to their assigned values. Intermediate temperatures are determined by a precisely specified interpolation procedure that is chosen to reproduce the true Kelvin scale as accurately as possible. The present version of the scale is IPTS68, adopted in 1968 by international agreement, covering temperatures from the triple point of hydrogen (13.81 K) upward.* Table B.3 gives the assigned temperatures for IPTS68.

In the range between 13.81 K and 903.89 K, which is the melting point of antimony, a platinum resistance thermometer is used as the interpolating instrument. In the range from 903.89 K to 1337.58 K, the melting point of gold, a platinum-platinum/rhodium thermocouple is used. Above 1337.58 K black body radiation is used.

Below 13.81 K no precisely defined procedure has been agreed. In the range between 5.2 K and 13.81 K various scales based on the vapor pressure of hydrogen are in practical use. Below 5.21 K, the critical point of ^{4}He, down to about 0.3 K, the 1958 and 1962 helium scales* are widely used as de facto extensions of IPTS68. The 1958 ^{4}He scale relates the vapor pressure of ^{4}He to the temperature T; the 1962 ^{3}He scale uses the vapor pressure of ^{3}He.

As the accuracy of primary temperature measurements improves, errors in practical scales such as IPTS become uncovered, leading eventually to revision of the practical scales. Table B.3 lists some errors now believed to exist in IPTS68.

---

* See, for example, *American Institute of Physics handbook*, 3rd ed., McGraw-Hill, 1972; Section 4: Heat, M. W. Zemansky, editor. Contains complete original references.

Table B.3   Assigned temperatures of the International Practical Temperature Scale of 1968

Equilibrium point		T_{68}, in K	$T_{68} - T$, in K
Substance	Type		
hydrogen	t	13.81	
hydrogen	b (250 torr)	17.042	
hydrogen	b	20.28	
neon	b	27.402	
oxygen	t	54.361	
oxygen	b	90.188	
water	t	273.16	exact
water	b	373.15	0.025
tin	f	505.1181	0.044
zinc	f	692.43	0.066
silver	f	1235.08	
gold	f	1337.58	

NOTE: Except for the triple points and the 17.042 K point, all equilibria are those at a pressure of one standard atmosphere, $p_0 = 101,325\,\mathrm{Nm^{-2}}$ ($=760$ torr). The 17.042 K point is the boiling point of hydrogen at 250 torr. The notations t, b, and f in the second column refer to triple points, boiling points, and freezing points.* The last column contains estimates of errors known to exist, from Physics Today **29**, No. 12, p. 19 (Dec. 1976).

* All data from the *American Institute of Physics handbook*, 3rd ed., McGraw-Hill, 1972; Section 4: Heat, M. W. Zemansky, editor.

Appendix C
Poisson Distribution

The Poisson distribution law is a famous result of probability theory. The result is useful in the design and analysis of counting experiments in physics, biology, operations research, and engineering. The statistical methods we have developed lend themselves to an elegant derivation of the Poisson law, which is concerned with the occurrence of small numbers of objects in random sampling processes. It is also called the law of small numbers. If on the average there is one bad penny in a thousand, what is the probability that N bad pennies will be found in a given sample of one hundred pennies? The problem was first considered and solved in a remarkable study of the role of luck in criminal and civil law trials in France in the early nineteenth century.

We derive the Poisson distribution law with the aid of a model system that consists of a large number R of independent lattice sites in thermal and diffusive contact with a gas. The gas serves as a reservoir. Each lattice site may adsorb zero or one atom. We want to find the probabilities

$$P(0), P(1), P(2), \ldots, P(N), \ldots,$$

that a total of $0, 1, 2, \ldots, N, \ldots$, atoms are adsorbed on the R sites, if we are given the average number $\langle N \rangle$ of adsorbed atoms over an ensemble of similar systems.

Consider a system composed of a single site. It is convenient to set the binding energy of an atom to the site as zero. The identical form for the distribution is found if a binding energy is included in the calculation. The Gibbs sum is

$$\mathfrak{Z}_1 = 1 + \lambda , \tag{1}$$

where the term λ is proportional to the probability the site is occupied, and the term 1 is proportional to the probability the site is vacant. Thus the absolute probability that the site is occupied is

$$f = \frac{\lambda}{1 + \lambda}. \tag{2}$$

The actual value of λ is determined by the condition of the gas in the reservoir, because for diffusive contact between the lattice and the reservoir we must have

$$\lambda(\text{lattice}) = \lambda(\text{gas}), \tag{3}$$

by the argument of Chapter 5. The evaluation of $\lambda(\text{gas})$ for an ideal gas was given in Chapter 6.

We now extend the treatment to R independent sites. Then

$$\mathfrak{Z}_{\text{tot}} = \mathfrak{Z}_1\,\mathfrak{Z}_2 \cdots \mathfrak{Z}_R = (1 + \lambda)^R. \tag{4}$$

By the argument used in Chapter 1 we know that the binomial expansion of $(\bigcirc + \bullet)^R$ or $(1 + \lambda)^R$ counts once and only once every state of the system of R sites. Each site has two alternative states, namely $\bigcirc$ for vacant or $\bullet$ for occupied, which corresponds in the Gibbs sum to the term 1 for λ^0 and the term λ for λ^1.

In the low-occupancy limit of $f \ll 1$ we have $f \cong \lambda$, whence

$$\langle N \rangle = fR = \lambda R \tag{5}$$

is the average total number of adsorbed atoms. The Poisson distribution is concerned with this low-occupancy limit. We can now write (4) as

$$\mathfrak{Z}_{\text{tot}} = \left(1 + \frac{\lambda R}{R}\right)^R = \left(1 + \frac{\langle N \rangle}{R}\right)^R. \tag{6}$$

Next we let the number of sites R increase without limit, while holding the average number of occupied sites $\langle N \rangle$ constant. The Poisson distribution is concerned with infrequent events! By the definition of the exponential function we have

$$\lim_{R \to \infty} \left(1 + \frac{\langle N \rangle}{R}\right)^R = \exp\langle N \rangle , \tag{7}$$

so that

$$\mathfrak{Z}_{\text{tot}} \simeq \exp\langle N \rangle = \exp(\lambda R) = \sum_N \frac{(\lambda R)^N}{N!}. \tag{8}$$

The last step here is the expansion of the exponential function in a power series.

The term in λ^N in $\mathfrak{z}_{\text{tot}}$ is proportional to the probability $P(N)$ that N sites are occupied. With the Gibbs sum as the normalization factor we have in the limit of large R:

$$P(N) = \frac{\lambda^N R^N}{N!} \cdot \frac{1}{\mathfrak{z}_{\text{tot}}} = \frac{\lambda^N R^N \exp(-\lambda R)}{N!} , \tag{9}$$

or, because $\lambda R = \langle N \rangle$ from (5),

$$\boxed{P(N) = \frac{\langle N \rangle^N \exp(-\langle N \rangle)}{N!}.} \tag{10}$$

This is the **Poisson distribution law**.

Particular interest attaches to the probability $P(0)$ that none of the sites is occupied. From (10) we find, with $\langle N \rangle^0 = 1$ and $0! = 1$,

$$P(0) = \exp(-\langle N \rangle); \qquad \log P(0) = -\langle N \rangle. \tag{11}$$

Thus the probability of zero occupancy is simply related to the average number $\langle N \rangle$ of occupied sites. This suggests a simple experimental procedure for the determination of $\langle N \rangle$: just count the *systems* that have no adsorbed atoms.

Values of $P(N)$ for several values of $\langle N \rangle$ are given in Table C.1. Plots are given in Figure C.1 for $\langle N \rangle = 0.5$, 1, 2, and 3.

Table C.1 Values of the Poisson distribution function $P(N) = \dfrac{\langle N \rangle^N \exp(-\langle N \rangle)}{N!}$

	$\langle N \rangle$									
	0.1	0.3	0.5	0.7	0.9	1	2	3	4	5
$P(0)$	0.9048	0.7408	0.6065	0.4966	0.4066	0.3679	0.1353	0.0498	0.0183	0.0067
$P(1)$	0.0905	0.2222	0.3033	0.3476	0.3659	0.3679	0.2707	0.1494	0.0733	0.0337
$P(2)$	0.0045	0.0333	0.0758	0.1217	0.1647	0.1839	0.2707	0.2240	0.1465	0.0842
$P(3)$	0.0002	0.0033	0.0126	0.0284	0.0494	0.0613	0.1804	0.2240	0.1954	0.1404
$P(4)$		0.0003	0.0016	0.0050	0.0111	0.0153	0.0902	0.1680	0.1954	0.1755
$P(5)$			0.0002	0.0007	0.0020	0.0031	0.0361	0.1008	0.1563	0.1755
$P(6)$				0.0001	0.0003	0.0005	0.0120	0.0504	0.1042	0.1462
$P(7)$						0.0001	0.0034	0.0216	0.0595	0.1044
$P(8)$							0.0009	0.0081	0.0298	0.0653
$P(9)$							0.0002	0.0027	0.0132	0.0363
$P(10)$								0.0008	0.0053	0.0181

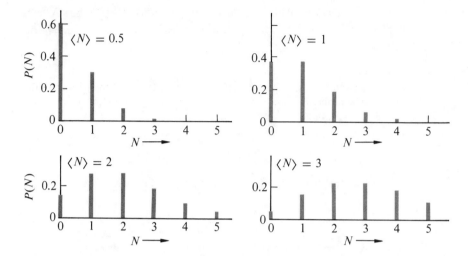

Figure C.1   Poisson distribution, P versus N, for several values of $\langle N \rangle$.

Example: Incorrect and correct counting of states.   (a) The Gibbs sum for the R sites is *not*

$$\mathcal{Z}_{tot} = 1 + \lambda + \lambda^2 + \lambda^3 + \cdots + \lambda^R. \tag{12}$$

Why not?

(b) The correct sum is

$$\mathcal{Z}_{tot} = (1 + \lambda)^R = 1 + R\lambda + \frac{R(R-1)}{2!} \lambda^2 + \cdots + \lambda^R = \sum_{n=0}^{R} g(R,N)\lambda^N , \tag{13}$$

where

$$g(R,N) = \frac{R!}{(R-N)! \, N!}$$

is the binomial coefficient. Note that $g(R,N)$ is the number of independent states of the system for a given number of adsorbed atoms N. The Gibbs sum is a sum over all states.

Example: Elementary derivation of $P(0)$.   Let a total of R bacteria be distributed at random among L dishes. Each dish is viewed as a system of many sites to which a bacterium may attach. The L dishes represent an ensemble of L identical systems. The average number of bacteria per dish is

$$\langle N \rangle = R/L. \tag{14}$$

Each time a bacterium is distributed, the probability that a given dish will receive that bacterium is $1/L$. The probability the given dish will not receive the bacterium is

$$\left(1 - \frac{1}{L}\right). \tag{15}$$

The probability in R tries that the given dish will receive no bacteria is

$$P(0) = \left(1 - \frac{1}{L}\right)^{R} \tag{16}$$

because the factor (15) enters on each try.

We may write (16) as

$$P(0) = \left(1 - \frac{\langle N \rangle}{R}\right)^{R}, \tag{17}$$

by use of $\langle N \rangle = R/L$. We know that in the limit of large R,

$$\exp(-\langle N \rangle) = \lim_{R \to \infty} \left(1 - \frac{\langle N \rangle}{R}\right)^{R}, \tag{18}$$

by the definition of the exponential function. Thus for $R \gg 1$ and $L \gg 1$ we have

$$P(0) = \exp(-\langle N \rangle) \tag{19}$$

in agreement with (11).

PROBLEMS

1. Random pulses.　A radioactive source emits alpha particles which are counted at an average rate of one per second. (a) What is the probability of counting exactly 10 alpha particles in 5 s? (b) Of counting 2 in 1 s? (c) Of counting none in 5 s? The answers to (a) and (b) are not identical.

2. Approach to Gaussian distribution.   Show that the Poisson function $P(N) = \langle N \rangle^N \exp(-\langle N \rangle)/N!$ closely approaches a Gaussian function in form, for large $\langle N \rangle$. That is, show when N is close to $\langle N \rangle$ that

$$P(N) \simeq A \exp[-B(N - \langle N \rangle)^2] \, ,$$

where A, B are quantities to be determined by you. *Hint*: Work with $\log P(N)$; use the Stirling approximation. In the Gaussian form both A and B are functions of $\langle N \rangle$; in the development of the Poisson function you may find A, B are functions of N, but the two forms of A, B are closely equivalent over the range in which the exponential factor has significant values.

Appendix D
Pressure

Let a pressure p_s be applied normal to the faces of a cube filled with a gas or liquid in quantum state s. By elementary mechanics (Chapter 3) the pressure is equal to

$$p_s = -dU_s/dV \ , \tag{1}$$

where U_s is the energy of the system in the state s. We can also write the pressure as

$$p_s = -(dU/dV)_s \ , \tag{2}$$

where $(dU/dV)_s$ denotes the expectation value* of dU/dV over the state s at volume V. It is important that we can calculate p by (2) which is at a fixed volume with no ambiguity about the identity of the selected state s, whereas (1) involves following the state through two volumes, V and $V + dV$, with some possible doubt whether the state remains the same. The ensemble average pressure p is the average of p_s over the states represented in the ensemble:

$$p = \langle p_s \rangle = -\langle (dU/dV)_s \rangle \tag{3}$$

Because the number of states in the ensemble is constant, the entropy is constant, so that the derivative is at constant entropy. We may therefore write

$$p = -\left(\frac{\partial U}{\partial V}\right)_\sigma . \tag{4}$$

The result (4) uses the energy of the system expressed as $U(\sigma, V, \ldots)$; that is, as a function of the volume V and the entropy σ—not the temperature τ. It is the entropy and not the temperature that is to be held constant in the differentiation.

---

* The equivalence of (1) and (2) is an example of the Hellmann-Feynman theorem of quantum mechanics, according to which the derivative of the hamiltonian $\mathcal{H}$ and energy eigenvalue U with respect to a parameter λ are related by $dU/d\lambda = \langle |d\mathcal{H}/d\lambda| \rangle$. The derivation may be found on p. 1192 of C. Cohen-Tannoudji, B. Diu, and F. Laloë, *Quantum mechanics*, Wiley, 1977; see also E. Merzbacher, *Quantum mechanics*, 2nd ed., Wiley, 1970, p. 442.

Appendix E
Negative Temperature

The result of Problem 2.2 for the entropy of a spin system as a function of the energy in a magnetic field is plotted here in Figure E.1. Notice the region in which $(\partial\sigma/\partial U)_N$ is negative (Figure E.2). Negative τ means that the population of the upper state is greater than the population of the lower state. When this condition obtains we say that the population is inverted, as illustrated in Figure E.3.

The concept of negative temperature is physically meaningful for a system that satisfies the following restrictions: (a) There must be a finite upper limit to the spectrum of energy states, for otherwise a system at a negative temperature would have an infinite energy. A freely moving particle or a harmonic oscillator cannot have negative temperatures, for there is no upper bound on their energies. Thus only certain degrees of freedom of a particle can be at a negative tempera-

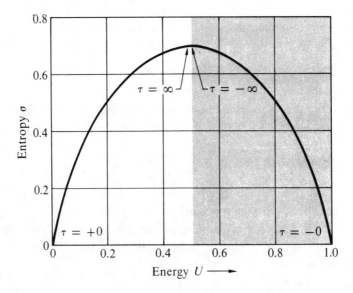

Figure E.1 Entropy as function of energy for a two state system. The separation of the states is $\varepsilon = 1$ in this example. In the left-hand side of the figure $\partial\sigma/\partial U$ is positive, so that τ is positive. On the right-hand side $\partial\sigma/\partial U$ is negative and τ is negative.

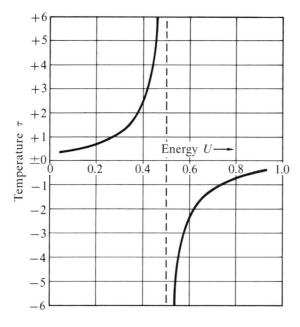

Figure E.2 Temperature versus energy for the two state system. Here

$$\tau = \frac{1}{\left(\dfrac{\partial \sigma}{\partial U}\right)_N} = \frac{1}{\log\dfrac{1-U}{U}}.$$

Notice that the energy is not a maximum at $\tau = +\infty$, but is a maximum at $\tau = -0$.

ture: the nuclear spin orientation in a magnetic field is the degree of freedom most commonly considered in experiments at negative temperatures. (b) The system must be in internal thermal equilibrium. This means the states must have occupancies in accord with the Boltzmann factor taken for the appropriate negative temperature. (c) The states that are at a negative temperature must be isolated and inaccessible to those states of the body that are at a positive temperature.

The ordinary translational and vibrational degrees of freedom of a body have an entropy that increases without limit as the energy increases, in contrast to the two state or spin system of Figure E.1. If σ increases without limit, then τ is always positive. The exchange of energy between a system at a negative temperature and a system that can only have a positive temperature (because of

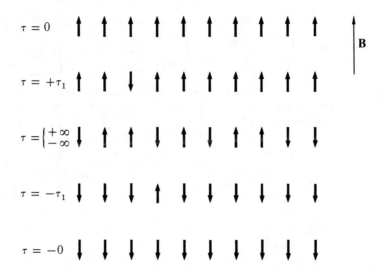

Figure E.3 Possible spin distributions for various positive and
negative temperatures. The magnetic field is directed upward. The
negative spin temperatures cannot last indefinitely because of weak
coupling between spins and the lattice. The lattice can only be at a
positive temperature because its energy level spectrum is unbounded
on top. The downward-directed spins, as at $\tau = -\tau_1$, turn over one
by one, thereby releasing energy to the lattice and approaching
equilibrium with the lattice at a common positive temperature. A
nuclear spin system at negative temperature may relax quite slowly,
over a time of minutes or hours; during this time experiments at
negative temperatures may be carried out.

an unbounded spectrum) will lead always to an equilibrium configuration in
which both systems are at a positive temperature.

Negative temperatures correspond to higher energies than positive temperatures. When a system at a negative temperature is brought into contact with a
system at a positive temperature, energy will be transferred from the negative
temperature to the positive temperature. Negative temperatures are *hotter*
than positive temperatures.

The temperature scale from cold to hot runs $+0\,\text{K}, \ldots, +300\,\text{K}, \ldots, +\infty\,\text{K}$,
$-\infty\,\text{K}, \ldots, -300\,\text{K}, \ldots, -0\,\text{K}$. Note that if a system at $-300\,\text{K}$ is brought
into thermal contact with an identical system at $300\,\text{K}$, the final equilibrium
temperature is not $0\,\text{K}$, but is $\pm\infty\,\text{K}$.

Nuclear and electron spin systems can be promoted to negative temperatures
by suitable radio frequency techniques. If a spin resonance experiment is

carried out on a spin system at negative temperature, resonant emission of energy is obtained instead of resonant absorption.* A negative temperature system is useful as an rf amplifier in radio astronomy where weak signals must be amplified.

Abragam and Proctor[†] have carried out an elegant series of experiments on calorimetry with systems at negative temperatures. Working with a LiF crystal, they established one temperature in the system of Li nuclear spins and another temperature in the system of F nuclear spins. In a strong static magnetic field the two thermal systems are essentially isolated, but in the Earth's magnetic field the energy levels overlap and the two systems rapidly approach equilibrium among themselves (mixing). It is possible to determine the temperature of the systems before and after the systems are allowed to mix. Abragam and Proctor found that if both systems were initially at positive temperatures they attained a common positive temperature on being brought into thermal contact. If both systems were prepared initially at negative temperatures, they attained a common negative temperature on being brought into thermal contact. If prepared one at a positive temperature and the other at a negative temperature, then an intermediate temperature was attained on mixing, warmer than the initial positive temperature and cooler than the initial negative temperature.

FURTHER REFERENCES
ON NEGATIVE TEMPERATURE

N. F. Ramsey, "Thermodynamics and statistical mechanics at negative absolute temperature," Physical Review **103**, 20 (1956).

M. J. Klein, "Negative absolute temperature," Physical Review **104**, 589 (1956).

---

* E. M. Purcell and R. V. Pound, Physical Review **81**, 279 (1951).
† A. Abragam and W. G. Proctor, Physical Review **106**, 160 (1957); **109**, 1441 (1958).

Index

ISBN 0-7167-1088-9

9 780716 710882

EAN

90000 >